The Irish Scientist

Editorials
An exciting time to be a scientist 7
Sequence of articles
 in the 1999 Year Book 7
Ex-Directory 7

Articles
Booklet on North/South research
 opportunities
 by Gerry McKenna 8
Universities key to unleashing
 the "Economic Tiger"
 by George Bain 9
Sunset or new dawn?
 by Ian Hughes 10

People
Denis Weaire, F.R.S. 11

Page 11

New President of UCC
 – Professor Gerry Wrixon 12
New Vice-Chancellor of UU
 – Professor Gerry McKenna 13
Jane Grimson, first woman President
 of the Institution of Engineers of
 Ireland 14

Winners
Ophelia Blake *(RIA Biochemistry
 Writing Competition)*
 The prostate - "a man's thing" 15
Oliver Blacque
 (Merville Lay Seminars)
 Understanding how cancer
 cells spread 16
Michael Sharkey
 (Merville Lay Seminars)
 The importance of discovering
 novel proteins 17
Lorraine Maher
 (Merville Lay Seminars)
 Inflammatory bowel disease
 & immune wars! 18

Eamonn O'Donnell
 (RDS Science Communication Forum)
 Age-related neuronal damage
 & its reversal by vitamin
 supplementation 19

Some Science-Based Careers
John Connolly 20
Lillian Cromie 20
Frances Doyle 21
Susan Harrington 21
Nuala Kerley 22
Barbara McCartney 22
Michelle McGarraghy 23
Aileen Moore 23
Jean Moran 24
Liam O'Dwyer 24

Page 23

Education
Department of Education (NI)
 Science in Northern Ireland
 Schools 1989-1999:
 past achievements & future
 challenges 25
**Department of Education
 & Science (RofI)**
 Improving the participation
 levels in the physical sciences 26
The Higher Education Authority
 Higher Education Institutions to
 become world-class Research
 Centres 28
 Allocations announced in July 1999 29
**National Centre for Technology
 in Education**
 Initiatives from the grassroots
 – Schools IT 2000 supports
 innovation 30
**National Council for
 Educational Awards**
 Skills shortages in science-based
 industries in Ireland 30

Science Centres
 & History of Science
Awake! O Leaders and Press
 of Ireland 31
Birr Castle
New Galleries of Discovery 32
St Patrick's College, Maynooth
Well worth a visit 32
AGB Scientific Limited
The Scientific Museum at AGB
 – forging success in the future
 with respect for the past 33
***The ecos - millennium
 environmental centre***
A new approach to promoting
 sustainability for the new
 Millennium 34

Reports from Scientific
 & Technical Societies
***Association of Clinical Biochemists
 in Ireland*** 35
***Industry Research &
 Development Group*** see page 85
The Institute of Physics in Ireland 35
 see pages 200, 208
***Institute of Chemistry
 of Ireland*** see page 203
Institute of Petroleum see page 209
***Institution of Engineers
 of Ireland*** see pages 14 and 205
***Irish BioIndustry
 Association*** see page 80
***Irish Pharmaceutical &
 Chemical Manufacturers'
 Federation*** see page 81
***Irish Research Scientists'
 Association*** 36
***Irish Science Teachers'
 Association*** see page 210
Irish Statistical Association 38
Teachers' Union of Ireland see page 211
Women in Technology & Science 38

Reports from Private Organisations
Royal Dublin Society
The RDS Irish Times
 Boyle Medal Award 39

Page 39

THE IRISH SCIENTIST YEAR BOOK 1999

Short list for the centenary
  Boyle Medal   39
Age-related neuronal damage
  & its reversal by vitamin
  supplementation   see page 19
Royal Irish Academy
Awards & Fellowships available
  to Irish scientists   40

Reports from Forfás & ICSTI
Forfás
Science, Technology & Innovation
  Awareness Programme continues
  through 1999   41
Knowledge as a Development Factor
  – Technology Foresight Ireland   44
**Irish Council for Science,
Technology & Innovation**
Irish Council for Science,
  Technology & Innovation   42

Reports from Enterprise Ireland & Programmes in Advanced Technology
Enterprise Ireland
Campus Companies Programme   45
EU Fifth Framework Programme
  for R&D   45
Developing & delivering leading
  edge technologies for industry   46
AMT Ireland   see page 169
BioResearch Ireland
BioResearch Ireland   48
Materials Ireland
R&D - the future of the
  healthcare sector   52
The Polymer Development Centre
  – The Centre of Polymer Expertise   53
Materials Ireland Polymer Research
  Centre, Physics Department, TCD   53

Reports from Commercial State and Semi-State Organisations
Bord na Móna
Research & Development in Bord
  na Móna Horticulture Limited   54
On-line radiometric analysis
  of peat   see page 185
Coillte Teoranta
Satellite Imagery brought down
  to earth in Ireland's forests   55

Reports from State and Semi-State Agricultural, Fishery & Forestry Organisations
**Department of Agriculture
for Northern Ireland**
Scientific advances in food quality
  & safety – Food Science at
  Newforge Lane, Belfast   56
Silvopasture - a new land use   57
**Forest Service, Department of
Marine & Natural Resources**
Satellite Imagery brought
  down to earth in Ireland's
  forests   see page 55
Teagasc
Agri-environmental indicators   57
Food research - the key to
  competitiveness   58
Grass-fed beef:
  a natural health food!   61

Distribution pattern of
  supplementary concentrates
  for finishing cattle   62
Enhancing the texture & sensory
  quality of meat products – use of
  high pressure treament   63
Extracting residues from food
  with carbon dioxide   63
Eating quality of deep-water
  fish species   64
Atmospheric change studies
  at Oak Park   see page 149
**The Veterinary Laboratory
Service**
The Veterinary Laboratory Service   64

Page 66

Central Fisheries Board
Extending the salmon's range   66
Algal control in waterways
  using barley straw   67
COFORD
Forestry research in the assessment
  of tree health from nursery
  to field   65

Reports from State & Semi-State Bodies
An Bord Glas   see page 206
Environmental Protection Agency
Water quality in Ireland 1995-1997   68
Innovative Web site   68
Food Safety Authority of Ireland
The Prevention of E. coli O157:H7
  Infection – A Shared
  Responsibility   69
                              see page 202
Forensic Science Laboratory
Forensic Science Laboratory   73
Geological Survey of Ireland
The seabed project   76
Protecting our geological heritage   76
                              see page 212
**Geological Survey
of Northern Ireland**
Landscapes from stone   77
Health Research Board
HRB Research Project grants 1999   70
Prevalence of Hepatitis B virus
  in the Republic of Ireland   72

Depression in the community
  dwelling elderly: no longer the
  "common cold" of psychiatry   72
Susceptibility genes for
  Attention Deficit Hyperactivity
  Disorder   see page 79
Immunophysiology of inflammatory
  bowel disease   see page 165
**Information Society
Commission**   see page 208
**The Irish National Accreditation
Board**
The Irish National Accreditation
  Board   74
Marine Institute
Into deeper waters   77
Into the Age of the Ocean...   78
                              see page 207
**National Rehabilitation
Board**   see page 205
North Eastern Health Board
Prevalence of Hepatitis B virus
  in the Republic of Ireland see page 72
**Radiological Protection Institute
of Ireland**
The effect of Sellafield discharges
  on harbour porpoises   75
The Patents Office   see page 211
The State Laboratory
Genetically modified (GM) foods   73

Reports from Hospitals
Beaumont Hospital
Ultraviolet radiation & the skin   79
The Mater Hospital
The prostate
  – "a man's thing"   see page 15
St James's Hospital
Susceptibility genes for Attention
  Deficit Hyperactivity Disorder   79
See also
Depression in the community
  dwelling elderly: no longer
  the "common cold"
  of psychiatry   see page 72
St Patrick's Hospital
Depression in the community
  dwelling elderly: no longer
  the "common cold"
  of psychiatry   see page 72
St Vincent's University Hospital
Liver lymphocytes: new insight
  into liver disease   80

Reports from Industrial & Commercial Firms & Organisations
AGB Scientific Limited
The Scientific Museum at AGB
  – forging success in the future
  with respect for the past see page 33
Ashling Microsystems Limited
Pan-European R&D yields new
  developments, new opportunities
  in Microprocessor Design
  Equipment   86
Biotrin International Limited
The 4th RTD Framework
  Programme   87
Cadbury Ireland Limited   see page 21
Cell Media Limited   see page 170
Changing Worlds Limited see page 170
City Analysts Limited   see page 176

William Clark & Sons see page 84
Danu Industries see page 108
Foodlink R&D Europe see page 108
FSL Electronics see page 82
Golden Vale plc
Novel Trans-European project
 on fish oil incorporation into
 functional foods 87
Guinness Research & Development
Biosensing to ensure
 the perfect pint 88
Industrial Research
 & Technology Unit 82
Industry Research
 & Development Group 85
Intel California see page 24
Intel Ireland see page 203
Intelligent Biomaterials
 Limited see page 170
Irish BioIndustry Association 80
Irish Business & Employers'
 Confederation see page 208
Irish Pharmaceutical & Chemical
 Manufacturers' Federation 81
The Irish Times see page 39
LAKE Communications
Broadband urban rural based
 open networks 88
Loctite (Ireland) Limited see page 24
Card technology for a smarter,
 faster world 89
Magnetic Solutions see page 155
Merck Sharp & Dohme
 (Ireland) Limited
Merck Sharp & Dohme – A global
 company in a local community 90
MINEit Software Limited see page 189
Nanomat Limited see page 170
Norbrook Laboratories
 Limited see page 20
Nortel Networks
Nortel Networks are at the forefront
 of everything that's exciting in
 the fast-moving world of
 telecommunications 93
Secure File Transfer over
 the Internet (SFTI) see page 134
University of Ulster's Applied
 Research Centre see page 190
Nycomed Ireland Limited see page 21
Pfizer Pharmaceutical Production
 Corporation see page 22
Programmable Systems see page 108
Radiocontact Limited see page 82
RFT Vision Systems see page 108
Roche Ireland Limited see page 23
Rural Generation Limited see page 83
Scientific Systems Limited see page 45
Shaw Scientific Limited see page 209
SIFCO Turbine
 Components Limited see page 20
Laser cladding development for
 the gas turbine industry 91
Valpar Industrial Limited see page 84
WBT Systems Limited see page 170
Yamanouchi Ireland see page 198

Reports from Academic & Teaching Institutions

Armagh Observatory
Cannibal white dwarf creates
 trembling giant 95

Page 94

Dublin Institute for Advanced Studies
Forty years of experimental
 particle research at DIAS 94
*Greenmount College of Agriculture
 & Horticulture*
The Internet – learning for life 96
*Royal College of Surgeons
 in Ireland* see page 79
Psychosocial contributions
 to illness & healthcare 96

Reports from Institutes of Technology

Athlone
Ecotoxicology Research at
 Athlone Institute of Technology 98
Cleaning validation, how clean
 is clean? 98
Developing the Irish Business
 Excellence Model for an
 educational environment 99

Carlow
Biotechnological & Environmental
 Sciences at IT Carlow 100
Biochemistry & biotechnological
 applications of amino acid NAD(P)+-
 dependent dehydrogenases 100
Antibiotic resistant microorganisms
 in the River Barrow 101
Directed evolution of bacteria for
 the detoxification of pollutants 101

Cork
Sustainable development in Ireland
 – all talk and no action? 102
Probing the hearts of quasars 103
"Smooth & Shiny" 103
CIT develops virtual reality milling
 machine for interactive training 104
Electronic analysis & design
 for industry 104

Dublin
Pervaporation for waste
 management 110
Dublin Institute of Technology
 Spectroscopic Facility 110
Centre for Industrial
 & Engineering Optics 111
Optical sensor research 111
Power estimation at high levels
 of integrated circuit design 112

Identification & control
 of delayed systems 112
 see pages 205, 212
Dundalk
Research strategy & expertise at
 Dundalk Institute of Technology 106
Low fat poultry products 106
The behaviours of edge panels
 in reinforced concrete flat slab
 structures 107
A chilled fibre optic humidity
 sensor 107
Supporting knowledge-based enterprise
 in the North-East Region 108
Supporting entrepreneurship at
 Dundalk Institute of Technology 109

Dun Laoghaire see page 10
Science & Technology come to
 Dun Laoghaire 97

Galway-Mayo
Marine microalgae as a source
 of ω3 fatty acids 114
The periwinkle, a perfect model
 for biodiversity studies 115
Sewage detection in seawater
 – the detector needs a boost 116
Meat quality assurance in GMIT 116
Mathematical modelling in
 science & engineering 117
Minimax & applications 117

Letterkenny
Research Colloquium at Letterkenny
 Institute of Technology 105
The use of Middleware in
 E-Commerce applications 105

Limerick
Making sense of water pollution 113
A framework for migrating to
 distributed object computing 113
The diffusion of gases into
 polymer films see page 119

Sligo
Clean up of oil spillages
 using biosurfactant 99

Tallaght
Analytical probes for the future 118
Ultrasound propagation in wood
 – developing quality assessment
 systems 118

Tralee
Institute of Technology, Tralee
 – a Centre for Applied Research
 & Product Development 120
The SHaPE Centre
 – tomorrow's sports stars to-day 121

Waterford
The diffusion of gases into
 polymer films 119

Reports from Universities

Dublin City University
Initiatives from the grassroots
 – Schools IT 2000 supports
 innovation see page 30

Page 137

National Cell & Tissue Culture
   Centre   *see page 50*
Molecular recognition
   & chemical sensors   121
Rapid antibody-based analytical
   methods: tolls for detection of
   food, drug & environmental
   contaminants   122
Solitons in field theory   123
The chemistry of pure hydrogen   123
Creating a plasma   124
Electronic Engineering research
   highlights at DCU   124
Indexing and retrieval of
   digital video   125
Intelligent software project
   management   125
Sensing knee injury   126

National University of Ireland, Galway
National Diagnostics Centre *see page 48*
Martin Ryan Institute   *see pages 127-8*
The Centre for Health Promotion
   studies, NUIG   126
Measuring water clarity from space 127
Keeping an eye on the grass
   of the seas   127
Growing a new sustainable
   sea-vegetable in Ireland   128
Helping seabed life   128
Laser Centre expands to meet
   new needs of Industry   129
Physics research at NUI Galway   130
Biomedical Science & Biomedical
   Engineering at NUI, Galway   132
Biomaterials science
   – a challenge for the future   133
Air Quality Technology Centre   133
Secure File Transfer over the
   Internet (SFTI)   134
Rapid PDM   134
Geoscience probes mantle secrets   135

National University of Ireland, Maynooth
Well worth a visit   *see page 32*
The shape on the map   135
Faculty of Science, NUI Maynooth 136
Quarks, symmetry
   & cold electrons   137

Mathematics, modelling
   & medicine   137
Probing the Early Universe
   with the PLANCK Surveyor   138
Electrochemistry at NUI Maynooth 138
Parvovirus B19 infection
   – an underestimated problem   139

The Queen's University of Belfast
Universities key to unleashing
   the "Economic Tiger"   *see page 9*
Scientific advances in food
   quality & safety – food
   science at Newforge Lane,
   Belfast   *see page 56*
Silvopasture
   – a new land use   *see page 57*
The behaviours of edge panels
   in reinforced concrete flat
   slab structures   *see page 107*
Air turbine wave energy
   plants   *see page 186*
Research at The Queen's University
   of Belfast on reducing life
   cycle cost   139
150 years of Engineering
   excellence at Queen's, Belfast   140
Image & vision systems   141
Queen's chips in with new designs 142
The UK Multiphoton & Electron
   Collisions HPC Consortium   142
Chasing solar eclipses at Queen's 143
The QUILL Research Centre
   – rewriting the future of solvents 143
Oxidative metabolites or aromatic
   hydrocarbons: chemistry between
   QUB and UCD   144

Trinity College Dublin
Denis Weaire, F.R.S.   *see page 11*
Age-related neuronal damage
   & its reversal by vitamin
   supplementation   *see page 19*
Awake! O Leaders
   and Press of Ireland   *see page 31*
National Pharmaceutical
   Biotechnology Centre   *see page 50*
Materials Ireland Polymer
   Research Centre, Physics
   Department, TCD   *see page 53*
Depression in the community
   dwelling elderly: no longer
   the "common cold" of
   psychiatry   *see page 72*
Susceptibility genes for
   Attention Deficit Hyperactivity
   Disorder   *see page 79*
The TCD Enterprise Centre   145
The School of Pharmacy at TCD   145
The TCD Department of
   Psychology   146
Researching children's experiences:
   parental separation, homelessness
   foster care, & prevention of early
   school leaving   146
Medicinal Chemistry at TCD   147
Exercise ability with obesity   147
The Ocular Genetics Unit, TCD   148
Inflammation & cancer   148
Atmospheric change studies
   at Oak Park   149
New signal processing techniques
   for use in medicine   150

High Performance Computing
   & Bioinformatics   151
Mathematics & the knowledge
   economy   152
Origin of the proton's spin   153
Lattice QCD   153
TCD Department of Physics   154
Physicists take a look at foam   155
Attractive materials   155
Lasers the size of a photon   156
Geotechnical research at TCD   156
Research projects in Mechanical
   & Manufacturing Engineering
   at TCD   157

University College Cork
New President of UCC   *see page 12*
National Food Biotechnology
   Centre   *see page 49*
The effect of Sellafield discharges
   on harbour porpoises   *see page 75*
Silicon sensors and sensor
   systems at NMRC   157
Programming approaches subsumed
   by new model of computing   158
Institute for Non-Linear Science
   at UCC   159
Probing volcanic dust from the past 160
The Caledonian granites of Ireland
   & Scotland: where why & how? 161
Optimising the aquaculture
   production of turbot & halibut 161
Atmospheric chemistry at UCC   162
NMR spectroscopy for chemistry
   research & teaching   162
UCC's Department of Anatomy   163
Regulating cell survival & lifespan 163
Sensing membrane stress   164
Chinese takeaway   165
Immunophysiology of
   inflammatory bowel disease   165
Food Science & technology for
   the 3rd Millennium   166
Oral health, fluoride toothpaste
   & fluorosis: information based
   planning for Europe   168
Instrumental flavour research   169

University College Dublin
The prostate
   – "a man's thing"   *see page 15*
Understanding how cancer
   cells spread   *see page 16*
The importance of discovering
   novel proteins   *see page 17*
Inflammatory bowel disease
   & immune wars   *see page 18*
National Agricultural & Veterinary
   Biotechnology Centre   *see page 51*
Forestry research in the
   assessment of tree health
   from nursery to field   *see page 65*
Liver lymphocytes: new insight
   into liver disease   *see page 80*
Oxidative metabolites or aromatic
   hydrocarbons: chemistry
   between Queen's University
   & UCD   *see page 144*
Chinese takeaway   *see page 165*
AMT Ireland   169
University Industry Programme
   – support for innovation
   & technology transfer at UCD 170

Conway Institute of Biomolecular & Biomedical Research	171	
Broad spectrum of activities at UCD's Department of Zoology	172	
Medicinal chemistry & carbohydrates at UCD	173	
Genetic modification of wheat & food safety	173	
Nutrition & fertility: a dilemma for the modern dairy cow	174	
Endocrine disrupting compounds & reproductive problems	174	
Some beneficial effects of Yucca plant extracts in sheep and other domestic animals	175	
Cardiovascular disease – the missing link	176	
Water Quality: trouble in the pipe-line?	176	
The imperative to green design	177	
Centre for Water Resources Research	177	
Advanced internet applications at the Smart Media Institute, UCD	178	
Hypertext navigation on the world wide web	179	
Cortical software reuse: a model of large-scale cortical computation	179	

University of Limerick
Academia-Industry research links at the University of Limerick	180
Bioscience & Technology at UL	182
Materials & Surface Science Institute	183

Page 184

LiteFoot: a smart dance floor or a new musical instrument	184
Communications & electronics research at UL	185
On-line radiometric analysis of peat	185
Air turbine wave energy plants	186
Pore development in cold drawn PET fibres	186
Fundamental motor skill development	187
Equine science research: technical developments for an indigenous industry	187
Molecular technology in action – a tale of two ß-Lactams	188
Centre for Applied Mathematical Sciences, UL	188

University of Ulster
Booklet on North/South research opportunities	see page 8
New Vice-Chancellor of UU	see page 13
Biosensing to ensure the perfect pint	see page 88

Data mining made easy	189
University of Ulster's Applied Research Centre	190
Centre for Sustainable Technologies (CST)	192
Fire safety engineering research & technology transfer	192
UUTECH	193
The Northern Ireland Centre for Energy Research & Technology	193
Environmental toxicology at the UU – a new initiative	194
The Northern Ireland Centre for Diet & Health (NICHE)	194
Vision Science looks to the future in Coleraine	195
Probing molecular & cellular aspects of diabetes	196

Page 196

Biotechnology research at the University of Ulster	196
Investigating earthquakes	197
Maritime archaeology	197

Youth Science
In praise of our Young Scientists	198
An industry-school links programme	198
Cryptography - a new algorithm versus the RSA	199
Wild mushroom – *Hydnum repandum*	199
Parasites & propulsion – or when the experts get it wrong	200
A modified water barometer	200
Hydration practices in sport	201
Vegetricity – the power of the future	201
"Gaeilgeoirí nó déagóirí" – studying the students who attend Irish medium schools	202
Food borne diseases – are you at risk?	202
Curve representation of digitised objects	203
Slimeocity	203
Parental involvement in primary education	204
Operation deceleration: an inquiry into speeding – who speeds & why?	204

Alzheimers – cabhair teicheolaíocht	205
Programming & robotics for primary schools	205
Spotlight on *Elodea*	206
Natural food colourings extracted from carrots	206
Starfish, predators of shellfish	207
A microbiological analysis of Bundoran's bathing waters	207
Science thrills without the bills – a website for National Schools	208
Low cost 3-D scanner: virtual reality?	208
Mop it up!	209
Telephone home link	209
Heavy metal ions in plants	210
Decomposition in coniferous & deciduous woodland soil	210
The self-repairing wheel	211
Cow pat power	211
Geomorphology of river valleys in Louth	212
Mixing dynamics in Killone Lake	212

Schools
Abbey Vocational School, Donegal	199
Aquinas Grammar School, Belfast	200
Balinteer Community School	209
Belvedere College, Dublin	208
Bundoran Vocational School	207
Coláiste Choilm, Ballincollig	210
Coláiste Eoin, Stillorgan	205
Coláiste Ráithín, Brí Cualann	202
Coláiste Ris, Dun Dealgan	212
Coláiste Spioraid Naoimh, Cork	200
Coleraine Academical Institution	205
Convent of Mercy, Roscommon	204
Coolmine Community School, Dublin	198
Gael Choláiste an Chláir, Ennis	212
Kildysart Vocational School	204
Loreto College, St Stephen's Green, Dublin	208
Methodist College, Belfast	210
Mount Mercy College, Cork	202
Our Lady's College, Drogheda	203, 209
Pobalscoil Ghaoth Dobhair, Tír Chonaill	201
St Francis College, Rochestown	206
St Joseph's Community College, Kilkee	211
Salerno Secondary School, Galway	207
Sandford Park, Dublin	201
Scoil Mhuire Gan Smál, Blarney	199, 203
Sutton Park School	206
Terenure College, Templeogue	211

Page 203

THE IRISH SCIENTIST YEAR BOOK 1999

Samton books in print

  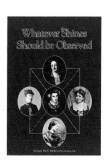 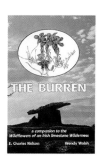 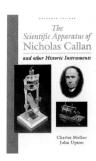

SAMTON HISTORICAL STUDIES

No. 1 – Campanology in Ireland
by Fred Dukes, ISBN 1 898 706 00 X (HB), £15, 256 pages, 1994. The result of a lifetime's study by the doyen of Irish bell ringing – a detailed inventory of the old bells of Ireland, with introductory chapters, list of bell founders, etc.

No. 2 – The Chemical Association of Ireland 1922-1936
By J. Philip Ryan, ISBN 1 898 706 07 7 (HB), £15, 240 pages, 6 black & white illustrations, 1997. The story of the unsuccessful attempt of Irish chemists to set up a Chemical Association with statutory powers in the Free State. Contains also a 38 page outline, by Charles Mollan, of the history of Irish chemistry from 1660 to 1936.

No. 3 – Whatever *Shines Should be Observed
By Susan McKenna-Lawlor, ISBN 1 898 706 14 X (HB), £10, 136 pages, 30 black & white illustrations, 1998. Biographies of Irish women scientists: Mary, Countess of Rosse, 1813-1885; The Hon. Mrs Mary Ward, 1827-1869; Agnes Mary Clerke, 1842-1907; Ellen Mary Clerke, 1840-1906; Lady Margaret Huggins, 1848-1915.

WILDFLOWERS

The Burren – A Companion to the Wildflowers of an Irish Limestone Wilderness
By E. Charles Nelson and Wendy Walsh, ISBN 1 898 706 10 7 (SB), £25, 343 pages, 103 colour and 32 black & white illustrations, 1997. The second edition of this highly acclaimed guide.

HISTORIC SCIENTIFIC INSTRUMENTS

The Scientific Apparatus of Nicholas Callan and other historic instruments
by Charles Mollan & John Upton, ISBN 1 898 706 01 8 (HB), 1 898 706 02 6 (SB), £25 (HB), £15 (SB), 304 pages, 385 entries, 390 photographs, 1994. Comprehensive catalogue of the important collection of historic instruments at St Patrick's College, Maynooth – Nicholas Callan invented the induction coil there in 1836.

The Mind and the Hand – instruments of science 1685-1932
by Charles Mollan, ISBN 1 898 706 03 4 (SB), £3, 64 pages, 144 illustrations, 1995. Highly illustrated catalogue of an exhibition held in the Colonnades Gallery (on the way to the Book of Kells) at Trinity College, Dublin, throughout 1995.

Irish National Inventory of Historic Scientific Instruments
by Charles Mollan, ISBN 1 898 706 05 0 (HB), £50, 5104 records, 501 pages, 1995 (only a few copies left). The result of ten years of travelling throughout the country (North & South) collecting and collating records – the first comprehensive National Inventory for any country in the World.

THE IRISH SCIENTIST

The Irish Scientist – 1994-1995
ISBN 1 898 706 13 1 (HB), 1997, £10, 32+32+64 pages (plus covers), 217 colour and 35 black & white illustrations. Hard bound edition of the first three issues of *The Irish Scientist*.

The Irish Scientist – 1996 Year Book
ISBN 1 898 706 08 5 (HB), 1 898 706 09 3 (SB), £6 (HB), £3 (SB), 104 pages, 190 colour and 11 black & white illustrations.

The Irish Scientist – 1997 Year Book
ISBN 1 898 706 11 5 (HB), 1 898 706 12 3 (SB), £8 (HB), £4 (SB), 136 pages, 254 colour and 10 black & white illustrations.

The Irish Scientist – 1998 Year Book
ISBN 1 898 706 15 8 (HB), 1 898 706 16 6 (SB), £10 (HB), £5 (SB), 180 pages, 360 colour and 23 black & white illustrations.

The Irish Scientist – 1999 Year Book
ISBN 1 898 706 18 2 (HB), 1 898 706 19 0 (SB), £12 (HB), £6 (SB), 212 pages, 430 colour and 26 black & white illustrations.

* *

PLEASE ADD POSTAGE COSTS

(Surface mail only – for Air Mail please enquire)
Ireland: First £10 – p+p £3;
More than £10 – p+p £4
GB: First £10 – p+p £4; £10-£20 – p+p £5;
More than £20 – p+p £7
Rest of World: First £10 – p+p £4; £10-£20 – p+p £5;
More than £20 – p+p £10

All available from Samton Limited, 17 Pine Lawn, Blackrock, Co. Dublin;
Tel: 01-289-6186; Fax: 01-289-7970; E-mail: cmol@iol.ie

An exciting time to be a scientist

The Irish Scientist was founded in 1994 in order to provide a vehicle for scientists to tell the world what they are doing. (I consider engineers and technologists to be scientists too – though not all seem to agree.) At the time scientists were feeling hard done by. They knew that they were doing useful work, and that they could do a great deal more and better with the right funding and infrastructure, but nobody seemed to be interested. What could be done to improve the situation? There were no media science correspondents, virtually all politicians and senior civil servants had been educated in an arts/humanities tradition, and the reaction to the confession "I am a scientist" was a glazed expression and some comment like "I'm afraid I don't know anything about science".

If you are in a commercial company, and you are convinced that you have a wonderful product which will transform peoples' lives, but it is not selling, what do you do? You tell the world. If the public still don't buy, then you probably have a duff product. But if your product is indeed excellent, it will sell once people know about it, and you will prosper. By analogy, there was a need for scientists to project a positive and dynamic image. How to do it? *The Irish Scientist*, which became a *Year Book* in 1995, was born.

Of course, many scientists did not recognise the need. There was confusion between publication and publicity. More and more, scientists were publishing their results in ever more specialised journals. These did, indeed, reach the select band of specialists in that particular area. But the specialist journals were not read by others outside that specialisation. Very few people knew about the quality of even the best of scientific research and development going on in the country. There was publication, but no publicity.

Fortunately an enlightened band of forward thinking scientists supported the idea inherent in *The Irish Scientist*, and the project got underway. (This is, of course, an unbiased comment!) From small beginnings, the idea is catching on *(see graph)*.

While *The Irish Scientist* can, and does, claim only some of the credit, there is no doubt that the climate for science has been transformed in the last five years. **It is an exciting time to be a scientist.** This is not to say that everything is perfect. But there is now widespread recognition that science and technology are critical to the economic and social welfare of the country. More funds are being directed to basic and applied research, and there are significant developments in collaborative research – between Northern Ireland and the Republic of Ireland *(see especially page 8)*, between Ireland and Great Britain, between Ireland and Europe, and between Ireland and the rest of the world. Just leaf through the pages of *The Irish Scientist 1999 Year Book* and pick out the number of co-operative projects. It is impressive and growing all the time.

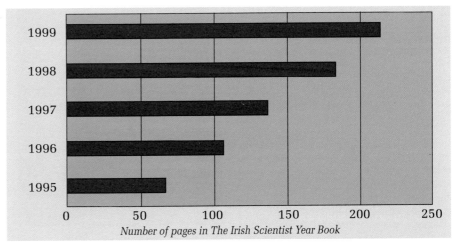

Number of pages in The Irish Scientist Year Book

There are still problems. For example, the editorial of *The Irish Scientist 1997 Year Book* highlighted the fact that fewer and fewer young people in both Northern Ireland and the Republic of Ireland were studying chemistry and physics in school. This problem is now receiving serious attention, as will be seen in several articles in this *Year Book (see especially pages 25-27)*.

The Irish Scientist continues to play its part. In a Special Feature, school pupils who entered projects in Esat Telecom Young Scientist and Technology Exhibition in January, and won prizes there, are featured *(pages 199-212)*. We hope this will encourage other school pupils to experience the excitement of scientific discovery. We also give brief biographies of young professional scientists who are enjoying the opportunities which their scientific training has presented to them *(pages 20-24)*. Our thanks are due to them and to the companies which supported their contributions. These biographies weren't collected in a very systematic way. Various people were asked to nominate suitable people. As will be seen, a surprising number suggested young and vigorous female PhDs. Is this a pointer to the future in Irish science?

Anyway, it is good to acknowledge with many thanks all those who have supported this issue. Pride of place must go to the **Forfás Science, Technology and Innovation Awareness Programme**, which, once again, provided generous sponsorship. Many others have given support, by sponsorship or participation or both. There has been a dramatic increase in contributions from Institutes of Technology in the Republic of Ireland, reflecting the increasingly impressive role they are playing in research, development and training both locally and nationally; and also a substantial increase in contributions from commercial firms engaged in research and development.

It is an exciting time to be a scientist. Let our young career scientists have the last word:

- "These are exciting times....What better time to get involved in science and have a part to play in the new innovations of the future" (Nuala Kerley – *page 22*).
- "It is now a very exciting time to be involved in a technological industry in Ireland. Irish graduates are respected globally as highly educated, competent people and are well positioned to contribute meaningfully well into the future" (Liam O'Dwyer – *page 24*).
- "This is a very exciting time for those involved in the pharmaceutical environment in Ireland. Irish graduates are.... competent people and have the confidence and ability to face the challenges of the future in such a competitive arena" (Lillian Cromie – *page 20*).

Sequence of articles in the *1999 Year Book*

In recent years, articles in *The Irish Scientist Year Books* have been arranged largely under subject content. However, this was becoming more and more difficult, as inter-disciplinary topics are very much the order of to-day's science. So this time, we are arranging the articles by institution or organisation. Feedback on whether this is an improvement or a dis-improvement would be welcome, as indeed would any helpful criticism about the publication.

Ex-Directory

Due presumably to "human error" (other less polite phrases spring to mind), the telephone and fax numbers for **Samton Limited**, publishers of *The Irish Scientist Year Book*, have been omitted from Business Listings of the *01 Phone Book 1999-2000*. They **were** listed in the 1998-1999 edition. (In case you might have wondered, Samton stands for "**S**cience **A**ncient and **M**odern, **T**echnology **O**ld and **N**ew" – the history of Irish science is of particular interest to the Company, as of course is the science of today.)

The omission from the telephone book is hardly helpful to a small business with ambitions to grow. (The telephone number *is* listed under Dr R. Charles Mollan in the Residential Listings.) So please make a special note of the Samton Limited contact details, in case you might wish to get in touch: **Tel: 01-289-6186; Fax: 01-289-7970; E-mail: cmol@iol.ie**

Booklet on North/South research opportunities

The University of Ulster has an outstanding record of collaboration at national and international level. This is reflected in its many joint research projects with leading research centres overseas, and the award of nearly £20 million in European Union funds for research (all involving collaborations) over the past five years. The University's commitment to technology and knowledge transfer is exemplified by its ten applied research centres ranging from biotechnology, bioengineering, and diet and health to energy, sustainable technologies, advanced materials, and knowledge-based systems. The University is a leading university in the UK in terms of participation in the Teaching Company Scheme which is designed to solve problems of direct relevance to specific companies. The University is no less committed to social and culturally relevant research, examples of which include centres in nursing, higher education, voluntary action, conflict resolution, and language and literature.

The University has a wide range of collaborative links with all the other Irish Universities and with a number of government agencies in the Republic. On learning of the commitment, in the Good Friday Agreement 1998, to closer co-operation between Northern Ireland and the Republic of Ireland in a number of specified areas, the University of Ulster anticipated that many of its existing links could form the basis for major collaborations. The University has recently published a booklet entitled *The Agreement: North/South Research Opportunities**.

The booklet outlines the University's current research interests, some of the existing North/South research collaborations, and scope for further collaborative activity in the six "areas for co-operation":
- Transport
- Agriculture
- Education
- Health
- Environment
- Tourism.

Looking ahead to the topics and issues likely to confront the cross-border "implementation bodies", UU has also outlined current and potential co-operation in the following areas, in which there exists substantial expertise:
- Arts/Culture
- Management Development
- Economics/Financial and Banking Services
- Conflict and Conflict Resolution
- Voluntary Sector Research.

Full details of UU's academic and research management structures are also given.

Professor Gerry McKenna, the University's Pro-Vice-Chancellor with responsibility for research**, intends the booklet to act as an aid to politicians and government officials and to be a source book of potential areas for further discussions.

Professor Gerry McKenna, Pro-Vice-Chancellor (Research) at UU**.

Some key collaborative activity in Science and Engineering

One of UU's areas of expertise is in nutrition, through the Northern Ireland Centre for Diet and Health (NICHE), the work of which features in the University's top rated Biomedical Sciences research group (1996 UK Research Assessment Exercise). NICHE is part of the Irish Universities Nutrition Alliance (IUNA), a formal association of the nutrition units at UU, TCD and UCC. The Alliance is committed to developing joint initiatives in education and research and a range of projects with implications for the production and marketing of healthy foods. For example, the North-South Dietary Survey of Ireland, funded by the Irish government and the food industry, is the first study to examine the food and nutrient intakes of people living in the island of Ireland.

Among the collaborative work outlined under the heading of "Agriculture", are major projects related to forest and pest management and silviculture, pointing towards scope for major R&D all-Ireland projects on forest plant protection.

Other Environmental Studies work involving north-south collaboration and potential is to be found in the areas of lake management and water quality, wildlife conservation, land cover change and biodiversity, the environmental effects of agriculture, waste management, quaternary sediments, sand and gravel, inland fisheries, and marine matters – that is coastal geomorphology, coastal dynamics, coastal zone management and maritime archaeology. UU has recently established a Centre for Maritime Archaeology, which presents exciting possibilities for all-Ireland research.

In the context of European Union funded projects, UU's work in the Northern Ireland Centre for Energy Research and Technology (NICERT) is well known, and offers potential for north-south collaborative projects to encourage the use of renewable energy and to improve energy efficiency.

In UU's Faculty of Engineering there are particular interests in energy use in buildings, innovative energy-saving building components, advanced glazing and fire structure systems, and solar energy systems, both thermal and photovoltaic. Work is already being undertaken in collaboration with NUIG, and potential exists for strategic alliances with other ROI universities.

Also in Engineering is the Transport and Road Assessment Centre (TRAC), which is an umbrella organisation for research groups in transport, logistics, mobility, utility technologies and highway engineering. The Highway Engineering Research Group has been involved in North/South co-operation with TCD in identifying sources and developing new products for infrastructural engineering purposes.

Waterways are of course an important topic of research from a variety of perspectives, and offer great potential for tourism both north and south of the border. In UU, research is ongoing in relation to hydraulic behaviour of rivers with flood plains and hydraulic characteristics of environmental features in rivers, as well as on the scope and impacts of tourism and recreation on waterways. There are opportunities for collaboration with ROI partners on shared problems relating to enhancement of the environment as well as to the development of tourism, while the sustainable development of river catchments is a crucial international issue.

As UU is well known as a major provider of healthcare education, it is not surprising that the research programme is particularly strong in this area – specifically in Biomedical Sciences, Nursing, Health Informatics including Biomedical Engineering, Rehabilitation Sciences, and Life and Health Technologies. A range of projects with ROI partners is ongoing – collaborations such as the Centre for Innovation in Biotechnology (involving UU, QUB and BioResearch Ireland), and the BEST Centre (that is, Biomedical and Environmental Sensor Technology) involving also DCU, UL and QUB, have now become well known as a focus for cross-border research.

Current activities could be developed even more widely across the full range of healthcare services North and South, and might even include Clinical Trials. In addition, expertise in Medical Informatics, North and South of the border, could be developed to mutual benefit.

---

*Any readers interested in obtaining a copy of the University of Ulster's booklet on North/South research opportunities should contact the University's Research Office at the Coleraine Campus; Tel: 01265-324124; Fax: 01265-324905; E-mail: nc.dallat@ulst.ac.uk. Likewise any enquires about possible collaboration may be directed to the Research Office or to the appropriate Faculty, School or Research Group (these are all outlined in the booklet and in the University's worldwide web site http://www.ulst.ac.uk).
**Prof McKenna has been appointed Vice-Chancellor of UU (see page 13)

Universities key to unleashing the "Economic Tiger"

The role of universities in society has prompted intense debate throughout their history, and increasingly so during the last 50 years.

In essence, this discussion has led to the evolution of two main philosophies. The first stresses learning for learning's sake, while the second contends that this knowledge should be applied in tangible form for the greater good of society.

These two viewpoints are not necessarily contradictory. It can quite easily be argued that the application of learning and research to the needs of society is a natural progression in the higher education process, and even a moral imperative.

There is no doubt that, while it is widely agreed that the primary duty of any university is to generate, transmit and advance knowledge through teaching and research, the focus of attention as we approach the 21st century has sharpened to centre on the relevance of universities to the wider society.

There are a number of reasons for this, but two of the most pertinent are the need for accountability to the tax-payer, and the sheer volume of the practical demands on higher education as the pace of technological change has quickened during the last few decades.

The leading American academic and writer, John Brubacher, crystallised the evolving role of universities in modern society, when he said:

> *Just to understand, let alone to solve the intricate problems of our complex society would be next to impossible without the resources of college and university. Government, industry, agriculture, labour, raw materials, international relations, education, health and the like – once solved empirically – now demand the most sophisticated expertise. The best place to procure such expertise and people trained in its use is in our higher institutions of learning.*

Certainly, the advent of the post-industrial age and the development of the knowledge-based economy means that universities are in a stronger position than ever before to contribute to the wider community by aiding economic development.

At Queen's, while we are well aware of this, we are also acutely conscious of the need for increased funding for basic research in Northern Ireland to sustain long-term economic development.

This is critical if we are to develop a high-technology-based economy, especially since we have a high number of small and medium-sized enterprises which have limited capacity to fund research and development.

Queen's has a long history of close co-operation with local business, industry and key government agencies to provide practical assistance in economic development and to help attract inward investment for the benefit of everyone in Northern Ireland.

Our research already has a positive impact on the local economy; for example, research excellence in the Faculty of Engineering has had a significant influence on the attraction of key inward investors such as Seagate in Londonderry and BCO Technologies in West Belfast.

But we want to build on this. We aim to make Queen's a networking university engaging in co-operation with public and private sector organisations. We want to continue to provide the cutting-edge technological support needed by inward investors to Northern Ireland, as well as catering for the needs of local fledgling companies.

There has been some good news in these respects recently. In March Northern Ireland won £2 million to help turn academic research into wealth-creating business opportunities.

The funding was awarded by the University Challenge Fund, a partnership set up last year by the British Government, the Wellcome Trust and the Gatsby Charitable Foundation. The award will be supplemented by a further £500,000 from Queen's and £250,000 from the University of Ulster.

The Northern Ireland bid, led by Queen's and in partnership with the University of Ulster, will provide the opportunity and the resources to exploit commercially the respective research strengths of both institutions across a broad scientific and technological base.

This award, which will help us keep world-beating ideas – and the wealth they create – in Northern Ireland, is a clear signal that an enterprise culture is thriving in the university sector.

Another important building block in Northern Ireland's economic regeneration is the plan for a science park to be established by Queen's, the Department of Economic Development and the University of Ulster.

The impact of this initiative will be felt right across Northern Ireland – not just in Belfast. The experience elsewhere is that science parks attract inward investment to the whole region, as well as stimulating home-grown technology-based businesses.

To date, Queen's has created 20 companies and some 350 jobs by recognising the potential of its research, and investing in it. We have already demonstrated that our strong research base, together with the spirit of enterprise, provides a perfect climate for economic growth. We have proved that we can make a difference.

But we must continue to increase our investment in high-quality fundamental and strategic research. The economic survival of Northern Ireland in a competitive world requires us to harness and exploit the top flight skills and knowledge available in our universities.

Research and development is the key which will release Northern Ireland's economic tiger from its cage.

---

* Professor George Bain is Vice-Chancellor of The Queen's University of Belfast.

Sunset or a new dawn?

Go to the people
Live amongst them
Start with what they know
And when the deed is done,
The mission accomplished
Of the best leaders
The people will say,
"We have done it ourselves."
                              Lao Tsu

The Voluntary Human Extinction Movement is calling for volunteers to forego procreation as a step towards the total voluntary phasing out of our species. Such a goal, its proponents believe, would be in the best interests of life on earth. This idea, attributed to the movement's founder Les Knight, has been acclaimed in a recent article in *The Economist* as one of the most revolutionary ideas in human history. Regardless of whether we are ready for such revolutionary action (or inaction) just yet, Knight's call raises a question of profound significance – Why should there always be a next generation?

Many, within academia and without, are asking this very question of our universities and other institutions of higher education. Can these institutions, outlived within Europe only by the Icelandic Parliament and the Catholic church, serve the needs of our present and future learning society? Or should they, in the spirit of Knight, and in the greater interest of education and society, forego procreation and voluntarily bring about their own demise?

In support of Knight's call, critics of our higher education system could point to analogies between humanity's adverse impact on our planet and the adverse impact of higher education on education and learning. The environmental damage being caused by humanity could be compared to the damage being caused by the educational system as a whole by outdated divisions between first level, second level, further and higher education; alarming population growth worldwide causing unbearable strains on limited resources is analogous to the impact of the growing numbers of students in higher education; and finally, demographic trends,

whereby an increasing proportion of the population approaching retirement are being supported by a decreasing population of working age, may be familiar to many within academia.

To this list could be added higher education's contribution to what historian of science I. Bernard Cohen called the fallacies of the public's perception of science. These include the fallacy of scientific idolatry ("believing scientists to be lay saints, priests of truth, and superior beings who devote their lives to the selfless pursuit of higher things"); the fallacy of critical thinking (the belief that an understanding of science necessarily provides this faculty, the disproof of which "may easily be demonstrated by examining carefully the lives of scientists outside the laboratory"); the fallacy of scientism (the belief that science is necessarily the best or even the only way to solve many sorts of problems); and the fallacy of miscellaneous information ("the belief in the usefulness of unrelated information such as the boiling point of water... the distance in light years from the earth to various stars... the names of minerals").

This gulf between science, and the public's perception of science, can be traced in part to the development of science as a profession. In their book *Science in Public*[1], Jane Gregory and Steve Miller chart the rise of the professional scientist, and the attendant separation of the scientific community from the public at large. They point to the establishment of the Royal Society in England in 1660 as the beginning of this process. The late eighteenth and early nineteenth centuries saw important changes within science itself with the establishment of the first dedicated scientific laboratories in which scientists worked. Then in 1847 the American Association for the Advancement of Science was formed with the central aim of drawing a clear line between professional and amateur science. Thus, by the end of the nineteenth century, science had become a separate profession. In our own century, the process has continued with the development of more and more specialties and subspecialties in science. As a result we have seen the scientific community, and scientific communication, further divided – between disciplines, and between science and the public.

As Dorothy Nelkin points out in *Selling Science*[2], this division between science and the public has very real consequences for society. On the one hand, public acceptance of science appears to be based on expectations of immediate applications: promising solutions for environmental problems, effective new therapies, and efficient ways to handle toxic wastes. On the other, the public mis-image of science perpetuates several convenient myths: that science can provide definitive answers about risk, that "facts" speak for themselves rather than being open to interpretation, and that decisions about socially acceptable risks are scientific rather

Sunset?

than political judgements. The current controversies surrounding genetically modified organisms and BSE attest to the very real human, political, and economic consequences of the myths and fallacies of science that higher education has helped to create.

Partly as a result of such public controversies, there has been an explosion of activities aimed at improving public understanding of science and technology in recent years. But is the scientific community within higher education capable of bridging the gap between science and the public; of correcting Cohen's fallacies of science as they exist in the public mind?

It is telling that the major initiatives in public understanding of science within higher education, with a few notable exceptions, are coming not from within departments of science or technology. Rather they are originating from within departments of sociology, history and philosophy of science, communications and journalism, and psychology. This would suggest that science itself has become the establishment, incapable of self-inspection and self-criticism. However, it is clear that it is no longer acceptable for the scientific community within higher education to continue to bolster the myths and fallacies of science which exist in the public mind. It would appear therefore that the voluntary extinction of the scientific community within higher education would indeed be in the best interests of both science and society. Fortunately there is still time for such revolutionary inaction to be approved by Academic Council in time for the new millennium.

---

* *Dr Ian Hughes is Head of the School of Science and Technology at Dun Laoghaire Institute of Art, Design and Technology; Tel: 01-214-4724; E-mail: ianhughes@tinet.ie*

References
1. *Science in Public*, by Jane Gregory & Steve Miller, Plenum Press, 1998.
2. *Selling Science*, by Dorothy Nelkin, W.H. Freeman and Company, 1995.

Denis Weaire, F.R.S.

Professor Denis Weaire, F.R.S.

What does a physicist do? If we rattle off as an answer, ...publishes history, runs conferences on finance, helps rebuild heritage sites, makes videos, rescues old laboratories, dresses up as a Victorian, writes essays on odd socks, uses comics to teach science, and supports the employment of graduates to build and sell magnetic instruments.... then we can only be talking about one person: Denis Weaire, recently elected in London a Fellow of the Royal Society. A rare distinction for an Irish scientist but one which continues a link that goes back to the times of Robert Boyle, it is an honour which has pleased Denis' friends.

Denis Weaire's career has ranged across the spectrum of high academic achievement in Ireland, the UK, the United States and Continental Europe. From schoolboy days in Belfast through to studies at Cambridge University, he moved on to the Mathematics and Physics community in the United States in California, Chicago, Harvard and Yale, and ultimately held professorships in Herriot-Watt, and UCD before coming, in 1984, to TCD where he still practises his craft. With pitiful local resources available for scientific research, Weaire and his physicist colleagues in Trinity grasped the lifeline offered by European Research Programmes: their Department in Trinity has been one of the most successful University Departments in Ireland in achieving international status in its research.

A succession of management functions – from Head of Department, through Dean of Science at TCD, to President of the European Physical Society – has not diminished Weaire's appetite for research and for innovation. If not busy promoting the now successful restoration of the Great Telescope at Birr Castle, then he is combating the indifference of scientists and literati alike to the memory of his predecessors like Prof. Francis FitzGerald by theatrical appearances in period costume on Bloomsday in Dublin. He has given his attention recently to the Oscar Wilde Centre in TCD, to planning the re-issue of a classical textbook of physics and to catering for modernists with the first European Physical Society Conference on Physics and Finance in TCD, July 1999.

A man who has clambered aboard a dumpster in a Californian University to salvage the library of a deceased academic, and who is a rescuer of scientific instruments from dissolving convents, deserves to be recognised for these contributions to cultural life. But in addition to these beachcombing forays, Weaire continues to be a researcher. Following a long period as an investigator of the structure of amorphous silicon (a candidate for the development of solar cells), his interest has turned in recent years to foams. Generations of scientists have peered past froth into their beakers (whether of beer or of chemicals), ignoring the delicate structures that are observable in this "head". Weaire and his students and collaborators have examined these structures as a problem in the economical partitioning of space. In applying modern computational techniques to these structures, they discovered a new basic structure which is more economical in packing the "bubbles" together than was known previously.

A quiet but human satisfaction with this achievement enters into Weaire's own account. As he describes how the new Weaire-Phelan structure inches past the century-old description that another Belfast man, Lord Kelvin, had achieved, he is careful to credit appropriately his associates and those who lent him tools for the job, emphasising the co-operative structure with which science now advances. The award of a prize by the international oil company Shell to a related paper written by Weaire's team illustrates the rapid connection between academia and the needs of industry for new knowledge, another working genre which Weaire has promoted and participated in for the past decade. A director of a campus company that imports Chinese raw materials and sells instruments across the globe, Weaire has always supported his academic colleagues in their endeavours to tempt the Celtic Tiger to master its own technological future.

One of the great strengths of basic sciences like Physics is that there are many areas of endeavour where a talented physicist can flourish. But it is rare to find a scientist who succeeds in so many dimensions of his vocation without losing focus on his research. If there is in Ireland somewhere a virtual well providing a cocktail conducive to enthusiasm, scepticism, application, criticism, innovative scientific thought and a humorous sensitivity to cultural environment, then Denis Weaire must have drunk copiously from this source. From wherever he draws this strength, it will surely fuel further achievement.

A representation of the structure discovered by Weaire-Phelan devised by John Sullivan.

UNIVERSITY COLLEGE CORK

DICK HOGAN*

New President of UCC

Professor Gerry Wrixon (courtesy of The Irish Times*).*

With the appointment of Professor Gerry Wrixon as President, UCC has entered a new era. Continued growth, further capital investment and more emphasis on research driven projects – already a huge source of revenue for the College, will be combined with the President's personal goals. His wish is to ensure that being a student at UCC will be a time to remember – a happy time – not only because of the ambience that UCC can provide, but because of the excellence of its teaching and courses. And he believes that with so much on offer in the education system at third level, young students will weigh up their choices and opt for what suits them best. In other words there is competition in the education market place, and universities like UCC must respond.

Professor Wrixon, who so successfully headed the National Microelectronics Research Centre (NMRC), knows the world of industry better than most presidents who have occupied the office. Therefore, he brings a unique presence to the President's chair. His presidency, he says, will be "hands on", but not overwhelmingly so. Effectively, he sees his new role as chief executive of a major industry in the education sector. To Cork, UCC is a major industry and a major contributor to the local economy. It now serves almost 12,000 students, has an annual budget of £85 million, and has a capital expenditure programme running at £11 million per annum. Its external research funding – £17 million each year – is the highest in the State, and supports more than 400 people on campus. UCC employs almost 1,800 staff members and it is estimated that its total economic value to the region is at least £150 million annually.

The President's vision for the university is to make it more accessible to the people of Cork, city and county. To explain itself more: to remove the perception that there is some arcane business afoot in the hallowed grounds of one of the largest universities in the State. Sometimes academics are great communicators when talking to one another, but sometimes, too, they are not good at explaining their purpose, role and vision to others not immediately involved in college life. "That must change", says Professor Wrixon, "UCC must try even harder to become part of the fabric of everyday life in Cork and Munster".

The new President's plan is to develop a strategy for UCC embracing plans for future development – plans that will see it expanding as the millennium approaches. The various departments and faculties in the college have been asked to become part of this process and submit proposals to the President on how and where they see future trends. The various strands will then be drawn together, providing UCC with a blueprint for the years to come. But the plan will not be written in stone – Professor Wrixon's idea is that it should be subject to revision reflecting the changing needs and demands of UCC. Part of the process will include a strengthening of UCC's management team in the areas of human resources, planning and development, and communications.

Professor Wrixon says he came to the job with great expectation and hope. UCC is an important asset to Cork and Munster, and must continue to be a centre of excellence to which students will be drawn for that very reason. He is keen to ensure that, in an age of dramatic technological developments, the humanities will not be forgotten at UCC. He recognises that the humanities are a major strength of UCC, comprising, as they do, almost 40% of the University. There is a need for a greater recognition of their worth and he is anxious that there be a humanities component in all science and technology courses.

Professor Wrixon and his wife Marcia have two sons and a daughter. Alexandra, 21, has just graduated in French and Spanish at UCC. Robert, 27, and Adrian, 25, were both born in the United States. Robert graduated in chemical engineering from Princeton and is working with the Mars Group in Sydney. Adrian is an electronics graduate of UCD and Berkeley in California and works with the Bell Corporation in New Jersey.

Professor Wrixon has been chairman of Éolas, the Science and Technology Agency. He is a director of Cork Opera House, and of Triskel Arts Centre in Cork and is convinced that UCC must continue to foster close links with the artistic community in Cork. He said recently that there must be a dramatic re-direction of investment to education and training, and a eventual end to "incentive-based" industrial policy. "The strategic question for those charged with industrial development planning was", he said, "how relevant current policy which incentivised companies with State grants, low rates of corporation tax and tax breaks, would be in five or ten years time".

Contact: Marie McSweeney,
Information Office, University College, Cork;
Tel: +353-21-902371; Fax: +353-21-277004.

---

* *Dick Hogan is a journalist with* The Irish Times.

New Vice-Chancellor of UU

The distinguished biomedical scientist, Professor Gerry McKenna, has been appointed Vice-Chancellor of the University of Ulster – the largest university on the island of Ireland. He will take up his post in October, on the retirement of Lord Smith of Clifton, the present Vice-Chancellor.

Professor McKenna, who is 45, was born in Benburb, County Tyrone, and attended St Patrick's Academy in Dungannon. He and his wife, Phil, have two sons, and live in Portrush, County Antrim.

Professor McKenna was himself a student at the University of Ulster, graduating in 1976 with a first class honours degree in Biology. After completing a PhD in Genetics at Queen's University Belfast he took up a lecturing post, in Human Genetics and Biology, at the then New University of Ulster at Coleraine in 1979. He became Senior Lecturer in 1984, the year the University of Ulster was formed by the merger of The New University of Ulster with the Ulster Polytechnic, and was appointed to a personal Chair in 1988.

From 1988 Professor McKenna was Head of the Department of Biological and Biomedical Sciences for six years, during which time the range of courses offered was developed to include Human Nutrition, Radiation Science and Biotechnology. He took on the role of Dean of the Faculty of Science from 1994-1997, when he led the research team in Biomedical Sciences which was awarded the highest possible 5* rating in the UK's 1996 Research Assessment Exercise.

It was therefore a logical progression for Professor McKenna to be appointed Pro-Vice-Chancellor with responsibility for research and technology transfer. In this role he reviewed the University's research strategy, looking forward to the next UK Research Assessment Exercise in 2001, while at the same time placing a renewed emphasis on technology and knowledge transfer, particularly for the economic, social and cultural wellbeing of Northern Ireland.

Achievements during Professor McKenna's period as Pro-Vice-Chancellor (Research) include the University's successful bids for location of sites of the NI Science Park at the Coleraine and Magee campuses, the successful bid (with QUB) for funding under the UK "University Challenge" initiative, attraction of funding for incubator units on campus, and transition to the second phase of the UK Science Enterprise Challenge (again with QUB), on which a final decision is expected later this summer. He also led the establishment of UUTECH, the University's technology transfer company (see page 194). Seven new campus companies have been spun out via UUTECH in the past year.

Professor McKenna's academic research interests are in DNA repair, mutagenesis and nucleotide pool balance, areas where he has published over 200 scientific papers, supervised 25 DPhil theses, and attracted over £6 million in research grants. However during his career he has developed strong interests in the worlds of politics, industry and commerce. Indeed the role of Pro-Vice-Chancellor (Research) has afforded insights into the range of subject areas in which the University of Ulster conducts research, and has brought him into contact with opinion formers in a range of fields, including the life and health sciences, the arts, sport, education, public and social administration, as well as industry and commerce.

Professor Gerry McKenna.

> While Professor McKenna is well known in the academic world in the UK – he has been appointed the Chair of a Research Assessment Exercise Panel for the 2001 Exercise – he has always devoted attention to North-South affairs. It is anticipated therefore that collaborative links throughout Ireland, involving the higher education sector and other major players, will be consolidated and developed during his Vice-Chancellorship.

Under his leadership, the University of Ulster recently published a comprehensive summary of research areas related to the "areas for co-operation" in the Good Friday Agreement where there exists, or is potential for, North-South collaboration (see page 8).

In recent years Professor McKenna has lent vocal support to the strategic importance for the region of investment in R&D funding. In 1996, when government R&D funding in Northern Ireland was cut, with colleagues from QUB, Professor McKenna campaigned for its reinstatement. This campaign is likely to continue in the new administrative and governmental structures in Northern Ireland.

Turning to the role Professor McKenna is to take on from October 1999, the challenge is significant. The University of Ulster has more than 20,000 students, 3,000 employees and an annual budget of over £110 million sterling. It operates across four campuses – two in the greater Belfast area (York Street, Belfast and Jordanstown), one in Coleraine and one in Derry. Plans are also progressing, with the Belfast Institute of Further and Higher Education, for the educational village at Springvale in west Belfast.

Professor McKenna will have regard to the University's performance and reputation internationally, within the UK, and in an all-Ireland context, as well as in serving the needs of the community in Northern Ireland. One pressing matter will be to ensure that the varied work of the University which impacts on wealth creation is structured in a way which best facilitates the University's contribution to regional development in Northern Ireland.

Above all, in the University's research, teaching and reach-out activities, the new Vice-Chancellor has indicated that he will continue to base his approach on his personal philosophy of the importance of quality in higher education – that is, quality of education, quality of research and technology transfer, quality of the learning and working environment, and quality of service.

First woman President of the IEI – Professor Jane Grimson

Professor Jane Grimson, President of the IEI.

The current President of the Institution of Engineers of Ireland (IEI) is Prof. Jane Grimson. She is the first woman to be elected IEI President in its 165-year history. She follows her father, William Wright, into the position. He was President in the late 1970s.

Jane is Dean of Engineering and Systems Sciences at Trinity College Dublin and she is also Associate Professor and Co-Chair for Health Informatics at the College.

Prof. Grimson attributes much of her interest in engineering to her father.

> "He regularly brought me with him to his workplace. At the time he was doing research on a large-scale model of Southampton Water and that provided lots of opportunities for fun!"

Later she attended Alexandra School, whose ethos was that no career was off-limits for women. Jane studied Mathematics, Physics and Chemistry at Honours level.

> "The fun and excitement of engineering passed on to me by my father and the strong belief Alexandra generated that anybody can follow the career they choose, irrespective of sex, provided me with two of my projects for my Presidency. The slightly wackier one is to have a *Bring your Daughter (or niece or granddaughter) to Work Day*! Let them see what engineers do."
>
> "The second project is an *Engineering Day* when we will let young people, their teachers and their parents get to know engineering, see the excitement, see the huge variety of work engineers do – building computers, designing software, biomedics, genetics, bridges, roads, improving the environment – the list goes on."
>
> "Ireland must continue to produce a stream of highly qualified engineers in every discipline. Government has a big responsibility in this regard, particularly in ensuring that there are sufficient trained teachers and sufficient equipment to allow young children develop an interest in scientific subjects."

Prof. Grimson was the first female graduate in Engineering from Trinity College, Dublin, in 1970. She did an MSc in Computer Science at the University of Toronto before returning to Scotland, the land of her birth, to do a PhD at the University of Edinburgh.

Apart from her normal lecturing duties, Jane is heavily involved in research, particularly in databases and health informatics. She says that the next decade will see major changes in the way healthcare is delivered, for example by booking appointments with consultants at hospitals *via* the Internet. Research is playing a vital role in preparing us for these changes, and Prof. Grimson feels that an increasing amount of national endeavour, by Government, educational establishments and industry must be devoted to generating opportunities for investment in research and development.

Prof. Grimson acts as consultant to a number of commercial, industrial and semi-state bodies and is an evaluator of various EU and national research programmes. She is a member of the visiting panel at the EU Research Centre at Ispra in Italy. She is Technical Manager of TCD teams in a variety of externally funded projects under COST 11-bis, MAP, DELTA, STAR, Esprit, AIM, RACE, ACTS and Telematics programmes of the EC, industry and under the National Strategic Research Programme. She is currently Project Manager of the SYNAPSES Project involving 26 institutions from 14 different countries, funded under the Health Telematics Programme.

Professor Grimson at TCD.

Contact: Professor Jane Grimson, President, Institution of Engineers of Ireland, 22 Clyde Road, Dublin 4; Tel: 01-668-4341; Fax: 01-668-5508; E-mail: iei@iol.ie

The prostate - "a man's thing"

The incidence of prostate diseases is on the increase in the West. The reason for this is complex and may be due to longer male life expectancy and increased public awareness of the problem. In the US, screening for prostate cancer has led to a dramatic increase in the detection of this cancer at a stage when it can be treated with a high probability of cure. In Ireland, which has no screening program, most patients are diagnosed at a late stage. The debate, concerning the screening and treatment of prostate cancers, remains one of the most controversial subjects in Medicine.

Since the 1980s a blood test which detects a protein produced in the prostate gland has been used to diagnose prostate disease. High concentrations of this protein indicate inflammation of the prostate. Scientists were so sure of its specificity that they called the protein Prostate Specific Antigen (PSA). Like so many certainties, this has proven to be wrong, but the name PSA has persisted.

The gene that codes for PSA in the prostate is located on chromosome 19, which is also present in women. PSA has been detected in tumours unconnected with the prostate, including female breast tumours. Hence this protein, which was originally thought to be valuable only in the male, is now under examination in female breast tumours.

tumour, many men who develop the disease die from other causes. This fact has generated an oft-quoted truism that "prostate cancer is a disease that older men die *with* rather than *of*". Prostate cancer detected while still within the confines of the gland has a good prognosis. Modern X-ray, ultrasound and biopsy techniques have improved the ability to diagnose tumours. Early detection is essential for successful treatment.

Frequent urination, inability to urinate or a burning sensation when urinating are the symptoms which indicate it is time to seek medical help. These symptoms are due to an enlarged prostate gland but are coincidentally associated with prostate cancer only if the tumour is large. Most cancers of the prostate are initially too small to cause any symptoms.

Other diseases of the prostate can cause the same symptoms. The more common prostate disease is Benign Prostatic Hyperplasia (BPH) and has nothing to do with cancer. Over half of men in their sixties and over 90% in their seventies or older will have symptoms from the prostate gland enlarging as men grow old. This enlargement is as much a part of ageing as grey hair. The prostate gland, which is part of the male reproductive system, is doughnut shaped and is wrapped around the urethra. As the prostate enlarges, it constricts the urethra, thereby impeding the flow of urine.

concentration of Insulin-like Growth Factor in blood was found to be an effective indicator. Proteins, similar to PSA but under separate genetic control, are now being evaluated for the diagnosis of prostate disease. Methods of measuring the amounts, ratios and absolute production of these proteins are the subject of current research.

The discovery of PSA in the female has renewed interest in this protein. PSA is normally produced in some women during pregnancy and in breast milk. Not all breast tumours produce PSA, but those that do seem to have a favourable outcome. At present, it is not known if the PSA produced in females is identical to that produced in males. In Ireland there is little fundamental research being carried out in prostate disease despite its increasing incidence.

Some workers in this field have described prostatic disease as "the male epidemic". Last October the theme for Europe Against Cancer Week was "Men and Cancer". This raised public awareness of prostate cancer. Every effort must be made to encourage men with symptoms to seek medical advice. This is the single most effective action that can be taken. Ideally men above the age of 50, or younger if symptomatic, should have an annual prostate evaluation as part of a health check-up. Other issues raised include the cost of maintaining funds for research into cancer, and the ever-increasing costs of financing a health service.

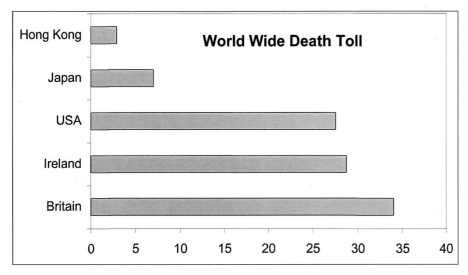

Number of prostate cancer deaths per 100,000 men per annum.

In Ireland, prostate cancer is the most common internal cancer in men. It represents 11% of all cancers, and 16% of non-skin cancers in men. There are about 1,000 new cases diagnosed each year. Although the statistics for this disease are similar to those of female breast cancer, it receives considerably less publicity or health resources.

Risk factors for developing prostate cancer are complex - but include family history, increasing age, racial origin and dietary intake. As this cancer is a very slow growing

This is successfully treated by surgery, drug or laser therapy.

The introduction of the PSA assay has had a stimulating effect both on the public and on health care workers. Even as the shortcomings of the test became apparent, more effort was put into its improvement. PSA is a serine protease belonging to the kallikrein family. In the circulation it is bound to protease inhibitors. This has contributed to some methodological problems associated with this assay. Other blood tests, either alone or in combination, are being investigated. The

Surely there should be more effort put into the research and management of prostatic disease in Ireland? A chauvinistic view? Hardly, since the author is female but her husband, father, brother and half her friends are males and they all deserve the best!

---

* This is a shortened version of the article which won the Royal Irish Academy Biochemistry Science Writing Competition 1999 - sponsored by Yamanouchi Ireland and The Irish Times.

Understanding how cancer cells spread

Cancer takes the lives of approximately 7,500 people each year in Ireland. This represents 20% of all deaths and is second only to cardiovascular disease. With respect to breast cancer, it is estimated that one in eight women will develop some form of the disease within their lifetime.

The vast majority of deaths from cancer are caused by secondary tumours which form when the disease state shifts from benign to malignant. These secondary tumours arise when cells split away from the original primary tumour and spread to other parts of the body by a process called metastasis. Tumour cell metastasis is a highly complex and dynamic process involving many events including: (a) detachment from the primary tumour, (b) migration through the tissue and entry into the circulatory system and (c) exit from the vasculature back into another tissue to form a secondary tumour.

Metastasis, as with almost all cellular processes, is driven and controlled by proteins. These molecules are the workhorses of the cell which interact with each other in specific ways in order to carry out all the dynamic functions that allow the cell, and hence the organism, to survive. A central goal therefore is to elucidate and understand the protein interactions involved in cancer processes such as metastasis.

How do we identify these proteins? There are many ways of tackling this question, but one method is to compare the protein patterns of normal and cancerous cells and look for any changes. Maspin, or mammary serine protease inhibitor, is a protein that was discovered in such a way in 1994. This protein was found to be made in the normal epithelial cells of mammals, particularly in the breast, but markedly absent from such cells once they became cancerous. Furthermore, it has been shown that maspin possesses the ability to reduce the metastatic potential of tumour cells. These exciting observations have led to the speculation that maspin may function as a natural tumour suppressor molecule in cells. Consequently, either directly or indirectly, maspin may have beneficial therapeutic effects.

Unfortunately, little is known about how maspin suppresses metastasis. Given that our laboratory is dedicated to research involving the serpin superfamily of proteins, of which maspin is a member, we designed experiments to shed light on the molecular mechanism(s) of maspin action. The question we specifically asked was – what protein(s) does maspin physically interact or associate with?

To do this we used a relatively new genetic based system called the yeast two hybrid interaction trap. This technique allows the user to take a protein of interest, called the bait, and test it for interaction with many thousands of other human proteins, called the targets. The two hybrid procedure has a number of advantages over conventional methods. First and foremost, the protein interactions are detected in yeast which is a living organism, and not within the relatively artificial confines of the test tube. Secondly, identification of the interacting target protein is comparatively straight-forward and rapid. This system has been used successfully to describe a wide range of protein-protein interactions and has provided valuable information on the intricate details of how many proteins work at the molecular level.

In our yeast two hybrid experiment, maspin was used as a bait to screen target human proteins derived from fibroblast tissue cells. The setting up of this system was a complex process and involved a variety of recombinant DNA techniques to introduce the bait and target genes/proteins into the yeast cells. The result was a large population of genetically engineered yeast, where all of the cells contained the maspin bait protein but each one had a different target protein. To identify cells harbouring a protein interaction, a screening process was employed using reporter genes also genetically engineered into the yeast cells. Activation of these genes occurs only when the maspin bait interacts with the target protein *(Figure 1)*. Positive interactor yeast cells were isolated and the maspin interacting target protein identified by genetic techniques.

From our screen we found maspin to potentially interact with another protein called collagen. This observation has since been backed up in our laboratory using purified maspin and collagen proteins in isolation. A key question arising from this work is – how would a protein interaction with collagen help to explain the putative anti-metastatic role of maspin? Collagen itself is one of the most abundant proteins in the body, making up a significant proportion of bone and skin. Most importantly with regard to metastasis, collagen is a critical component of the extracellular matrix (ECM), a mesh like structure of proteins surrounding cells in the tissues. The ECM serves to hold normal tissue cells in place through an elaborate array of protein-protein interactions between the ECM and the cell. Interestingly, maspin is often located on the outer surface of cells thereby placing it in an ideal location for interaction with the surrounding collagen. It is entirely possible that this contact between maspin and collagen helps in some way to prevent normal cells from migrating *(Figure 2)*. Future work will attempt to further describe the maspin-collagen interaction in an effort to elucidate the importance of this association with regard to the prevention of metastasis.

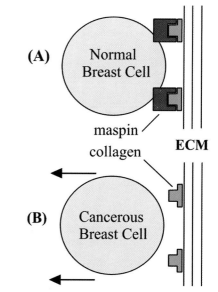

Figure 2. Interaction of maspin and collagen – A role in metastasis? (A) Maspin present on the outer surface of normal tissue cells may interact with the surrounding collagen in the ECM to aid in keeping the cell in place.
(B) Cancerous cells which have reduced maspin levels could therefore lose this mode of attachment to collagen, resulting in the cancer cell having an increased ability to spread.

Contact: Oliver Blacque, Department of Biochemistry, University College, Dublin 4; E-mail: oblacque@macollamh.ucd.ie

Oliver Blacque was one of the winners in the Merville Lay Seminars held in UCD in February 1999. This is a summary of his presentation.

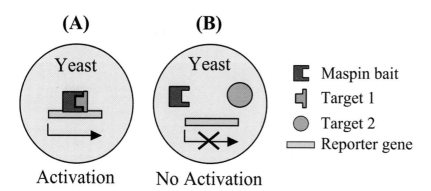

Figure 1. General scheme of the yeast two hybrid interaction trap. Yeast cells containing an interaction between the maspin bait and target protein are detected using a reporter gene system. Cells containing an interaction show activation of the reporter gene (A), whereas cells lacking an interaction demonstrate no such activation (B).

The importance of discovering novel proteins

A primary goal of biomedical science is the discovery and study of all of the proteins the body is capable of making. Some 60,000 to 80,000 proteins are thought to be encoded in the human genome, and a large number of these have yet to be characterised.

Many diseases occur as a result of the failure of the body to properly manufacture a given protein. One such disease is classical diabetes, which arises when there is a deficiency in the production of the protein insulin. Insulin is one of the most extensively studied human proteins, and diabetes can now be treated with genetically engineered human insulin. Given that many human proteins have yet to be discovered, it is likely that some may be as centrally important to certain diseases as insulin is to diabetes. Finding these proteins and identifying their roles in maintaining a healthy state will undoubtedly lead to intervention therapies in the future.

Our work focuses on a class of proteins known as Serpins (**ser**ine **p**rotease **in**hibitors) which may be ideally suited to biomedical intervention. Proteases are enzymes which can cleave or digest other proteins and are involved in many physiologically important events such as blood coagulation, complement activation and tissue remodelling. As their name implies, most serpins are inhibitors of a class of proteases which have a serine amino acid residue in their active site. Individual serpins control the activity of their target protease by irreversibly binding to the active site or "business end" of the protease *(Figure 1)*. This has the effect of permanently blocking the protease and preventing it from working. An exposed peptide loop on the surface of the serpin binds the protease, and the amino acid sequence of this loop is largely responsible for the specificity of the interaction.

Serpins have a track record in the treatment of certain disease states. Of the more than two hundred members of this protein family that are known to exist, at least two have been successfully developed into treatments for genetically inherited deficiency diseases. These are Alpha$_1$-Proteinase Inhibitor and Antithrombin III which are marketed by Bayer under the trade names "Prolastin" *(Figure 2)* and "Thrombate III". They are administered to treat conditions such as emphysema and thrombosis respectively, where the disease arises as a result of a congenital insufficiency of the serpin. In Alpha$_1$-Proteinase Inhibitor deficiency, the unregulated target protease elastase causes destruction of lung tissue, and in Antithrombin III deficiency the target protease thrombin causes excessive blood clotting. This underlines the importance of serpins and the need to continue research into finding more members of this protein family.

The classical way to search for novel proteins has been to isolate the protein directly, but this is often difficult if the protein is unstable or not abundant. In recent years, attention has focused on the DNA which codes for the protein. DNA is highly stable and easy to purify, and techniques have been developed over the years that allow for the re-creation of the encoded protein in the laboratory. A major endeavour is currently underway to sequence the entire complement of human DNA (the human genome sequencing project). When this is completed, it should be possible to identify all of the genes and predicted proteins encoded in the genome. Alongside this undertaking, other research groups are examining different tissues to determine and sequence what genes are specifically expressed in them. This has led to the generation of DNA sequence information of actively transcribed genes known as Expressed Sequence Tags (ESTs).

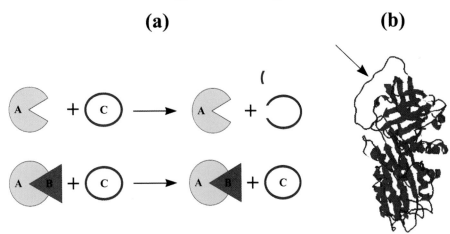

Figure 1. **(a)**: *A typical proteinase (A) is prevented from cleaving its substrate protein (C) by the inhibitory action of a serpin molecule (B) which irreversibly binds to the proteinase's active site.* **(b)**: *Structure of Alpha$_1$-Proteinase Inhibitor. The position of the reactive site loop is indicated by an arrow.*

In an effort to further add to the serpin superfamily, we examined EST library databases for novel serpin-like sequences. ESTs are generated by preparing a complementary DNA (cDNA) library from a given cell/tissue type, and partially sequencing a random selection of the resultant clones. The contents of an EST library should, therefore, reflect a range of proteins being produced by the tissue/cells from which the cDNA library was made. Since many ESTs are derived from as yet uncharacterised genes, EST databases are a good starting point for the identification of novel genes and, therefore, novel proteins.

One EST sequence expressed in activated immune system T-cells was of particular interest to us. This entry in the database was a gene fragment which showed striking similarity to a region of typical serpins near the critical reactive site loop. Using a Polymerase Chain Reaction (PCR) based screening procedure, we isolated the full length gene from a library of thousands of DNA molecules. We have sequenced the entire insert and have identified a region within the gene which encodes a serpin-like protein. Utilising a bacterial expression system, we have since created the recombinant protein and are in the process of elucidating its role as a potential protease inhibitor.

The most immediate benefit of this basic research is that it will add to the current picture of how proteins interact in maintaining normal tissues. This will ultimately lead to a greater understanding of how diseases develop when the biochemical balance is disturbed, and most importantly to development of novel therapies.

Figure 2. "Prolastin" – a treatment for inherited Alpha$_1$-Proteinase Inhibitor deficiency.

Contact: Michael Sharkey,
Department of Biochemistry, UCD, Dublin 4;
E-mail: msharkey@ucd.ie

Michael Sharkey was one of the winners in the Merville Lay Seminars held in UCD in February 1999. This is a summary of his presentation.

Inflammatory bowel disease and immune wars!

The environment in which we live is hostile. Evidence to support this statement includes the fact that, if skin is physically damaged, systemic infection by micro-organisms is quite likely to occur. Similarly, if the epithelium lining the intestine is damaged or functionally impaired, micro-organisms can enter the blood stream. For these reasons, peptic ulcer disease was, in the pre-antibiotic era, very much more dangerous than now.

Not all micro-organisms are harmful. Some multicellular animals, for example all ruminants, depend on the presence of commensal bacteria – bugs which normally inhabit the intestinal lumen and digest cellulose into products which are the nutrients for the "host" animal. It is perhaps not surprising that this situation has evolved, since bacteria have been around in such large numbers throughout mammalian evolution. If you consider that one gram of intestinal contents can contain up to one trillion bacteria, you can see how these barrier (epithelial) cells, which line the intestine and separate the body from the external environment, are integral to our health. It is a common misconception that we feed our gut when in actual fact it feeds us.

Certain micro-organisms are very harmful. The most common cause of death in children worldwide is dehydration due to intestinal infections with micro-organisms such as cholera, E. coli, campylobacter and salmonella, which are other common bacteria which evoke secretions leading to diarrhoea. Curiously, diarrhoea is a defensive response, designed to rid the host of these parasites. Tragically this is such a powerful mechanism that it can lead to fatal dehydration if water and electrolytes are not replaced.

Protective responses to harmful micro-organisms may be excessive or inappropriate. Inflammatory bowel disease (IBD) refers to a range of clinical conditions (e.g. Crohn's disease and ulcerative colitis) which display similar symptoms that can range from mild intestinal disturbances to chronic diarrhoea and weight loss. Inflammation, a common feature of these diseases, is defined as how the immune system responds to injury. Ideally inflammation allows the body to heal damage that has occurred in response to a noxious stimulus. However, in IBD, non-resolving or uncontrolled inflammation is a feature. IBD is a young person's disease, affecting a growing number of people.

Why does chronic inflammation occur? Causes of IBD are, as yet, unknown, with suspicion raised at everything from infection to diet. In addition, recent information strongly implicates a genetic component. However, what is clear is that the immune system is a major player in this disease. The majority of the time the immune system is a "good thing" and plays a crucial role in keeping us healthy. However, like most effective systems, the immune system can malfunction, with consequent collateral damage due to "friendly fire". In this case, the body's immune system becomes out of control and turns on itself, damaging our own tissues and cells.

Science may help understand disease processes. Three main areas being investigated in our laboratory are:
i) *How does the immune system function in the gut?*
ii) *Why does it malfunction in diseases like IBD?*
iii) *Can we develop new ways to interrupt or even prevent inflammatory disease processes?*

For those who suffer from IBD, and their families, the last question may seem the most important. However, as scientists, we must recognise that it is only by understanding how a particular system works *normally* can we ever hope to discover why it fails and, therefore, how to stop this from happening.

So what does an immune reaction consist of? One of the most important events that occur is that immune cells respond to the stimulus (e.g. infection/autoimmune disease) by secreting chemical messengers or "mediators". These mediators are chemicals with which cells of the body communicate with each other. Designated cells of the immune system orchestrate this complex crosstalk. For example, locally released chemical mediators act on epithelial (barrier) cells lining the gut and stimulate water loss; the mechanism underlying diarrhoea. Another function of a specific sub-group of chemical mediators is to recruit immune cells from the blood to the site of infection to amplify protective responses to infection (or sustain an inappropriate or excessive inflammation). Thus cells will migrate toward sites of infection. Research carried out in many laboratories, including ours, focuses on trying to characterise these pro-inflammatory mediators which are so important in amplifying the immune response, as we consider these to be targets for novel anti-inflammatory drugs.

Nature supplies "off" switches. Usually, wounds heal. Such resolution of damage is the ideal outcome of a disease process or response to injury. Thus, in addition to preventing synthesis or release of pro-inflammatory mediators, we can begin to consider amplification of normal curative processes. An opportunity has recently been identified by the discovery of a family of molecules which are anti-inflammatory - the lipoxins. Lipoxins are enormously powerful endogenous anti-inflammatory molecules which appear to limit or reverse inflammation as part of natural healing.

The dog that didn't bark. Inflammation occurs when necessary and shuts down when no longer needed, and so normal control systems must be continuously operating. Currently, we are investigating such internal braking signals in intestinal tissue with a view to harnessing natural mechanisms to combat disease processes. Only by identifying the weapons being deployed in "immune wars", which are a feature of virtually all chronic diseases, can we ever hope to identify rational targets for novel and effective therapies for use in IBD.

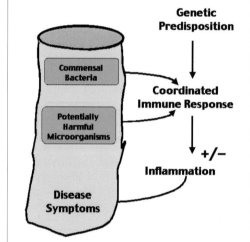

Figure 1: Crosstalk mechanisms involved in host response to micro-organisms.

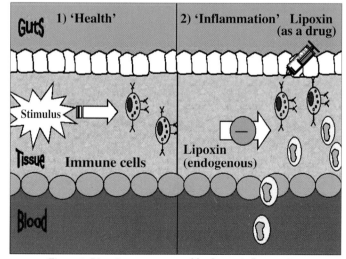

Figure 2: Lipoxins act as natural brakes to inflammation.

This work was supported by Forbairt (now Enterprise Ireland).

Contact: Lorraine Maher & Alan Baird, Department of Pharmacology, UCD, Fosters Avenue, Blackrock, Co. Dublin;
E-mail: lmaher@macollamh.ucd.ie

*Lorraine Maher was one of the winners in the Merville Lay Seminars held in UCD in February 1999. This is a summary of her presentation.

TRINITY COLLEGE DUBLIN   EAMONN O'DONNELL*

Age-related neuronal damage and its reversal by vitamin supplementation

Aging is a universal process and yet it is a remarkably difficult phenomenon to define. The human body theoretically should be able to regenerate itself indefinitely as it has numerous processes that repair and renew time dependent damage. And yet we age. In the Department of Physiology at Trinity College we are interested in the processes that underlie cellular aging and in ways in which these processes can be delayed.

All organisms fundamentally operate on a cellular level. Cells co-operate and co-ordinate to form tissues and organs. To do this they must produce energy, and it is this process, it is hypothetised, that creates the conditions that "cause" aging.

To respire, a mammalian cell must combine oxygen and glucose and, from this reaction, "energy" is released and stored in a chemical form. However life evolved in an oxygen-free environment over a billion years ago, and it is imperfectly suited to handling the very high energy potential present in molecular oxygen. It is this leakage of semi-reacted but highly energetic oxygen or reactive oxygen species (ROS) from respiration sites within the cell that is hypothesised to cause aging. For these moieties rip through the delicate cellular architecture, distorting and inactivating proteins, damaging DNA and hardening membranes. Over a life time the destruction wrought by these agents overwhelms the cells reparitive capacities. Muscle cells whither, skin cells lose flexibility, and brain cells decline. An organism grows "old".

The cell has evolved an elaborate defence mechanism against these toxic by-products of respiration. This system comprises enzymatic and non-enzymatic components. Both arms function to neutralise the ROS before they can interact and damage cellular components.

In the laboratory of Dr Marina Lynch in the Physiology Department of Trinity College, we have been studying a portion of the brain, the hippo-campus, believed to be involved in memory formation. The brain is composed of an entangled tapestry of specialised cells called neurons. These cells are unable to regenerate themselves and appear to be especially sensitive to the constant barrage of ROS. Over time, the capabilities of these neurons erode, and we believe this is due to a parallel decline in the antioxidant defence system of these cells.

The nonenzymatic portion of the defence system, which is comprised mainly of the vitamins C and E, deteriorates significantly in the aged hippocampus. But also, in aged hippocampal neurons, the activity of a key enzyme involved in trapping ROS is paradoxically increased. This over-activity has deleterious consequences. The enzyme, superoxide dismutase (SOD), converts the toxic ROS into another toxic agent, hydrogen peroxide. This molecule can cause cellular damage if it is not instantly removed, and the enzymes that break down this molecule struggle to cope with its overproduction. The resultant combination of SOD over-activity and the decrease in the concentrations of the vitamins C and E is believed to significantly contribute to the "aging" in these neurons.

It is known that a particular peptide released during infection in the brain, interleukin-1ß (IL-1ß), is also present in high concentration in aged hippocampal neurons. We hypothesised that ROS-mediated neuronal damage causes this peptide to be released and that this in turn stimulates the activity of SOD – thus creating a destructive feedback loop. We postulated that, if this cycle could be stopped by artificially boosting the non-enzymatic portion of the defence system and thus limiting the extent of ROS mediated damage, perhaps the hippocampal neuronal injury witnessed in aging could be deferred?

To test this hypothesis, young and aged rats were fed a diet rich in the vitamins C and E to boost their antioxidant defence system. It was found that there was no increase in concentration in IL-1ß in these animals compared to corresponding controls, SOD activity was not increased, and oxidative damage was also significantly decreased. Furthermore, in a key experiment designed to test hippocampal neuronal function, it was found that aged cells from the diet group recovered their responsiveness to a specific electrical stimulus, and the extent of that response was similar to that of hippocampal neurons from young rats.

We have shown that long-term vitamin supplementation can reverse certain age related changes in hippocampal neurons and that time dependent decline in neuronal function in this portion of the brain can be inhibited.

Mrs Jeanne Calment (122 years old) – can antioxidants delay age-related cellular damage?

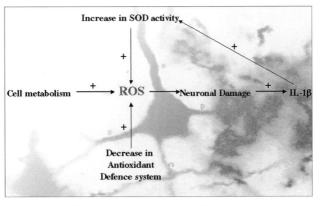

Effect of age on hippocampal neurons.

* Eamonn O'Donnell won first prize and the RDS Medal in the Science Communication Forum in November 1998. This is a summary of his winning presentation.

Contact: Eamonn O'Donnell, Department of Physiology, Trinity College, Dublin 2; E-mail: odonneec@mail.tcd.ie

---

Message from Mr Chris Shouldice, Chairman, Royal Dublin Society Committee of Science and Technology:

Congratulations to Eamonn O'Donnell on his impressive presentation of an important research project. The subject of the Public Benefits of Contemporary Science Education will be addressed at a conference to be held in the RDS on Wednesday 1 December 1999. Enquiries to Carol Power, Development Executive, Science & Industry, RDS; Tel: 01-668-0866; E-mail: carol.power@rds.ie

SOME SCIENCE-BASED CAREERS — SIFCO TURBINE COMPONENTS, CARRIGTWOHILL, CO. CORK

John Connolly

I was born in Cork in 1967, and attended Douglas Community School, where I took Irish, English, Maths, French, Physics, Chemistry and Business Organisation in the Leaving Certificate. I attended Cork Regional Technical College (Now Cork Institute of Technology) from 1985 to 1990, and obtained the degree of Bachelor of Engineering in Mechanical Engineering with a 2.1 Honour.

I took a very worthwhile year out from 1988-1989 as a Student Mechanical Design Engineer at Krups Engineering in Limerick, and from 1990 to 1992 joined SIFCO Turbine Components as Process Engineer. I transferred to Shannon Turbine Technologies as Process Engineer and Supervisor of Coating Processes from 1992-1994, and then returned to SIFCO Turbine Components, where I was Manager of Engineering at the Blackrock Facility between 1995 and 1998. Since then I have been Manager of Product Development at SIFCO's Carrigtwohill Facility.

The SIFCO Turbine Group provides a gas turbine hot section component overhaul and material support service to airlines and airmotives, worldwide. Established in 1913 in Cleveland, Ohio, the company grew and diversified into turbine component repair and materials technology, and is now an independent supplier equipped with the technologies of General Electric Aircraft Engines, SNECMA, Rolls Royce and Pratt and Whitney.

My present job responsibilities include leading a team of engineers to develop and productionise new product repairs. We prepare production instructions in great detail to ensure the repairs are applied as per OEM (Original Engine Manufacturer) requirements. We carry out pre production repair substantiations, and obtain OEM approval. We are responsible for developing fixturing and tooling, equipment and production processes by working with suppliers and OEMs, to successfully and efficiently repair engine run turbine blades and nozzle guide vanes.

Products repaired are all Nickel based superalloys; poly crystalline and directionally solidified. The repairs are designed to return the part to a blue print condition, dimensionally and metallurgically. Quality Assurance is integral to the repair development.

The scope of engineering processes is very broad and "leading edge". The following are three samples of the processes we develop:

5 Axis CNC laser drilling
5 Axis CNC electro discharge machining and laser drilling equipment is programmed and used to generate small diameter straight and shaped holes for component film cooling.

Hydrogen Fluoride cleaning and high temperature vacuum furnace brazing
Used to heal cracks in components. The components are first cleaned using HF and Hydrogen mixture at 950+°C. Then superalloy braze powder suspended in a binder is applied to the cracks. The component is heated to around 1200°C in a vacuum furnace (10E-4 mbar) to melt and flow the alloy into the cracks, followed by a diffusion heat treat.

Oxidation and temperature resistant coatings
These are applied using inert atmosphere aluminising heat treatment and robotic plasma spray processes.

SOME SCIENCE-BASED CAREERS — NORBROOK LABORATORIES LIMITED, NEWRY

Lillian Cromie

I was born on April 21st 1966 and grew up in Newry, Co. Down. I attended a local national school and followed on by attending the Sacred Heart Grammar School in Newry. I completed my O'Levels in 1982 and took Biology, Chemistry and Mathematics at A'level. The results I obtained enabled me to study Biomedical Science at the University of Ulster (Coleraine campus). I completed my Honours Degree in 1987 and chose to continue my studies for a PhD in Biomedical Research with a particular emphasis towards Immunology. I spent three years researching the effects on the Immune Reponse on Cellular aspects of the cancer process, and obtained my PhD in 1991.

A summer placement at Norbrook Laboratories Limited, Newry, in the Regulatory Affairs Department encouraged me to undertake a course of Legal Studies at Queen's University, Belfast, which facilitated my understanding of the Regulatory process. This course of study was undertaken on a part-time basis, whilst in the employ of Norbrook, spanning a period of three years, during which I studied topics as diverse as Law of Business Organisation, Evidence, Contract, Constitutional Law, Torts and Equity.

Norbrook Laboratories Limited researches, develops, manufactures and markets novel and generic veterinary pharmaceutical products and, as a young and energetic company in the 1970s, has grown rapidly to become a leading Veterinary Pharmaceutical Company distributing to more than 110 countries worldwide, with its customer base including 75% of the world's multinational companies. We are currently expanding into the Medical Pharmaceutical arena.

Within the Research and Development facility, staff come from a wide range of disciplines within the scientific field – including Veterinary Science, Pharmacy, Biomedical Science, Biochemistry, Chemistry, Microbiology, Food Science, Agriculture, with more than 80% of our staff Irish Graduates.

My current role within Norbrook, where I have worked for eight years, is as Manager of the Research and Development Department, which offers me an opportunity to marry my scientific and legal strengths. Working in such a vibrant, intellectually challenging environment and in such an innovative Company provides for a satisfying and enjoyable career.

This is a very exciting time for those involved in the pharmaceutical environment in Ireland. Irish graduates are respected globally as highly educated, competent people and have the confidence and ability to face the challenges of the future in such a competitive arena.

Moving towards the next millennium, the veterinary pharmaceutical industry faces a more competitive market coupled with changing regulatory and manufacturing environments. These challenges, which are being presented to Norbrook, are being met by a continuous investment in our Facilities and our People, with the success and high standards achieved a direct result of the expertise and dynamism of our highly trained work force.

SOME SCIENCE-BASED CAREERS — NYCOMED IRELAND LIMITED, CARRIGTWOHILL, CO. CORK

Frances Doyle

I was born in London in 1966 and came to live in the seaside village of Garryvoe in East Cork at the age of four. The bright lights of London soon forgotten, I attended primary school in Shanagarry and completed my Leaving Certificate in Midleton at St Mary's high school. I took science for what was then called the Inter Cert and chose Biology for the Leaving Cert.

Sitting the Leaving Cert in 1984 now seems like a different era: there was still quite a lot of pressure but not the same gateway of opportunity, which I feel exists now. "Jobs for life" is what we were seeking; the emphasis was on Job Security – i.e. Civil Service Mentality.

I completed a secretarial course in the Collage of Commerce in Cork and started work in production at an electronics company in Cork City. I worked there for six years.

I suppose it depends on whether you see life as a treadmill or a window of opportunity. I stepped off my treadmill when I left the electronics company to join Nycomed Ireland in 1993.

Nycomed Ireland produces contrast media for use in the diagnosis, treatment and monitoring of soft tissue disease. In order to enhance the contrast between healthy and diseased tissue, radiologists inject these media into a patient's circulatory system or central nervous system before making an X-ray or magnetic resonance (i.e. CT Scan) examination.

Working within Nycomed taught me the entire process and, armed with this knowledge, I was promoted to Team Leader within the Sterile Production area.

This required not only an overall understanding of the process but also the skill to instil this information to others. Hard work, perseverance and ability to get on with people are just some of the requirements of a Team Leader. Last year I was promoted to Good Manufacturing Practice (G.M.P) Officer here at Nycomed. GMPs are the guide lines or rules which Nycomed and all other pharmaceutical companies must follow.

Nycomed has offered me the chance to enhance my career and academic knowledge. I have attended and participated in a number of courses both internally and externally. These courses have helped me to broaden my knowledge and gave me the ability to support, motivate and guide the operators within the production area.

When I did my Leaving Cert in '84 I may not have taken "the road less traveled by", but I have enjoyed the journey.

To those making choices on subjects for the Leaving Cert, perhaps it's all about vision, innovation and risk taking. My choices served me well in the real word: I enjoy what I do and would recommend my career path to anyone. Do not worry, you are not locked into your first choice.

SOME SCIENCE-BASED CAREERS — CADBURY IRELAND LIMITED, DUBLIN 5

Susan Harrington

I was born in Ferbane, Co. Offaly, in 1969, and attended secondary school at La Sainte Union de Sacre Couer, Banagher, Co. Offaly. I took Biology, Physics/Chemistry, Accountancy, French, English, Irish and Maths in the Leaving Certificate in 1986.

I graduated with a Bachelor of Science Degree (Biochemistry/Microbiology) at University College Galway in 1989. Since then I have also gained the following diplomas:

1990:	Diploma in Industrial Biology – Austin Waldron Regional Technical College
1993:	Diploma in Management for Scientists & Engineers – Trinity College Dublin
1997:	Diploma in Training – Irish Management Institute
1998:	Introduction to Public Relations – University College Dublin.

In 1990 I joined Cadbury Ireland Limited as Process Analyst, where I was involved in Sensory Evaluation Panels, Quality Control, and Laboratory/Plant Equipment Calibration. The following year I became Process Technologist, a job which included Plant Commissioning/New Product Development, ISO9002 Implementation, Plant Optimisation, Preparation of Plant Operation Manuals, Training Plant Operators, and Assessment/Monitoring of raw materials.

Since 1997 I have been Training Officer, in which position I:

- Manage Production Training Instructors
- Liaise with production managers regarding plant training
- Maintain and update ISO9002 Training Manual
- Manage training records
- Develop/update training manuals
- Manage interactive computer training centre
- Manage Business and Education Links Programme
- Develop yearly training plan/training budget with the Human Resources manager
- Source companies for internal/external training
- Liase with all departments regarding training needs
- Deliver Food Hygiene Programme to staff.

During my career I have also attended training courses in Pest Control, ISO9002, Interpersonal Skills, Problem Solving Skills, Influencing Skills, Hygiene for Food Handlers, and Manual Handling. I have also attended the Instituto Cervantes for the last three years studying Spanish.

I hope to apply for a Masters in Business Administration with Spanish in the near future.

SOME SCIENCE-BASED CAREERS — PFIZER PHARMACEUTICAL PRODUCTION CORPORATION, RINGASKIDDY, CO. CORK

Nuala Kerley

We are now entering the new millennium. This is a time of continuing change and technological advancement. Science has become part of our everyday lives. The benefits of success in science can be seen all around us, from the widget in a beer can to the major medical advancements in treatment of diseases such as cancer and heart disease. It has become increasingly important to have an understanding of science, with issues such as genetically modified foods (GM) and the green house effect hitting the headlines every day.

My interest in science stemmed from a love of all things practical, and I still to this day enjoy the practical side of science. I completed my leaving certificate in 1987 at St Vincent's Secondary School Dundalk having studied seven honours including maths, chemistry and biology. With a determination to pursue something in the science field, I accepted a place at the University of Ulster at Coleraine to study Biological Chemistry. This degree combined pure chemistry with biology and biochemistry. After a thoroughly enjoyable three years I graduated in 1990.

With a sound basis in chemistry, I decided to study for a PhD at Queen's University Belfast. This led me into the field of synthetic organic chemistry. I spent four years researching the biotransformation (using bacteria to carry out chemical reactions) of hydrocarbons (e.g. benzene) and their role in synthesis. A brief spell in Scotland with Zeneca (now Astra-Zeneca) was the stepping stone to my current job at Pfizer in Ringaskiddy.

Pfizer is a global healthcare company. It develops, manufactures and markets medicines for the treatment of a wide variety of illnesses, including heart disease (Norvasc), fungal infections (Diflucan), bacterial infections (Vibramycin), and anti-depressants (Lustral), among others. Pfizer has become a household name since the launch of Viagra – its drug used for the treatment of MED (male erectile dysfunction). The active ingredient in this drug is only one of eighteen finished products manufactured at the Pfizer plant in Ringaskiddy which is one of the major manufacturing facilities within the entire company world-wide.

My role in Pfizer enables me to put into practice many of the skills I learned at college. As a development chemist I work closely with my colleagues at the Pfizer central research site in Groton Connecticut USA. The job involves evaluating new synthetic routes to drugs in early phase clinical trials. The development process involves studying yield improvements, analysing scope for recycling solvents, and also finding alternative cheaper and easier ways of carrying out a process. All process development must take consideration of the ultimate goal of scale up to full production without compromising the very high quality of the product. Each new drug for the treatment of widely differing illnesses - from diabetes to migraine to cancer - can yield its own set of chemical problems to overcome.

These are exciting times for us in Pfizer: there are many new challenges to be met. Just as there are in many other technology industries. What better time to get involved in science and have a part to play in the new innovations of the future.

For further information http://www.pfizer.ie

SOME SCIENCE-BASED CAREERS — RANDOX LABORATORIES LIMITED, CRUMLIN, CO. ANTRIM

Barbara McCartney

I was born in November 1972 and grew up in Newtownards, Co. Down. I went to Regent House School and for as long as I can remember was fascinated by different aspects of science. At A-level I studied Biology, Chemistry and Maths, and then chose the Biomedical Sciences course at the University of Ulster's Coleraine campus. The course is a four year sandwich course and for my third year placement I went to Randox Laboratories, Co. Antrim.

Randox Laboratories was established in 1982 and is the only manufacturer of clinical chemistry reagents in the United Kingdom. It is a progressive and rapidly expanding company with 85% of sales being exported throughout the world. The major product ranges include quality control serum, clinical chemistry reagents and diagnostic tests for a wide range of parameters such as infectious diseases, specific proteins, thyroid hormones, fertility hormones and immunoglobulins.

During my placement year I worked in the Raw Materials Quality Control Department. This involved visual and weight checks of raw materials and running assays on the enzymes used in Production to check their activity. Randox manufactures some of its own raw materials and part of my job was to test preparations at different stages of the manufacturing process. The practical experience which I gained in this year greatly helped my final year studies and also reassured me that I wanted a career in this area of science.

I returned to Randox after graduating and entered one of the Research and Development Laboratories where I worked on immunoturbidimetric assays for Anti Streptolysin-O (ASO), C-Reactive Protein (CRP) and Rheumatoid Factor (RF). After 18 months I changed research laboratories and started to develop liquid stable reagents for a new range of clinical chemistry products. Through this work I have gained experience in the field of clinical chemistry and in using a wide range of clinical chemistry analysers.

I have continued working in research but 18 months ago I also started to work part-time in the Technical Support Department. In order to answer customer enquiries I have gained a detailed knowledge of the entire product range but also have used my technical experience from the laboratory to investigate customer queries. I continue to divide my time between research in the laboratory and technical support. One of the many advantages in working for an international company is that there are many opportunities to meet people from around the world, and I also recently attended the IFCC-Worldlab Conference in Florence, Italy.

To conclude, I really enjoy the opportunities and challenges of my job and find great satisfaction in being able to apply scientific knowledge to resolve a problem in research or for a customer.

Michelle McGarraghy

Born in Sligo in 1970, I was educated at the Mercy College in Sligo. Having developed an interest in science from the Inter Cert, I chose to study chemistry, physics and biology for the Leaving Cert, along with maths, English, Irish and French. While studying for the Leaving Cert I became very interested in chemistry and physics and considered studying these subjects further with the intention of eventually teaching them.

After finishing school in 1987 I went to U.C.D. where I studied for a Bachelor of Science degree. In first year I studied chemistry, physics, biology and maths; in second year I studied chemistry, physics and maths; and chose to major in chemistry in the final two years. At the end of third year in college students are given the opportunity to work in a pharmaceutical company for the summer. This experience gave me a valuable insight into the working of the pharmaceutical industry. After completing my B.Sc. in 1991 I decided to specialise in organic chemistry and completed a Ph.D. at U.C.D. over the next ca. 4 years. During this time I had the opportunity to study at an Italian university for three months as part of an Erasmus exchange programme.

Once I completed my studies, I worked as a research chemist for about a year with a small pharmaceutical company, Unimed plc., based in Dublin. The work there involved the laboratory synthesis and development of new products, their study, classification and analysis.

In early 1997 I moved to Italy where I took up the position of research and development chemist with the pharmaceutical company Bristol Myers Squibb which is based outside Rome. My work was very varied and was divided between the laboratory and the manufacturing plant. Some of my responsibilities involved the development and evaluation of a chemical process in the laboratory and then the scale up of this process for bulk production in the plant. This work involved travelling between the company's international sites, which are based in both America and Europe. I also provided technical support to manufacturing plants, including process trouble shooting, analysis and resolution of production problems.

After eighteen months I returned to Ireland and began working with Roche in County Clare. This work still involves process development and optimisation both at the laboratory and plant scale; however there is also a business aspect to the job with an emphasis on the evaluation, production planning and costing for certain processes.

Chemistry is a versatile area of study and has provided me with great opportunities for both personal and professional development.

Aileen Moore

I was born on January 14, 1969, in Antrim, and had the delight to be raised on a farm just outside the small town of Randalstown. I began my education at Creggan Primary School in 1973 and in 1980 attended St Mary's Grammar School in Magherafelt, Co Derry. My love of the farm and involvement with animals stimulated an interest in Veterinary Sciences from an early age. I always preferred the science subjects at school and chose Chemistry, Physics and Biology at A Level. My career options had now shifted to either Medicine or Biomedical Sciences.

In October 1998 I started a four-year BSc Honours degree in Biomedical Sciences at the University of Ulster (UU) in Coleraine. I was delighted with my choice as it offered such a wide variety of subjects - biochemistry, immunology, histology, molecular biology and haematology, as well as a Third Year Placement Programme with the additional award of Diploma in Industrial Studies.

I graduated in 1992 and never really contemplated anything other than further research. I began my DPhil in October 1992 in the School of Biomedical Sciences at Coleraine in the area of DNA damage and repair. My research was mainly biochemistry based although I had gained considerable experience in molecular biology techniques. I completed my DPhil in September 1996 and lectured part-time in various Science-based degree programmes on different campuses within UU. I thoroughly enjoyed lecturing and in my own time also learned some new research techniques within the Cancer and Ageing Research Group at the University.

Marrying and settling in the coastal town of Portstewart, Co. Derry, I was delighted to be offered the position of Development Officer of the newly established Life and Health Technologies (LHT) Partnership in February 1998, my current job.

The Partnership has been established to improve the level of networking and flow of information, e.g. through breakfast meetings, seminars, workshops, between individuals and organisations active within the LHT sector (in Northern Ireland). The need for such a non-exclusive forum within this expanding sector was identified and documented by the LHT Panel in the Northern Ireland Technology Foresight Exercise.

As one might expect, the LHT sector is extremely varied, with individuals from UU, the Queen's University of Belfast, the National Health Service, industry, and government agencies. Gaining an insight into the needs of each of these groups has been and still remains a tremendous challenge for the Partnership and myself. While my own research experience has provided me with an understanding and empathy with the university and hospital-based research point of view, I have enjoyed entering into the world of the industrialist as well as appreciating the role of government agencies within the LHT sector. My position within the Partnership has resulted in being invited to participate in various panels discussing the future strategy of the sector.

Since starting my DPhil in 1992, I always felt that I would make this side-ways step from research at some point in my career. My scientific background has been essential training for this job, but I am also getting the chance to appreciate the much wider applications of LHT in our world today.

Jean Moran

I was born in Portlaoise, in November, 1970. I grew up in Limerick and went to the Salesian Secondary School, Fernbank. I studied English, Irish, Maths, French, Physics, Chemistry and Applied Maths for the Leaving Cert, which I completed in 1989.

I studied Science in UCC and completed a B.Sc. with Chemistry as the major subject. I continued in UCC for my postgrad and obtained a Ph.D. in Electrochemistry, working with Prof. L.D. Burke.

I started working for Intel Ireland in August 1996, immediately after I finished college. Intel was planning the start up of its new Wafer fabrication factory in Leixlip and was adding several new analytical tools to its capability on site.

I was sent to Intel in Santa Clara, California, to learn SIMS (Secondary Ion Mass Spectrometry) – a surface analysis technique. Briefly, SIMS involves bombarding a surface with primary ions and using a mass spectrometer to collect the secondary ions that get ejected from the surface. This gives information about the composition of the sample.

In late 1997 I returned to Ireland to set up a SIMS system at the Ireland site in time to support the ramp up of the new manufacturing facility (Fab 14). SIMS is used to check the performance and cleanliness of production tools, which is especially important at the start-up phase.

In March 1999, I accepted a position as Surface Analysis Area Manager with Intel Corporation, in Santa Clara, California – back where I originally trained. I now manage a group of people working on a variety of analytical systems such as SIMS, AFM (Atomic Force Microscopy) and Thermomechanics tools. We support both the development and transfer of new Intel technologies and the manufacture of current processes. It is interesting and challenging to be involved in the development of such processes at an early stage and to watch them come to completion. As the critical dimensions of the chips get smaller, we also have to find ways to analyse ever smaller areas – so there is a constant requirement to find new or more refined methods of analysis. We work on these projects in co-operation with some U.S. universities and all the other U.S. sites, as well as Ireland and Israel.

As for the weather in California, well, you have to take the rough with the smooth!!

Studying science often leads to travel abroad! Here Jean Moran and Richard Meade enjoy the sights of Utah. From Ballyporeen, Co. Tipperary, Richard also studied Chemistry and Physics in school – Mitchelstown CBS. He graduated in Electrical Engineering in UCC, took a post-graduate diploma at the National Microelectronics Research Centre (UCC). He now works for Cypress Semiconductor in San Jose, California.

Liam O'Dwyer

I was born on March 15, 1967, and grew up in a rural area close to Fethard Co. Tipperary. I attended the small local national school and Fethard Secondary School.

I completed my Leaving Certificate in 1985 and took Mathematics, Chemistry, Biology, French and Economics at higher level. The results I obtained enabled me to study Science at University College Dublin. I deferred my college place for one year before starting at UCD in September 1986.

I completed an honours degree in chemistry in 1990 and decided to undertake further research for two reasons: I was very interested in chemistry and the idea of specialised research in the subject appealed to me and, secondly, the career opportunities in Ireland were very good for people with post-graduate qualifications in chemistry. I started my Ph.D. at UCD in September 1990 and found it thoroughly stimulating and enjoyable. I completed it in April 1994 and began working at Loctite in the General Industrial Product Development Department.

During the following three years I gained excellent experience in a wide range of adhesive technologies and learned the importance of taking account of the requirements of the end-user when developing a new product. This necessitated a substantial amount of international travel to visit the factories of Loctite's industrial customers such as Philips, Nokia etc. This experience was also invaluable as a way of learning to appreciate how different cultures work and do business.

Since 1997 I have been part of the team responsible for the development of new consumer products in the Super Attak or Super-Glue range. Our customer is now the individual consumer as opposed to a large company. This requires a different but equally challenging approach to the development of products which satisfy the user's needs. It is the teamwork which must exist between people from product development, sales and marketing, the accounts, planning, production and health & safety departments which is largely responsible for getting a new product into the market, and this is the aspect of my work which gives me the greatest satisfaction.

I believe ongoing learning is important in order to become more effective in the workplace and also as a means of personal improvement. I have started a two-year part-time M.Sc. in technology management at UCD. This course will give me an enhanced appreciation of how technology and innovation are important for a company's present and future business.

It is now a very exciting time to be involved in a technological industry in Ireland. Irish graduates are respected globally as highly educated, competent people and are well positioned to contribute meaningfully well into the future.

Science in Northern Ireland Schools 1989-1999: past achievements and future challenges

Since the introduction of the Education Reform (Northern Ireland) Order in 1989, science has become an integral and important part of the statutory curriculum of all pupils of compulsory school age in Northern Ireland (NI). To enable schools to implement the science curriculum, programmes of study for science in the NI Curriculum (NIC) were written for each of the key stages of compulsory schooling. The Department of Education for Northern Ireland (DENI) has provided, since 1990, some £140M to help improve the level and quality of accommodation for science and technology and design in post-primary schools in the Province: the curriculum and advisory support service of the Education and Library Boards has provided intensive training for science teachers.

The programmes of study for science have undergone a number of changes over the past decade. In February 1996, for example, following a review of the primary curriculum by the Northern Ireland Council for the Curriculum, Examinations and Assessment (CCEA), the programmes for primary science were amended to include important aspects of learning in technology and design and have become known as the programmes of study for science and technology in the NIC at key stages 1 and 2.

Before 1990, science in most primary schools in NI was limited to environmental studies. While this work provided the children with often valuable experiences in important aspects of biology, it did not represent a sufficiently broad or balanced science programme. A recent report by the NI Education and Training Inspectorate (the Inspectorate), entitled *Children and Their Learning – Primary Inspections* (1992-98) indicates that, since Education Reform, primary schools in NI have made good progress in implementing many aspects of the programme of study for science and technology, and that the teachers have worked hard to increase their own knowledge and to consolidate science within the curriculum. The report also indicates that the children's knowledge and understanding in science and technology were significantly better than their attainments in investigating and making: in addition, much remains to be done to ensure consistent progression in the children's learning and development of skills in technology and design. Further, over 50% of the primary schools inspected during the period 1992-1998 needed to introduce procedures to ensure the effective assessment of the children's learning in science.

The Inspectorate's report, *Secondary Education 1990* highlighted significant weaknesses in science provision in many of the secondary schools inspected over the period 1986-1989. These weaknesses included the complete absence of science from the curricula of some pupils during their final two years of compulsory education. The report also commented on the lack of adequate breadth and balance and gender stereotyping in the science programme followed by most pupils during these years. Since 1990, matters have improved significantly: in 1999, almost all pupils in post-primary schools in NI follow a balanced science programme in their final two years of compulsory schooling. In addition, the findings from inspection indicate that most pupils in post-primary schools in NI demonstrate a satisfactory knowledge of the content of their science courses. However, while their competence in the skills of scientific investigation continues to improve, the standards achieved in this important aspect of their work are lower than those reached in the knowledge and understanding of science. Other issues include the need for the pupils to undertake more practical work: science for many pupils in post-primary schools in NI is characterised by listening and copywriting, rather than by discussing and doing. While most pupils can record accurately their observations in science in the form of charts, tables and graphs, the standard achieved in written work is more variable, with only a minority able to express themselves clearly, concisely and precisely in connected prose. In addition, approximately one-third of the pupils at key stage 4 follow a limited programme of study for science leading to a limited grade single award General Certificate of Secondary Education (GCSE). The emphasis on the single award science course has clear implications for the adequate supply of technicians, scientists and engineers to serve the needs of the NI economy, and also for the levels of scientific literacy reached by a significant minority of young people in NI by the age of 16 years. In the sixth form, the uptake of courses in science at General Certificate of Education (GCE) is too low and, along with many other regions and countries, the numbers involved in science at 16+ continue to fall.

In conclusion, there have been a number of improvements in science education in NI schools in the last decade. In addition to those mentioned above, most pupils now have good science vocabularies, and can apply the concept of a fair test: when involved in practical work, they have satisfactory or good manipulative skills, and pay due regard to matters of safety. Challenges which remain for both our primary and post-primary schools include the lack of continuity and progression in the pupils' experiences in science across the key stage 2/3 (primary/post-primary) interface, and the limited influence of science in helping to develop the pupils' skills in literacy, numeracy and information and communication technologies (ICTs). These challenges, along with those outlined above, will need to be met and overcome if the standards in science education in NI are to continue to improve and thereby help sustain and support economic growth and development, and contribute to the health and well-being of our citizens. Those involved in the various aspects of science education in NI are aware of these matters and want to ensure that our young people can contribute and compete on an equal footing with their peers locally, in Europe and beyond.

Contact:
Department of Education, Northern Ireland,
Rathgael House, 43 Balloo Road,
Bangor, Co. Down BT19 7PR;
Tel: 01247-279279;
Fax: 01274-279100.

Improving the participation levels in the physical sciences

The importance of Physics and Chemistry in the context of Ireland's increasingly technological society and economy can scarcely be overstated. As stated in The Competitive Advantage of Nations: "..... there is little doubt from our research that education and training are decisive in national economic advantage. The nations that invest most heavily in education have advantages in many industries that could be traced in part to human resources".[1]

The World Competitiveness Report for 1998, which compared 46 of the most developed countries worldwide on a wide range of competitiveness factors, has ranked Ireland as number one in terms of its educational system meeting the needs of a competitive economy.[2]

Recently, IBEC has indicated that: *"the education system is 'unacceptably weak' on the number of students taking science subjects, particularly physics and chemistry ... It is crucial to Ireland's sustained economic development that a greater number of school leavers be attracted to these subjects"*.[3] A 1998 survey from Forfás revealed that 89% of top business professionals felt more school children need to be persuaded to study science and engineering subjects. This is against a background where 94% of business people believe that only by applying the most modern technology will our economy become more competitive.[4]

Various commentators in both the industrial and educational sectors have expressed opinions in a similar vein to those quoted above. Recent developments have shown that Ireland can no longer compete on a cost basis at the lower level of technology and so it must move up the value chain into more knowledge intensive manufacturing and into services, where skills and technology are even more important. An ability to do this is predicated to a large extent on the quality and experience of the students leaving second-level schools, and it is clear that physics and chemistry have a central role to play in this regard. More recently, analysis of the skills needs of science and technology based industries for the coming years has clearly indicated that, if the current pattern of participation in the study of the physical sciences at second and third level persists, the needs of industry will not be met.[5] This will severely hamper growth in the industrial sector and consequently the strength of our economy. It is therefore imperative that the declining interest in the physical sciences must not only be arrested but also reversed.

In the context of the above, it is clearly essential that everything possible is done to develop these subjects in schools. Issues, which have become of concern in recent times, must be dealt with as a matter of urgency. These concerns relate not only to the number of students taking these subjects as part of the Leaving Certificate Programme but also the delivery of these subjects in schools, and to the apparent difficulty of students in achieving high grades in these subjects. Indeed, it has been suggested that the three issues may be related. It is suggested that limited exposure of students to practical work has made the subjects unattractive and this, combined with the quest for high "points", are steering students towards subjects in which they believe it is easier to obtain higher grades.

The background to the current state of physics and chemistry in schools is summarised. In addition, the programme of measures which are being put into place aimed to increase participation levels in these subjects, as well as targeting resources at the development of the overall quality of learning in these subjects, are also outlined.

Participation in physics and in chemistry at Leaving Certificate level

Over the period 1986-87 to 1996-97, there has been a gradual but substantial increase in the number of candidates taking the Leaving Certificate examinations - from 104,897 to 123,973. This increase in the numbers sitting the Leaving Certificate (*circa* 19%) has to a large extent masked trends in the uptake in the physical sciences. Against this growing student population, there has been a decrease in participation rates in the physical sciences. In chemistry, 20,347 students took the Leaving Certificate examination in 1986-87 which represents 19.4% of the Leaving Certificate cohort. Whilst this number had decreased to 15,247 in 1996-97, it is perhaps more significant to note that this represents only 12.3% of the Leaving Certificate cohort. In physics, the numbers taking the Leaving Certificate have remained relatively unchanged, showing only a slight decrease from 21,130 to 20,362 over the same time period, but again this represents a significant decrease in participation levels from 20.1% to 16.4%. Participation levels in biology remain relatively stable, with an increase in numbers taking the subject from 54,366 to 62,559 over this period. This represents a slight fall in uptake from 51.8% to 50.5%. It is worth noting that the decrease in participation rates in the physical sciences experienced in Ireland is a phenomenon that is mirrored internationally. This was one of the issues explored in the international Colloquium on Attainment in Physics at age 16+, which was organized by the Department of Education and Science and held in Cork in September 1998.

It is interesting to note that, whilst the percentage of pupils taking the physical sciences has declined, the percentages of schools providing these subjects has increased slightly. In other words, at a time when more pupils are provided with an opportunity to study the physical sciences, fewer are actually taking up the option. Another interesting point is the differential take-up of the physical sciences between the sexes, which has shown significant changes over the period. The ratio of males to females in physics has fallen from approximately 3.5:1 to just over 3:1 in the 1986-96 period. The Department's scheme of Intervention Projects in Physics and Chemistry is primarily responsible for this change. Further improvements are clearly necessary in this area. In the case of chemistry, the ratio of males to females in this subject is currently approximately 1:1 compared to 1.6:1 in 1986/87.

The up-take of Leaving Certificate subjects is closely related to the teaching of the physical sciences in the Junior Certificate. While the numbers taking Junior Certificate science have increased somewhat over the period, there is concern in relation to:
- the physics and chemistry content of the syllabus;
- the standard of teaching, especially in relation to practical work;
- the preponderance of teachers whose primary qualification is in biology.

At Leaving Certificate level, a number of issues have been identified as effecting participation levels in the physical sciences:
- The percentages obtaining the various grades in the Leaving Certificate examination are generally reasonably consistent over the years. In Ordinary Level physics the percentage of candidates obtaining E to NG grades has averaged 20.8% between 1990 and 1998; in Ordinary Level chemistry the average percentage obtaining these grades over the same period was 19.2%. Concern has arisen in recent years at the numbers of candidates who do not achieve Grade D or higher in physics and chemistry, especially at Ordinary Level. In particular, it has been noted that the percentages obtaining the lower grades in these subjects is significantly higher than in many other subjects at this level.
- Laboratory facilities are in need of investment to support the most modern approaches to teaching physics and chemistry. The Department of Education & Science has recently carried out a survey of all second-level schools to determine the condition of their laboratory facilities.

The results of this survey are expected shortly, though initial information confirms the necessity for significant investment.
- Both domestic and international information suggests that many teachers require in-service support to help them develop their skills to encompass the most modern approaches to teaching physics and chemistry. This is particularly so as the majority of teachers will both have biology as their principal third-level qualification and will predominantly teach biology.

Investment in the teaching of the sciences

To address these issues, an investment plan valued at £15 million over three years has been put in place. This plan involves the modernisation of school science laboratories, the revision of different science syllabi, and the provision of extensive training for teachers. It also includes an annual payment per student for schools to provide materials for the teaching of physics and chemistry at Leaving Certificate level.

Capital Grants

Specific grant aid to assist the delivery of the sciences has been put in place.
- A special grant for the teaching of physics and chemistry of £10 per pupil in respect of pupils studying these subjects for the Leaving Certificate. This is in addition to the *per capita* grants provided to schools.
- The Department is currently undertaking a survey of science facilities in all second-level schools. Based on the data obtained from this survey, measures to address identified areas of need will be devised and put in place on a phased basis.
- The importance of the use of Information and Communication Technologies (ICTs) in all aspects of teaching and learning is increasing rapidly, and this is particularly so in the case of the physical sciences. It is intended to develop fully the use of ICTs in the teaching of the physical sciences and, in co-operation with the National Centre for Technology in Education, to incorporate these technologies in in-career development programmes. Schools will receive a grant for the purchase of ICT equipment specifically for use in the teaching of science subjects.

Capital expenditure in these areas will amount to in excess of £14m over the next three years.

Curricular change

A rolling curricular review as part of making the physical sciences more relevant to today's students is essential.
- At Leaving Certificate level current syllabi in physics and in chemistry have been in place since the mid-eighties. New syllabi, along with support materials, will be issued to schools at the beginning of next school year. Comprehensive reference handbooks for teachers will also be issued to schools in conjunction with the new syllabi. Teaching of the new courses will start in September 2000, and they will be examined for the first time in June 2002.
- In order to fully address the needs of Ordinary Level candidates, specific attention has been given to the Ordinary Level syllabi by the National Council for Curriculum and Assessment (NCCA).
- There can be little doubt that the health of the physical sciences at senior cycle is greatly influence by the junior cycle science programme. Accordingly, the NCCA has undertaken a review of the Junior Certificate science syllabus, particularly in respect of its physics and chemistry content. A restructuring of this syllabus is in train.

In-career Development

It is important to have a well-informed and vibrant teaching community to deliver the subjects.
- An extensive in-career development initiative for teachers of physics and chemistry at both Junior Certificate and Leaving Certificate will be launched in 1999. The initiative will consist of a comprehensive programme which will emphasise modern developments in teaching methods, including the role of practical work, in relation to these subjects. Changes in content in the new Leaving Certificate syllabi will also be addressed.
- At Junior Certificate level, these courses will be particularly targeted in the first instance at those teachers whose major subject qualification may not include physics or chemistry.
- It is intended that 480 teachers will receive additional training in their subject disciplines of physics and chemistry during 1999, with a target of providing such training to up to 1,500 teachers over the next three years if such an expanded programme is felt to be necessary.

Support for Teachers
- A panel of trainers will be put in place to support the in-career development programmes at both Leaving Certificate and Junior Certificate levels. These trainers, who will be regionally based, will have a primary role in the provision of courses for teachers of physics and chemistry.
- The trainers will also provide on-going support for teachers of physics and chemistry at both junior and senior levels in their region.
- In addition, the trainers will also have a role in supporting schools in identifying the appropriate laboratory resources required, and in the effective utilisation of school laboratory facilities.
- The Department of Education and Science is currently in discussion with a number of third level institutions in relation to both the pre-service and in-service needs of teachers of the physical sciences.

Assessment

The Department of Education and Science has decided to revise the style and layout of the Ordinary Level papers in Junior Certificate science and Leaving Certificate physics and chemistry. The purpose of this revision is, in general, to make the presentation of these three papers more appropriate for the cohorts of candidates for whom they are designed. In particular, it is intended to simplify the style and layout of the papers, make the questions more accessible in relation to structure and language, and change the emphasis in relation to the skills and knowledge currently tested.

Laboratory practical work is seen as an integral part of the new syllabi in physics and in chemistry. In 1997 the Department of Education and Science carried out a feasibility study into practical assessment in Leaving Certificate physics and chemistry. The Department commissioned further research on the results of this feasibility study in 1998-99. This research has clearly demonstrated that further work is necessary in the area of assessment of practical work.

Conclusion

It is intended that this Science Education Initiative will build on the excellent work and commitment of teachers in the physical sciences. These teachers have achieved enviable standards over the years, and there is little doubt that these measures will further assist them in continuing to advance the achievements of their students. There should be optimism that this initiative will encourage an increased up-take of the physical sciences in our schools, opening up increased opportunities for the students themselves and, in the process, enhancing Ireland's highly deserved reputation for having a well-educated and scientifically literate workforce.

---

1 *The Competitive Advantage of Nations, Michael Porter*

2 *World Competitiveness Report, 1998*

3 *Statement by Mr Padraig O'Grady, Assistant Director of IBEC, 29.12.98*

4 *Forfás study, 1998*

5 *The first Report of the Expert Group on Future Skills Needs, Forfás, 1998*

Higher Education Institutions to become world class Research Centres

Dr Don Thornhill, Chairman of the HEA.

The Ir£180m Programme for Research in Third-Level Institutions is the most important and exciting development that has ever taken place in the history of research in Ireland. The programme will provide first class international research facilities in the universities and institutes of technology. In addition, it dramatically improves the outlook for research, and for researchers, in Ireland.

Following the announcement by the Government of first cycle funding allocations in July 1999, eleven higher education institutions (see details on next page) will share just over Ir£162m in the period 1999-2001 which will go towards providing new laboratories, research centres, scientific equipment, extra staff and support for researchers.

In addition, under the Programme, these institutions will now be able to recruit about 500 researchers, thus enabling young talented academics to pursue a career path in the world of research.

Today, the driving forces behind economy growth and social progress owe more to the application of knowledge, learning and education than they do to the traditional sources of economic growth such as natural resources and geographic position. In a rapidly changing society, where the pursuit and application of knowledge are of critical importance, the research capacity of the higher education system is an important national resource.

International economic studies have shown that as much as 50% of economic growth is due to innovation and to the application of advances in knowledge. In other words, we now need to become a "knowledge economy".

The world is going through a new industrial revolution – the knowledge revolution – and universities and institutes of technology are key players in this process. The research activity to be undertaken in our third level institutions will play a major part in generating future jobs for this country, particularly in the high value-added industries such as informatics and computers, pharmaceuticals and biotechnology.

The programme embraces research not only in science and technology but also the vitally important areas of the humanities and social sciences. An internationally credible research capacity is also needed in these areas – for utilitarian as much as for "higher order" reasons. The pace of economic, social, scientific and technological change is now so rapid, and the effects on individuals, communities and societies so pervasive, that the traditional role of academics in the humanities and in the social sciences is of increasing importance in explaining, interpreting and evaluating the implications of the powerful forces which impact on our society.

But the Programme for Research will generate more than knowledge. It will also provide state-of-the-art-equipment, new methodologies, trained researchers and interpersonal networks needed to do high quality R&D. The development of new enterprises and industrial activity are also a likely follow-on.

Some have argued that this country should not be engaged in basic research activity and that instead we should avail of the expertise of other nations. This view stemmed from a perception that, as a small country with a relatively low income, we could not afford the investment necessary to make major advances in research and that, in any event, our small size meant that it was unlikely that research in Ireland could make a "significant difference". This is a flawed position.

We cannot become a knowledge economy without achieving research leadership. We will not be able to sustain our hard-won potential to join the leading wave of advanced economies without developing our research capacity. It is not possible to be a "free rider" on the basic science performed in the rest of the world.

Basic research is one of the building-blocks of innovations systems and of a good innovation infrastructure. It underpins technological and social progress and therefore economic and social development, and is a key factor in implanting a capability for continuous learning among graduates and, ultimately, among their employers.

The Higher Education Authority was requested by the Minister for Education and Science to manage the Programme, the objectives of which are:

i) To enable a strategic and planned approach by third-level institutions to the long-term development of their research capabilities, consistent with their existing and developing research strengths and capabilities.

ii) To promote the development of high quality research capabilities in third-level institutions, so as to enhance the quality and relevance of graduate output and skills.

iii) Within the framework of these objectives, to provide support for outstandingly talented individual researchers and teams within institutions, and the encouragement of co-operation between researchers both within the institutions and between institutions – having particular regard to the desirability of encouraging inter-institutional co-operation within the two parts of the binary system and within Ireland, the EU and internationally.

Proposals totalling over Ir£250m were received from 23 institutions and were considered by an International Assessment Panel, chaired by the Chairman (non-voting) of the HEA, Dr Don Thornhill. The submissions were evaluated on three criteria:

i) Strategic Planning Criterion (40 points)
ii) Research Criterion (35 points)
iii) Teaching Criterion (25 points).

The Ir£180m Programme for Research in Third-Level Institutions 1999-2001 is a successor to, and subsumes, the programme for research in science and technology which was funded in 1998. Funding of Ir£4m was made available in 1998 to support science and technology research programmes, as follows:

> **Athlone Institute of Technology – £475,000** – Polymer Technology and Toxicology – failure of angioplast biomaterials and methods of evaluating their short and long term behaviour in physiological environments.
>
> **Institute of Technology, Carlow – £232,000** – Biotechnological and Environmental Sciences.
>
> **National University of Ireland, Galway – £712,000** – Bioengineering Sciences.
>
> **Royal College of Surgeons in Ireland – £331,000** – Biopharmaceutical Sciences and Experimental Therapeutics – proposal to establish a Biopharmaceuticals Sciences Research Resource.
>
> **Trinity College Dublin – £750,000** – Molecular Cell Biology, Advanced Materials, Information and Communications Technology, and Environmental and Earth Sciences.
>
> **University College Cork – £750,000** – Four inter-related projects under the themes of Molecular and Cell Biology and Microelectronic Materials and Semiconductor Devices.
>
> **University of Limerick – £750,000** – Materials and Surface Science and Informatics.

A further Ir£50m will be available in Cycle Two of the Programme.

The institutions can now move forward with significantly increased resources and an enhanced sense of direction. The proposals from the institutions have been subject to intensive review and evaluation against the highest international standards. We can look forward with considerable confidence to securing quality outcomes.

Contact: John L. Hayden, Secretary/Chief Executive, Higher Education Authority, Third Floor, Marine House, Clanwilliam Court, Dublin 2; Tel: 01-661-2748; Fax: 01-661-0492; E-mail: jhayden@hea.ie

JOHN HAYDEN

Allocations announced in July 1999

Athlone Institute of Technology £1,775,000
The funding will go towards a programme in Biopolymer and Molecular Activity Research. Much of the work will centre on the development of medical device products such as catheter-type products and load-bearing prostheses. A key activity will be a toxic assessment of materials used in this area of medicine.

NUI, Galway, and the University of Ulster at Coleraine will collaborate in the programme.

Dublin City University £22,446,000
The funding will be used to establish a National Centre for Sensor Research; a National Centre for Plasma Science and Technology; and the Research Centre in Networks and Communications Engineering. Collaboration with the Institute of Technology, Tallaght, will feature in the areas of plasma science and technology, and sensor research.

Sensor technology is used in a variety of fields including medical diagnosis, more efficient industrial processes and in developing safer food, for example, by means of packaging indicators that can reveal whether the essential requirements of freshness are being fulfilled. Plasma technology is heavily used in manufacturing electronic components such as microprocessors and for specialist coatings on lenses.

Dublin Institute of Technology £8,200,000
The funding will go towards the provision of an Optical Characterisation and Spectroscopic Facility for measurement and testing work in science, food science and engineering.

Institute of Technology, Carlow £948,000
The support provided will be used to fund a research programme in the areas of Biotechnology and Environmental Science. Among the areas to be explored under the programme are the treatment of toxic pollutants in soil and water; the environmental impact of current agricultural practices; and biochemical work in the area of cereal crops and the conversion of waste raw materials from the malting, brewing and distilling industries into valuable products.

National University of Ireland, Galway £15,406,000
The funding will support the establishment of a Centre for Biomedical Engineering Science. The programme includes collaboration with Galway-Mayo Institute of Technology, UCC, Athlone Institute of Technology, and Institute of Technology, Sligo.

The Centre will combine the expertise of engineers, medical graduates and scientists to research aspects of major human illnesses such as cancer, heart disease, bone diseases, kidney disease and strokes.

National University of Ireland, Maynooth £8,870,000
The funding will support a programme in Bioscience research in immunology, bioengineering and agroecology. NUI, Maynooth currently has the largest group of immunologists in Ireland and already acts as a knowledge base for immunology, providing expertise to other third level institutions, industry and hospitals.

The University also has an international reputation in the bioengineering/agroecology field and will engage in research activity in such areas as the reduction of chemical pesticides in food and the environment. Collaborative arrangements are in place with the Institute of Technology, Carlow.

The Royal College of Surgeons in Ireland £8,662,000
The allocation will be used to support biopharmaceutical sciences research. Collaboration with UCD and TCD is being explored. Among the areas of emphasis will be the development of a gene therapy programme availing among others of the expertise of staff who have taken a lead in research in cystic fibrosis and cardiovascular disease.

Trinity College, Dublin £18,616,000
The funding will go towards an Institute for Advanced Materials Science; an Institute for Information Technology and Advanced Computation Research; research programmes in Mediterranean and Near Eastern Studies, Irish-Scottish Studies and a National Political and Social Survey; Molecular Cell Biology research and Neuro-degeneration research. Research in neuro-degeneration and mental decline in the process of ageing is of key importance to every person on the planet.

A joint programme with UCD in the social science area will feature in the programme. Collaborative links have also been established with RCSI in the molecular cell biology area.

University College Cork £21,113,000
The investment will support a Biosciences Institute and Nanofabrication facility; Food and Health Science Research; and Humanities Research under the themes *History* and *Society*. Collaborative work with the Cork Institute of Technology will feature in the nanoscale science and technology initiative.

Nanoscale science involves reducing materials to the smallest possible measurement, often to just a few atoms, allowing scientists to exert greater control over chemical, electronic, biotechnological and other processes. Nanoscale science has been hailed as the science of the new millennium, with enormous commercial applications from microelectronics to healthcare.

The investment in research in History and Society is equally vital and will build on the experience of the University in such areas as Early and Medieval Irish, Law, Philosophy, Ancient Classics and Hispanic Studies.

The University has a long tradition in the Food and Health Science areas and among the areas of proposed research are Food Safety and Toxicology, and Consumer Analysis.

University College Dublin £24,535,000
The funding will be used to support the Conway Institute for Biomolecular and Biomedical Research; a new Institute for the Study of Social Change and a National Social Science Archive. Collaborative links have been secured with RCSI in relation to the work of the Conway Institute and, in the area of social science, with TCD.

The Conway Institute is named after Prof. E.J. Conway, one of Ireland's most distinguished scientists. Among the areas to be explored by the Institute are the study of human and animal diseases, e.g. inflammatory bowel disease, diabetic kidney disease, infectious diseases such as HIV and Hepatitis C, and rheumatoid arthritis.

The new Institute of Social Change will bring together research programmes in a range of areas in the social sciences – including economics, sociology and politics, with a particular focus on the impact of political change in Irish-British relationships, EU relations and global change.

University of Limerick £11,406,000
The funding will go towards supporting the Materials and Surface Science Institute. The Institute will be engaged in research activity in materials design for use in information storage and processing; transportation; healthcare; and environmentally sustainable industrial chemical processing.

Collaborative partnership arrangements have been made with Waterford Institute of Technology, and with UL's partners in the Atlantic University Alliance – NUI, Galway and UCC.

Research Resource £20,300,000
In addition to the programmes outlined, the Assessment Panel approved funding of just over £20m for a research resource comprising a library and information centre in Trinity College, Dublin. This confirms the national and international significance of the Trinity College Library. The new funding is to ensure that the Trinity Library will have the capacity to serve the whole of the Irish research sector. This will be an important resource for all areas of scholarship but particularly in the humanities and social sciences.

NATIONAL CENTRE FOR TECHNOLOGY IN EDUCATION — ANNE PHELAN

Initiatives from the grassroots - Schools IT 2000 supports innovation

Mr Micheál Martin, TD, Minister for Education and Science, with Ciara Galvin of Scoil Bríd, Celbridge, Co. Kildare, at the launch of the Schools IT 2000 Schools Integration Project. Ciara's school is participating in a project entitled "Primary Science and ICT".

Last year saw the introduction of Schools IT2000 in primary and post-primary schools throughout Ireland, and the establishment of the National Centre for Technology in Education (NCTE) to implement the project. Through its four different initiatives, which promote the development of technical infrastructure, training provision, on-line services and pilot project development, this three-year project has stimulated great interest among teachers, parents and pupils alike.

A forum for involvement of all interested parties was launched this year in the form of the Schools Integration Project (SIP) initiative. Under SIP, 228 schools were selected to take part in 48 pilot projects focussing on technology integration in schools, and in the curriculum in particular. Of the 48 projects, 17 have a Science, Mathematics or technical orientation. Mr Micheál Martin, TD, Minister for Education and Science, launched the projects on March 30, 1999.

SIP is about bringing creativity and participation, co-operation and innovation to action research and development projects which will serve as learning models for the future, more widespread integration of technology in all schools. Teachers, students and other partners in education have brought exciting ideas from an embryonic concept to project implementation stage in the course of a few months. Significant commercial sponsorship and participation has been instrumental in facilitating the projects' development and will continue to provide an important industrial link.

ScoilNet *(www.scoilnet.ie)*, the website for Irish schools, will host much of the information about, communication for and dissemination on the outcomes of SIP projects. ScoilNet is another major Schools IT2000 initiative, which provides information, educational content, interaction platforms and on-line services to schools.

Other achievements of Schools IT2000 during its first year of implementation were:
- £15 million distributed to Irish primary and post-primary schools for purchase of information and communications technology (ICT)
- All schools provided with Internet access courtesy of Schools IT 2000 partner Telecom Éireann
- Training facilities provided to Education Centres
- Training courses provided to over 35,000 teachers
- ICT planning, advice and information circulated to schools and available through ScoilNet
- Partnerships developed with a range of commercial companies to enhance the impact and effectiveness of Schools IT2000
- Acceptable use of the Internet policy guidelines for schools distributed
- Appointment of regional ICT advisors.

Further information:
National Centre for Technology in Education, Dublin City University, Dublin 9;
Tel: 01-704-8200; Fax: 01-704-8210;
Helpline: 1850-704040;
E-mail: info@ncte.ie; Web: www.ncte.ie and www.scoilnet.ie

NATIONAL COUNCIL FOR EDUCATIONAL AWARDS — MARIAN O'SULLIVAN

Skills shortages in science-based industries in Ireland

In January, the National Council for Educational Awards (NCEA) held a very successful conference entitled *Skills Shortages in Science-based Industries in Ireland – Myth or Reality?* at the Institute of Technology Tallaght. It was attended by over 60 delegates from the Institutes of Technology, private colleges, industry and public sector bodies.

The conference was officially opened by the NCEA Chairman, Dr Mary Upton, and the keynote address was given by Dr Chris Horn, Chairman of the Expert Group on Future Skills Needs. Dr Horn's paper highlighted the findings of the Expert Group in relation to the future demands for technologists within the Information Technology sector. Papers were also presented on skills needs in the Pharmaceutical and Chemical Industrial sector, the Physics/Instrumentation sector and the Computing sector. Mr Matt Moran, Director of IBEC's Irish Pharmaceutical and Chemical Manufacturers Federation (IPCMF), reported that results of a survey carried out on a sample of 33 companies revealed that employment will grow by 34% in the sector in the next three years, and that companies forecast difficulties in filling 28% of these

Dr Columb Collins, Director, Institute of Technology Tallaght, Mr Matt Moran, IBEC, Prof. Lee Harvey, Centre for Research into Quality, and Dr Marian O'Sullivan, attending the NCEA Science conference.

positions. Ms Jennifer Condon, Director, National Software Directorate *(http://www.nsd.ie)* gave an overview of the growth and dynamism of the Irish Software Industry. She reported that Ireland has almost 700 software companies and is the second largest exporter of software in the World after the United States. Ms Condon highlighted the importance of the Industrial and Educational sectors in working closely together in adequately addressing future skills requirements. Ways of increasing the skills pool include flexible approaches to learning, improvement of third level completion rates, re-training and in-company training. Professor Lee Harvey, Director, Centre for Research into Quality, University of Central England *(http://www.uce.ac.uk/crq)* presented research findings which suggested several ways in which higher education and employers could help students be successful at work. Mr Pat Timpson, Head of the School of Science, Institute of Technology Sligo, presented a paper on the trends of student intake into third-level science courses. The paper reported the serious decline in the number of first preferences for science (chemistry, physics, biology) courses in Institutes of Technology, DIT and Universities. In the former, first preferences have declined by 21% in the past three years.

The conference concluded with a lively plenary discussion and debate. The importance of co-operation between the policy makers, industrialists and educational and training groups was highlighted.

Further information from Dr Marian O'Sullivan; E-mail: mosullivan@ncea.ie

Awake! O Leaders and Press of Ireland

It is not every day one gets a chance to unveil one's own memorial....It is a sobering experience. But I have never turned down a chance to harangue the Irish public on my favourite themes!

I am tempted to spend my time telling you of the great need for properly financed research in applied science, so that we could for example generate electrical power with the peat of our boglands, or technical education to provide the skilled people for such things – which the government would be better employed in pursuing than setting up grandiose schemes like the Royal University in Earlsfort Terrace, in an attempt to settle the Irish University question.

My life has been torn between passions for scholarship and the advocacy of science, and I have not always had time to work out my ideas about physics in full. When I had the idea of the contraction of a moving body, while sitting talking to Lodge in Liverpool, I was, as usual, in too much of a hurry to do much with it and I sent it off to a journal in America that nobody paid much attention to, and forgot all about it, until Lorentz came along with the same thing.

Indeed I should have done more to work these things out, I suppose. Pontificating to the British Association every few years isn't enough.

But I can be proud of my close circle of friends, who pool their efforts: Lodge, Thomson, Heaviside, Larmor and the rest. And my students – Thomas Preston, John Sealy Townsend. And then there's Lyle, over in Australia....

Arrived down here from Coleraine one day – says in a quiet sort of a way he'd like to try his hand, or his foot, at Football, that is to say, Rugby Football. Next thing we know, that same season he's playing for Ireland. Goes on to be champion sprinter.... Then advises the powers that be on the best means of lighting lighthouses, is it to be gas, oil or electricity? ....a good offer comes from Melbourne and he's gone too.... but he's set up the first physical Laboratory south of the equator!

God, if we've done anything for society, we've lightened its darkness with all the marvellous forms of lighting we've dreamed up, even if we haven't lightened the darkness in the minds of politicians – or the Board of TCD!

I try all the time to bring such new inventions here for people to see – microphones, telephones, every description of electrical machine, photometers, calorimeters, spectroscopes, batteries – even the Lilienthal glider that I flew in College Park – we must open the eyes of the people to practical science.

The people did I say? No, it's the ruling class that is at fault, that won't give them a proper technical education. It is all very well

G.F. FitzGerald unveils his memorial plaque at 7 Ely Place on 10 June 1999.

to complain that the industrial classes are not industrious, that they are not cleanly, that they are fond of loafing.... this may all be quite true; but who is it who sets them the example of being content with what their forefathers did? Who sets them the example of refusing to change with the times? Who sets the examples of behaving like Red Indians and Australian savages who cannot change with the times and are consequently being exterminated?

- It is those who should be their leaders!
- It is the authorities of the University of Dublin.
- It is the Board of Intermediate Education.
- It is the Board of National Education.
- It is those gentry who think more of sport than of industry, who have left it to the nuns to teach the people to clear away the dirt from their houses and the manure-heaps from their front doors....

Awake! O Leaders and Press of Ireland.... Before it is too late – before the people of Ireland are swept away like the Red Indian and the Maori by the competition of a people with competent leaders. The people are starving for bread and you have given them educational stones. And their blood cries out against you.

Sometimes I smile when I read what I said, back at the beginning, on the subject of Maxwell's theory: that it might free us from the thralldom of the material ether and be full of new possibilities for science.... True enough! Now we have **wireless telegraphy**, based on the waves that we predicted. As I said to the British Association:

"It was a great step in human progress when man learnt to make material machines, when he used the elasticity of his bow and the rigidity of his arrow to provide food and defeat his enemies. It was a great advance when he learnt to use the chemical action of fire, when he learnt to use water to float his boats and air to drive them, when he used artificial selection to provide himself with food and domestic animals.

"For two hundred years he has made heat his slave to drive his machinery. Fire, water, earth and air have long been his slaves. But it is only within the last few years that man has won the battle lost by the giants of old, has snatched the thunderbolt from Jove himself, and enslaved the all-pervading – ether!"

So let that be my epitaph.

---

* *By a miraculous physical process, yet to be explained by science, Professor George Francis FitzGerald F.R.S. (1851-1901), the eminent TCD Physicist, appeared on the doorstep of 7 Ely Place, Dublin 2, on 10 June 1999 to unveil his memorial plaque. This is the speech he made on that memorable occasion. Any superficial resemblance to George's successor in the Erasmus Smith's Chair of Natural and Experimental Philosophy at Trinity (Professor Denis Weaire F.R.S.) is entirely co-incidental.*

The plaque.

ST PATRICK'S COLLEGE, MAYNOOTH — CHARLES MOLLAN

Well worth a visit

St Patrick's College, Maynooth, the only Pontifical University in these islands, has an important place in the history of science, for it was there in 1836 that Reverend Professor Nicholas Callan (1799-1864) invented the induction coil. An experimental genius, Father Callan also invented batteries and patented a means of protecting iron from rusting. He made use of "volunteer" seminarians to test his "high tension electricity", and, in the process, rendered unconscious a future President of the College, later Archbishop of Dublin, William Walsh. Another future President of the College, Charles Russell, had to spend time in the infirmary following the Professor's experiments.

The current President of St Patrick's College – Monsignor Dermot Farrell – a graduate of physics and mathematics, was not subjected to Callan's electrical attentions. But it was his initiative to celebrate the 200th anniversary of Callan's birth by completely renovating and refurbishing the Museum at Maynooth. The College has, as might be expected, a fine collection of ecclesiastical artefacts, but it also has a remarkable collection of historic scientific instruments. The main focus of the new "National Science Museum" is these instruments, but it also exhibits many of the finest of the ecclesiastical items.

Supported by a generous grant from KELT (County Kildare Leader II Company – funded by the EU), and aided by the donation of electronic equipment from Intel Ireland and the Institute of Physics, the Museum has been transformed by the team which mounted *The Mind and the Hand* Exhibition at Trinity College in 1995, and *The Irish Innovator* Exhibition at the Aer Lingus Young Scientists' Exhibition at the RDS in 1996 – Dr Charles Mollan of Samton Limited, Martin Murray of Tony Mullan Architects, and Denise Bradbury-Byrne of JDB Design Limited.

In specially designed and illuminated cases, you can see not only the finest of ecclesiastical vestments, crosses, chalices and medals, but the oldest Irish signed scientific instrument in the country (a surveyor's compass dated to 1688), many of Callan's original

Part of the scientific instrument display at the National Science Museum, Maynooth.

instruments and inventions, and major collections of physical apparatus in electricity, magnetism, surveying, navigation, sound and light. It is by far the finest public display of historic scientific instruments in the island of Ireland. Many of the items exhibited were used by staff and students at Maynooth, but the collections were dramatically increased over the years – mainly due to the efforts of the late Reverend Professors P.J. McLaughlin and Michael T. Casey. The current curator, Dr Niall McKeith, is continuing this tradition – most recently through the acquisition of a collection of instruments kindly donated by Loreto Abbey at Rathfarnham. Computers allow the use of CD ROMs and connection to the Internet to add a modern touch to the historical artefacts.

Certainly well worth a visit.

Contact: Dr Niall McKeith, Curator; Tel: 01-708-3780;
Opening Hours: Tuesday & Thursday 2.00-4.00 pm,
Sunday 2.00-5.00 pm. Entry: £1.00.

IRELAND'S HISTORIC SCIENCE CENTRE, BIRR, CO. OFFALY — ALICIA PARSONS

New Galleries of Discovery

Ireland has long been known as a land of dreamers and schemers, but it is only now with the explosion of technological advancement that the Celtic Tiger has come of age. However this success did not come overnight. For many decades the imagination and intellect of Irish men and women have played a part in preparing the ground for this generation to profitably expand our Celtic imagination. Nowhere can this be better understood and enjoyed than at Birr Castle Demesne where young scientists of the future can voyage with the inventors, thinkers and dreamers who went beyond the possible to push back the boundaries of discovery.

The recently opened Galleries of Discovery are dedicated to the significant achievements of members of the Parsons' family and to their contemporaries in the fields of Astronomy, Photography, Engineering, Microscopy, and Botany.

The Galleries use videos and scaled moving models to explain how the third Earl of Rosse built one of the greatest telescopes in the world in the middle of Ireland during the nineteenth century. He did this using local labour and machines which he designed and built to overcome each challenge as it emerged. Further Galleries focus on the observations and discoveries made with what became known as the "Leviathan of Parsonstown".

The wife of the third Earl was a pioneer photographer. Ironically, it was photography which sounded the death knell for the great telescope, as it was not suitable for celestial photography. The science of photography was in its infancy in the mid nineteenth century. This meant that photography was more akin to chemistry than art. Photographers were forced to prepare their own sensitised plates. Mary, Countess of Rosse was equal to the challenge.

The fourth Earl of Rosse's achievements in astronomy also feature, as he designed and constructed at Birr an apparatus which made the first accurate estimate of the temperature of the Moon.

The engineering galleries focus on Charles Parsons, the inventor of the steam turbine, which revolutionised marine transport. His invention, though, was not easily accepted by a sceptical admiralty in the 1890s. His experimental vessel, the "Turbinia", powered by his steam turbine and twin screw propellers, performed figures of eight around Queen Victoria's fleet in 1897. The point made, the Admiralty set about employing his technology, which then became standard for both naval and merchant vessels.

Part of the photographic display in the new Galleries.

In the Demesne of Discovery outside the Galleries can be found: the "Leviathan" now fully restored; a water turbine which adequately met the whole town of Birr's electricity requirements at the start of this century; a fountain powered by simple gravity; and a suspension bridge built nearly two hundred years ago that has stood the passage of time, as well as the passage of six generations of the Parsons family.

Contact: Ireland's Historic Science Centre,
Birr Castle Demesne, Co. Offaly;
http://www.birrcastle.com.
Open from 09.00-18.00, Adults £4.00,
Child £2.50, Student & OAP £3.20,
Guided tours £25 must be pre-booked.

The Scientific Museum at AGB - forging success in the future with respect for the past

In AGB Scientific, we are very aware that we are scientists dealing with scientists. There is a conscious effort to employ scientifically competent, qualified people in the sales and engineering staff, so that everyone speaks the same language. Communication is the key and, if we can communicate effectively, time is saved for all concerned. Communication is as old as time itself and, through the ages, scientists, inventors and free-thinkers have communicated effectively with each other and the scientific journals of the day to produce many valuable inventions which have revolutionised the lives of millions. In many, well-documented cases, serendipity lent a helping hand and, without unintentional events during the experimentation processes, we would have had to wait longer for the discovery of penicillin; vulcanised rubber; the structure of DNA; and even Post-It™ notes. In other cases, the progression from prototype to the final product has been long and painstaking. In the seventeenth century, John Harrison, inventor of the Marine Timepiece had to endure many setbacks, not only in the design and functionality of his invention, but also in the face of harsh opposition and incredulity from the so-called best minds of the day. Eventually, though, his persistence paid off and, with the intercession of the King and a head-to-head contest with the Astronomer Royal, Harrison succeeded in proving that his invention was the only way to accurately plot longitudes at sea.

While AGB Scientific is quite rightly seen as a forward thinking company, we retain our links with the past in a very special way - we have a museum of scientific artefacts at our Dublin headquarters. It is fascinating to see the origins of today's scientific instruments. Precision balances used today can weigh to the nearest microgram. In our museum, we have balances which date back to the eighteenth century, when coin balances were used to verify currency. Then, anyone claiming a balance had such mystical power would have been burned as a witch!

The measurement of temperature has been vital for thousands of years. It is only relatively recently, with the advent of commercial electronics, that the design of glass thermometers has been improved. In 1861, James Joseph Hicks from Roscarberry, Co. Cork, made the first clinical thermometer, and we have an example of a Hicks marine thermometer in our museum. It is an alcohol in glass thermometer, with graduations from 90 to 105°F.

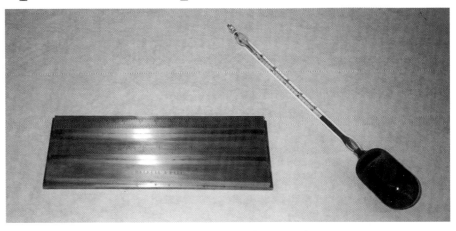
The Wedgwood pyrometer and Hicks marine thermometer.

Also on display is a pyrometer, manufactured in Paris in the nineteenth century and based on a 1782 design by Josiah Wedgwood for the measurement of high temperatures in his pottery kilns. According to the paper subsequently published in the *Philosophical Transactions of the Royal Society* of London in 1782 (Vol. LXXII, part 2), the principle by which the temperature was measured was a simple one. The pyrometer was made of brass, forming two graduated and converging channels. A piece of clay of a prescribed size was put into the oven during the firing process and was then slid along the brass channels as far as it could go. The clay,

The three Ugentarium bottles and Daniell hygrometer.

as described in later papers, was Cornish porcelain clay, which shrinks in a uniform manner when heated. The graduation on the brass channel of the pyrometer corresponded to the temperature to which the piece had been subjected. Obviously, these pyrometers had to be calibrated before use, to make sense of the graduations. Although, as SI units were only proposed in 1960, there was no ILAB accredited calibration to national standards back then.

The measurement of dew point has been important for many years, and is used to evaluate relative humidity. In our museum, we have a hygrometer designed in 1820 by J.F. Daniell. Ether is poured over the muslin-covered bulb. The ether evaporates, cooling the liquid inside the tube, causing water vapour from the ambient air to condense on the black surface of the second bulb. The temperature inside the black bulb is read off the integral thermometer and compared with the temperature of the ambient air. Tables were then used to determine the relative humidity. Some of today's hygrometers work on a similar principle, although the condensation of moisture from the air is brought about by electronic cooling devices and detected by a very accurate infrared sensing device, in a chilled mirror hygrometer.

Given our origins as a glass manufacturer, it is hardly surprising that we have some historically interesting glass pieces in our collection. The Ugentarium glass bottles on view date from the second century AD, and were used by the Romans to collect tears for medical diagnosis. Obviously, Associated Glass Blowers Limited would have been superfluous to requirements in those days, if the glassware was all built to last for two thousand years!

Visitors to our Scientific Museum are always welcome: why not pop in, the next time you are at AGB Scientific?

Contact:
Maureen McLoughlin,
AGB Scientific Limited,
Dublin Industrial Estate, Dublin 11;
Tel: 01-882-2222;
Fax: 01-882-2333;
E-mail: maureen.mcloughlin@agb.ie;
Website www.agb.ie

A new approach to promoting sustainability for the new Millennium

Situated on the outskirts of Ballymena town, the **ecos – millennium environmental centre** is a 150 acre, £10 million development jointly funded by the Ballymena Borough Council and the Millennium Commission. The aim of the centre is to promote the principles of sustainability through interpretation and providing examples of various approaches used on site.

There are a range of facilities being provided, but they are all based on the two main elements of a public park designed to encourage biodiversity, and an interpretive facility occupying a building designed for energy efficiency and obtaining two thirds of its energy from renewable sources produced on site.

The site consists largely of semi-improved wet grassland subject to regular flooding and providing rough grazing. As a result of past management, it has a reduced diversity of wildlife with a preponderance of rushes to the exclusion of other species. Improvement of the site for biodiversity has already begun and is involving a range of approaches:

- Improving the grassland management through introducing a grazing and mowing regime.
- Planting approximately 17 acres of native woodland in conjunction with the Woodland Trust and using seedlings grown from seed collected within a 15 mile radius of the site.
- Creating a lake to improve habitat diversity and improve visual amenity.
- Increasing marginal vegetation by planting reed beds on one side of the lake using reeds available elsewhere on site to seed the bed.

These actions will improve the biodiversity of the site gradually as the management actions take effect and as newly created habitats develop.

The 3,000m² landmark building has been designed as a thermally massive concrete structure with high levels of insulation, and incorporates a central tower of steel clad with copper *(see illustration)*. Throughout the design process, materials and design have been examined to try to achieve the best sustainable result taking into account scheme purpose and budget considerations. PVC elimination where possible, under-floor heating, and a sewerage treatment reed bed plant to treat waste water are some of the sustainable approaches taken.

In addition, approximately two thirds of the energy required by the facility will be produced by a combination of willow coppice fuelled combined heat and power production, a 30kw wind turbine, over 100m² of photovoltaic cells, and 20 solar water heating panels.

The important factor in all of the above systems is that they are all significant energy producers for the facility and will be monitored to provide data on energy production by renewable energy technologies at a usable scale in the Northern Ireland situation.

It will be equally important that the management regime reflects sustainable principles, and these will impact on all aspects of the ongoing management. This will range from sales materials stocked in the shop outlet, through the menu selection for the café, to the administrative systems and waste policies implemented on a day-to-day basis.

View of the model building.

Facilities being Provided

- Within the park a series of walks is being created through the improvement of an existing public path and creation of new paths, which will give a range of walks between 1km and 5km. Some of the paths will be combined cycle and pedestrian paths linking the Ballykeel estate to the town centre and connecting with developing cycle routes.
- In association with the paths, a series of sculptural icons is being constructed to relate aspects of the park to the interpretive centre. These will be designed to incorporate a play element and will combine with a separate play area to encourage use of the entire park area by children.
- A small 17 site touring caravan site is being incorporated in the scheme to provide a facility not currently available in the area and directly on route to the north coast.
- As part of the sustainable approach to the development, an organic market garden production area has been incorporated since the inception of the scheme. This will be operated by a local community company and will provide opportunities for employment and training, as well as tying in with the interpretive aspects of the development.
- The interpretive centre, with its associated café and shop, is the major user of the main building and utilises participation and involvement as a means to stimulate interest and hence learning. The facility aims to encourage asking the right kind of questions rather than providing definitive answers.
- Associated with the interpretation function is an education room providing facilities for school and other group use. The centre will provide a very useful venue for school groups, and information will be targeted toward the curriculum and provided in an appropriate form for different key stages.
- A 120 seater conference theatre has also been incorporated to provide all the facilities needed for conferences or lectures and, because of the surrounding facilities, is likely to prove a popular venue for environmental and other conferences.

The Northern Ireland Environmental Information Centre

- In addition to the facilities above, the building will also be home to the Northern Ireland Environmental Information Centre. The need for this facility has been recognised for a number of years but funding has not been available. Now the premises have been incorporated in this building and the centre will be operated by an independent charitable trust with the remit to make a wide range of environmental information on Northern Ireland available to the public in a "one stop shop" facility.

The above article briefly describes the ecos - millennium environmental centre facilities and approach. For the design team certainly one of the most important lessons has been that there are no right answers. However there are a range of options for every problem and ultimately the solution chosen in any particular situation will depend on a combination of purpose, available options and budget. The ecos - millennium environmental centre will not provide easy answers to questions about sustainable development, but it will stimulate thought and hopefully progress the debate about a sustainable approach to life in the new millennium.

Contact:
Mr Billy Reid, Project Officer, ecos,
Ballymena Borough Council,
80 Galgorm Road, Ballymena BT42, 1AB;
Tel: 01266-660300; Fax: 01266-660400;
E-mail: townclerk@ballymena.gov.uk

Association of Clinical Biochemists in Ireland

Introduction
The biochemical milieu of the human body is tightly regulated under normal circumstances, but biochemical abnormalities occur in many disease processes. Clinical Biochemistry deals with the measurement of the biochemical constituents (ranging from ions to complex proteins) of body fluids and tissues, for use in diagnosing disease, monitoring the course of disease and the response to treatment, and also in disease prevention.

Activities
The Association of Clinical Biochemists in Ireland (ACBI) was established in 1963 to represent the Profession in academic and professional matters. Today, the Association promotes the advancement of the practice of Clinical Biochemistry by increasing awareness of developments in medicine, biochemistry, technology and management, and by encouraging research and development. This is achieved through a programme of scientific meetings held throughout the year, a two-day annual conference, publication of four issues per year of *Clinical Chemistry News* and of guidelines on aspects of clinical biochemistry practice. The ACBI has standing committees for professional affairs, scientific affairs, education, registration, the annual conference and the newsletter, and it also participates in joint working groups on quality assurance, laboratory accreditation and information technology. The 120 members include graduates in science and medicine and corporate members.

The ACBI is the national society for clinical biochemistry, the Irish member society of the International Federation of Clinical Chemistry and Laboratory Medicine (IFCC) and of the European Communities Confederation of Clinical Chemistry (EC4). Members of the ACBI have been appointed as associate members on IFCC Scientific Committees and Working Groups dealing with plasma proteins, calibrators in clinical enzymology, markers of bone disease, advanced technology, prostate specific antigen (PSA), markers of cardiac damage, patient/sample identification, and archives. The ACBI is also the advisory body to the EC4 European Register for Clinical Chemists, which sets education and training standards appropriate for Clinical Biochemists at the higher professional level.

Right: Delegates at the ACBI Annual Scientific Conference 1998, during one of the poster presentation sessions.

ACBI '99 Annual Conference, October 29th - 30th, 1999
Major Themes:
Session 1: Guidelines in Clinical Biochemistry: application of evidence based medicine (Acute Coronary Disease, Therapeutic Drug Monitoring, Thyroid Function)
Session 2: Bone Biochemistry (Metabolic Bone Disease and Biochemical Markers)
Session 3: Applications of Information Technology in Laboratory Medicine (Bioinformatics, Decision Support Systems, Intranet in Medical Laboratories)

Contacts: ACBI Chairman - Dr Sean Cunningham, Clinical Biochemistry Department, St Vincent's Hospital, Dublin 4; Tel: +353-1-209-4430; E-mail: sean.cunningham@ucd.ie
Hon. Secretary - Dr Nuala McCarroll, Department of Biochemistry, St James's Hospital, Dublin 8; Tel: +353-1-453-7941, ext 2934; E-mail: nmccarroll@hotmail.com
ACBI Website: www.iol.ie/~deskenny/acbi.html (with links to websites of IFCC, EC4, and other national clinical biochemistry associations).

The Institute of Physics in Ireland

The Institute of Physics is a learned society and professional body for physicists in Britain and Ireland. Membership is open to all with an interest in physics, and includes categories for students, graduates, members, fellows, associates and affiliates. Qualified members are offered routes to two charted designations – Chartered Physicist (CPhys) and Chartered Engineer (CEng). In Ireland alone, both North and South, there are approximately 1200 members.

The origins of the Institute go back to 1874, when the Physical Society was formed to provide a framework for the discussion of new ideas and knowledge in physics through meetings, conferences and publications. Thus, in 1999, we celebrate our 125th anniversary and look forward to entering the new millennium with continuing enthusiasm.

Physics is a very important branch of science and culture which has played and continues to play a major role in our understanding of the nature of the world around us and of the universe beyond. The discoveries of physics represent one of the greatest achievements of mankind and they have had a very significant impact on our daily lives.

The Institute of Physics in Ireland organises a very extensive programme of lectures throughout Ireland from October to April each year.

Many schools enjoy the benefit of the Institute's Schools' Affiliation Scheme, which provides physics magazines, newsletters and other material outlining resources available to schools, along with information on courses and careers in physics.

For third level students, there is a higher education representative on each campus to encourage students to be involved in Institute activities. Students may become members of NEXUS, a network of student physics societies, with worldwide affiliations.

On the industrial side, the Institute operates a network of over a hundred industry representatives throughout Britain and Ireland. An Industry subgroup for members in Irish Industry encourages closer links with Institute activities. The Institute of Physics Publishing (IOPP) is one of the world's leading publishers in physics.

The Chairman of the Institute of Physics in Ireland (1999-2001) is Professor Denis O'Sullivan, Astrophysics Section, Dublin Institute for Advanced Studies, 5 Merrion Square, Dublin 2, with Dr P. Goodman, DIT, Dublin 8, Secretary, and Dr J. Costello, DCU, Dublin 9, Treasurer.

Outgoing Chairman Dr Bob McCullough with some committee members at the 1999 Spring Weekend Meeting.

For further information see web site at: http://www.tcd.ie/IOP/

IRISH RESEARCH SCIENTISTS' ASSOCIATION

Can or should Irish research have a future?

Irish Science and Technology need a new vision if they are ever to serve Irish Society to their full potential.

This year the Irish Exchequer will have a surplus of funds the like of which has never been seen before. At the same time the country needs a bold, radical vision of where it would like to be in the next 15 or 20 years. How can we use one to buy the other?

In the 1980s the language of retrenchment and belt-tightening was appropriate for an economy that had been on its knees virtually since the first oil crisis in the early 1970s. The time is now appropriate and the facilities and resources are available to move to a new economic lexicon. The single word "Future" must inform all of our thinking, but a future that exploits the opportunities we presently have, rather than a future that is constantly looking over its shoulder, is the future we need.

The Irish Council for Science, Technology & Innovation, Foresight Report recommends a major investment in intellectual capital, of the order of £500 million pounds over five years in identified areas of Irish expertise: this is part of the vision, but there must be another strand that supports constant development of new ideas. We feel that our proposed "Millennium Innovation Trust" provides the foundation for an Irish Research, Technology, Development and Innovation System that can assure Ireland's place in the future.

Over the last thirty years Ireland has developed its education system to the point where its graduates are equipped with the very skills that make this a place where overseas investors feel they can successfully invest. The growing group of Irish high-tech companies are also finding that Irish graduates can meet the needs of these companies, but the very success of our work force means that our expectations of what our life style should be are rising all the time, and that simply assembling goods with no intellectual input is unsustainable.

We are moving from a manufacturing centred economy to a services and, critically, a knowledge-based economy: if we make the correct choices now we can rely on the Celtic Tiger to continue to produce jobs for all her cubs in the future. In making that choice there are some things we should keep in mind:

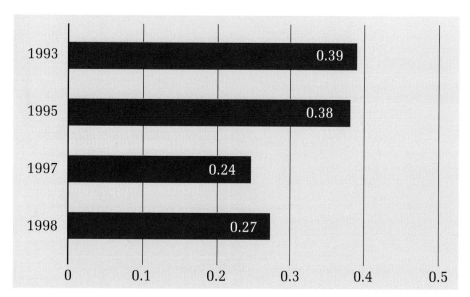

R&D as a percent of GDP (source, Forfás, CSO).

- S&T is driving more than 60% of our economic growth: that's 600 out of every 1000 new jobs, more than all other sectors combined.
- Jacques Delors, former President of the European Commission once said that fully 80% of all goods and services that we will be using in 2010 have not yet been invented.
- Several recent reports from the Government and its agencies all point to a "knowledge economy", and knowledge itself has been described as the "raw material" of the next millennium.
- Our competitors are making this investment now, knowing that they must run simply to stand still in the modern knowledge economy.
- In the IT industry, 70% of revenue is coming from products that didn't exist two years ago.

The inescapable conclusion is – **Brain Power is the future**. Human intellectual capital, contained within the skulls of the work force, is the single must potent factor in the future development of this economy.

One of the primary things that differentiates Ireland from the countries that we would emulate – the small rich nations like Denmark, Netherlands, Finland, and Norway, and the larger rich nations like Germany and Canada – is our pathetically small level of investment in the research base of the economy. The USA invests (privately and publicly) about 3.3% of GDP in research; Finland and Denmark slightly less than 3%; Germany and Norway about 2.5%. Our figure is about 1.5% (a little more than Turkey, Greece or Portugal), and remains one of the lowest figures in the OECD: but of even more concern, the fraction of that 1.5% that is spent on generating new intellectual capital (research) has fallen by almost 50% since 1993! *(see figure)*. If we are serious about our future, this is a game we cannot win.

Though the business expenditure on R&D is growing very quickly here, such research is largely confined to a very small number of large companies in a few areas. By and large, the multinationals present in Ireland do no research work here. Production lines are easy to move, but research facilities are not, as Galway discovered some years ago when Digital closed its production line but not its research facility.

Because of this absence of investment in "normal" science (either in industry or in public), we do not have the things that other normal countries take for granted – a thriving postgraduate and postdoctoral (fourth-level) training system, a thriving and innovative scientific culture, large-scale and ongoing interactions between third-level institutions and industry at all sizes, from fledgling campus enterprises to large multinationals. Normal in the sense that all of competitor countries have such a system. Normal in the sense that investment in science is recognised by our competitors as perhaps the most important investment that they can make in their futures. Normal in the sense that the best work is funded on the basis of a transparent, peer-reviewed system driven by excellence and innovation. Normal in the sense that our competitors accept that such investment pays off dramatically over time scales that are unrecognisable by the exigencies of the next election. Normal in the sense that they accept that such investment costs money, but is worth it. Instead, we have scientific groups which are run on shoestring funds, and agencies who disburse tiny S&T budgets; budgets which would be absorbed *in toto* by a single medium-sized research group abroad. In essence, without a proper investment in public S&T, Ireland is conducting a huge experiment with its future. It is the wrong experiment and there can be only one outcome – and that outcome will happen, not in ten years or five years but as soon as our worth as a manufacturing base is exhausted. We will have nothing to replace it if we do not take action now.

The Massachusetts Institute of Technology conducted a survey of the impact of research carried out by its staff and concluded that:

> If the companies founded by MIT graduates and faculty formed an independent nation, the revenues produced by the companies would make that nation the 24th largest economy in the world.
>
> The 4,000 MIT-related companies employ 1.1 million people and have annual world sales of $232 billion.
>
> That is roughly equal to a gross domestic product of $116 billion, which is a little less than the GDP of South Africa and more than the GDP of Thailand.
>
> MIT: The Impact of Innovation,
> http://web.mit.edu/newsoffice/founders/TofC.html

It is no coincidence that MIT should be so successful: it invests $1.2 billion in research every year. It is no surprise that the bulk of the world's great research universities sit squarely in the richest regions of the world's richest country. This is also true of the other smaller but wealthy countries that we would like to emulate.

So how do we pay for this investment? We can no longer rely on Brussels and so must climb up on to our own two feet. No matter how difficult we find this initially, standing on our own two feet, without the begging bowl of old, will be an exhilarating position to be in. The answer is simple – we must invest from our own resources, and this will cost us money. If we accept that this investment is a necessity, forced upon us by our decision to move to a knowledge-based economy with the demands on intellectual capital that this makes, then we have to make radical choices about resource allocation now. We must implement programmes of investment in research that maximise human intellectual potential. The future payoffs will be enormous and unpredictable. It is as likely to come from developments in the brain and cognitive sciences, as it is from investment in nanochemistry. Just think, for example, of the world markets that exist for rational drug treatments for common diseases that are currently without a cure, such as Alzheimer's Disease.

> **We propose that the Government, with the proceeds of the Telecom Éireann sale, establish a "Millennium Innovation Fund".**

There is widespread acceptance across the political spectrum that the role of the Government in respect of state enterprise is changing – the government can either be owner or be regulator of such enterprises, but not both. Neither can these enterprises be protected as monopoly service providers, as occurred in the past. Telecom Éireann has been sold, yet the "public" debate about what we will do with the proceeds of such privatisations or decentralisations of state-ownership has not really started. Is paying off the much diminished National Debt the only viable option?

We think not and suggest here that a permanent gift to our future would be to use the proceeds of the privatisation of Telecom Éireann (say five billion pounds – about 4% of GNP) to invest in a Millennium Innovation Fund. This fund, perhaps to be managed by the National Treasury Management Agency, would have several purposes. It would generate income, it would secure a permanent investment in brain power through public basic and strategic science programmes, broadly conceived from the social through the biological to the physical sciences. Such a fund would work rather like the Wellcome Trust in the UK (which incidentally funds more biomedical research in Ireland than we do ourselves) or a large pension fund; generating and sustaining its own capital base and income stream.

At the moment Ireland's public S&T effort is largely funded through transfers from the EU. The Germans, French, Danes etc. invest in Irish science; Ireland, by and large, does not. This fund will be used to replace the funding of science that will be lost to the country on the loss of EU structural funding. This fund will provide a future-proofed means of permanently secured research funding indefinitely into the future, a true gift from this generation to its own grandchildren and their grandchildren.

Over a period of a few years, the fund will grow, and the increased income stream can be used to further enhance the productive, innovative and research capacity of the economy. We will be able to retain our best and brightest whom we educate to degree level but who are lost to us when they go to seek intellectually stimulating postgraduate opportunities abroad. We will be able to attract the best and brightest from other countries (as proposed by the Irish Council for Science Technology and Innovation Foresight Exercise) to our third-level institutions and benefit from their presence in our economy – something that occurs in our normal competitor countries (who, incidentally, are benefiting from the presence of Irish graduates in their economies).

Does Ireland want to be a leader or a follower? We think we have been living in the dust of others for long enough. It is time to build the road ahead.

Contact:
Dr John Donovan,
Irish Research Scientists' Association;
Tel: 01-295-0630;
E-mail: secretary@irsa.ie
http://www.irsa.ie

Irish Statistical Association — Catherine Hurley

Conference on Applied Statistics 1999

The Conference on Applied Statistics in Ireland (CASI) is an annual conference held under the auspices of the Irish Statistical Association. This year, a very successful nineteenth conference was organised by Professor Gilbert McKenzie of Keele University, UK, and Mr Mike Stevenson, of The Queen's University Belfast, and held at the Dunadry Inn Hotel, near Templepatrick, Northern Ireland, from May 26-28.

The conference had about 80 attendees, who were informed and entertained by a total of 31 oral presentations, 11 poster presentations, drinks receptions and conference dinners. There were four major sessions led by invited speakers: Dr Sheila Gore of the Medical Research Council – Biostatistics Unit, Cambridge, spoke on BSE and nvCJD; Professor David Balding of Reading University talked about Modelling Genes; while Dr John Fox of the UK Office of National Statistics discussed Official Statistics and Data; and Professor Dennis Conniffe, Economic & Social Research Institute, introduced the General Statistics session with a discussion on Likelihood and its Variants. More details are available from http://www.maths.may.ie/staff/churley/isa/prog.html.

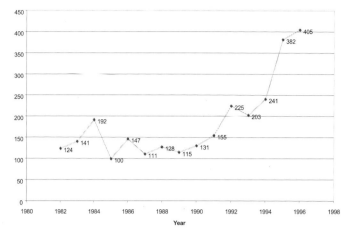
Notification of Bacterial Meningitis, Ireland, 1982-1996.

This year, a prize was given for the best poster presentation. The winner was Ms Gloria Crispino O'Connell, who presented a poster on the Dynamics of Meningoccal Meningitis in Ireland. Gloria began this work while at the Institute of Technology, Tallaght. At present, she is completing a Ph.D. at the Department of Mathematics, National University of Ireland, Maynooth, under the direction of Dr Catherine Comiskey.

According to Gloria, meningococcal meningitis is the most common type of bacterial meningitis in Western European countries, and it is an increasing problem in Ireland. From the 1980s to the present, the number of yearly cases of bacterial meningitis has jumped from less than 150 a year to over 400. In her poster, she formulated an epidemiological model that explained the transmission dynamics of the infection, based on set of differential equations. She presented a mathematical analysis of the model dynamics and the epidemiological impact of disease transmission. For her results, she used computer simulations to show how the disease has spread for the last twenty years and, finally, predicted its spread in future decades.

For more information on the Irish Statistical Association, contact Dr Catherine Hurley, Department of Mathematics, National University of Ireland, Maynooth, Co. Kildare.
E-mail: churley@maths.may.ie

Women in Technology & Science (WITS) — Helen Hughes

£30,000 for WITS Role Model Project

In November 1998 WITS (Women in Technology and Science) received confirmation from the Equality Committee of the Department of Education and Science that it had been successful in securing funding for an important new Role Model Project. Mr Micheál Martin, Minister for Education and Science, met Jacqueline Allan and Helen Hughes from WITS to launch the project.

As has been evident in the Irish media recently, Minister Martin is seriously concerned about the low numbers of students choosing science, engineering and technology (SET) subjects at school and in further education. By supporting this project, the Department is recognising the contribution of WITS in ensuring equality of opportunity in SET.

Jacqueline Allan (left), former Chairperson of WITS, with Minister Martin and Helen Hughes, Chairperson of WITS, at the launch of the Role Model Project.

WITS was set up in 1990 to actively promote women in science, technology and engineering in Ireland through initiatives at industrial, school, higher education, national policy and European levels. As a wholly voluntary organisation, WITS draws its members from all technical and scientific areas: research personnel, industrial scientists, technicians, journalists, engineers, administrators, policy analysts, teachers and lecturers, computer experts and consultants.

Since 1993 a thousand senior cycle girls from all over Ireland have benefited from the WITS Role Model Days. Responses to these, which include an introduction to science and technology career planning and themed workshops, have been enthusiastic.

This project consists of two Phases:

Phase 1 is to produce a revised Role Model Booklet with 40 biographical career sketches of women working in a scientific, engineering or technological career. The booklet will replace the WITS Booklet *Suitable Jobs for Women* which was produced in 1994 and will reflect the scope and flexibility which a scientific or technological training provides. This booklet will be circulated by the Department of Education and Science to all secondary level schools in the country during the academic year of 1999/2000.

In addition, a Role Model Pack will be produced for colleges wishing to organise their own Role Model Seminars. Several colleges have been running such Seminars with some assistance from WITS over the last few years and WITS would like other colleges to organise them. WITS sees this Role Model Initiative to work with educators to increase the profile and outline opportunities in science and technology as an effective way of fulfilling one of the organisation's primary aims. Several research studies have shown that the lack of role models for girls is one of the reasons for the under-representation of girls in SET careers.

Phase 2 of the project consists of running pilot Role Model Seminars in six colleges all around Ireland in the academic year 2000/2001. WITS will provide advice, support and assistance to these colleges in the running of their first Role Model Seminar. Dissemination of the Role Model concept to many other colleges with the aid of the Role Model Pack, is also a vital element of Phase 2.

WITS is grateful to the Department of Education and Science for the funding of this very worthwhile project.

Further information is available from WITS at PO Box 3783, Dublin 4;
E-mail: wits@iol.ie

The RDS Irish Times Boyle Medal Award

It was in 1800 that the Royal Dublin Society inaugurated a medal for scientific research of exceptional merit carried out in Ireland. The Medal is named after Robert Boyle (1627-1691) "the son of the Earl of Corke and the father of Chemistry". Robert was born in Lismore, Co. Waterford, and in 1661 introduced the notion of elements and compounds, almost single-handedly wrestling chemistry from the clutches of the alchemists, and propounding the experimental method on which all modern science is based.

The RDS can truly claim to be among Ireland's oldest scientific institutions, having been founded in 1731 "for the promotion of the Sciences, Husbandry, Manufactures and other Useful Arts". The need for such promotion is even more pressing in modern Irish life.

Today, the RDS Science and Technology Committee boasts an active programme. Annual activities, including the Youth Science and Arts Week, the Science Communication Forum, the Young Science Writers' Competition, along with conferences and a wide programme of lectures, aim to encourage an interest in science among young people and promote public awareness of scientific issues, while serving the scientific community.

The RDS in conjunction with *The Irish Times* have devised a new dynamic for its traditional Boyle Medal Award to distinguished Irish Scientists, initially awarded in 1899 to Prof. George Johnston Stoney who had first proposed the term "electron" for the elementary particle at an RDS lecture in 1891. This joint initiative is designed to direct attention to the Irish tradition of excellence in scientific initiative and innovation – a tradition which has not up to now received the recognition it deserves.

The centenary re-launch of the RDS Irish Times Boyle Medal Award in its new format will be inaugurated in 1999. The Medal and Bursary of £30,000 will be awarded to a scientist working in Ireland for research which has attracted international acclaim, to enable the recipient to employ a researcher for a period of three years to make significant further progress. The Medal in 2001 will be awarded to an Irish researcher working abroad. In 2003 the four year cycle will repeat. Much of the work of the RDS S&T Committee is directed to increasing scientific awareness in the national cultural context, and its membership includes representatives from most of the major scientific disciplines and institutions in Ireland.

Further details on the RDS S&T programme are available from Chris Shouldice, Chairman, S&T Committee, or Carol Power, Development Executive, Science & Industry, RDS, Ballsbridge, Dublin 4; Tel: 668-0866 ext. 217; E-mail: carol.power@rds.ie; WWW: www.rds.ie

Short list for the centenary Boyle Medal

The suggestion that there should be a new format, and a £30,000 research bursary, for the Boyle Medal was originally made in November 1996 by Dr Charles Mollan, former Science and Arts Officer of the Royal Dublin Society (RDS) in a submission to Professor Dervilla Donnelly, Past President of the RDS, Governor of The Irish Times Trust Limited, and Director of *The Irish Times*. Other key players in the development of the concept were Conor Brady, Editor, and Dick Ahlstrom, Science Editor, of *The Irish Times*, together with Carol Power, Development Executive, Science, at the RDS, and Aine Maguire and Maeve O'Meara of *The Irish Times*.

The three short listed candidates for the award were announced in *The Irish Times* at the end of June. The winner of the centenary Boyle Medal, and the £30,000 research bursary, will be announced in October.

The short listed candidates are: Professor Michael Coey of Trinity College Dublin, Professor Tom Cotter of University College Cork, and Dr Luke O'Neill of Trinity College Dublin.

The portraits of the short listed researchers, and abbreviated captions, are reproduced by kind permission of The Irish Times, *in which they were published on 28 June 1999.*

Professor Coey is professor of experimental physics in the Department of Physics, TCD. He is a leading international researcher in the area of permanent magnets, not the familiar iron versions but the much more powerful magnets based on novel combinations of elements. See also page 155.

Professor Tom Cotter is professor of biochemistry and head of the Biochemistry Department at UCC. His particular area of expertise is the natural and necessary process of programmed cell death, known as apoptosis.

Dr Luke O'Neill lectures in biochemistry in the Department of Biochemistry at TCD and is Director of the College's Biotechnology Institute. His main area of research is inflammatory disease, a typical example of which is rheumatoid arthritis. See also page 148.

Awards and Fellowships available to Irish scientists

It may not be well known that the Royal Irish Academy makes some fifty awards and fellowships each year in both the humanities and sciences. These are provided through the Academy's network of national Committees, exchange schemes with other countries, and some small commemorative funds. Those relevant to Irish scientists are summarised as follows:

European Exchange Study Visits and Fellowships

To promote academic exchanges between Ireland and Austria, France, Great Britain, Hungary and Poland. Senior scientific researchers working in Ireland must apply by **October 15th** each year. Awards include the cost of all living expenses and travel for a specified period, not more than 4-6 weeks in the case of study visits. The Royal Society (UK) scheme also supports long-term fellowships and joint projects.

Senior Visiting Fellowships

To enable a new scientific research technique or development to be introduced into Ireland from OECD countries. Senior scientific researchers must apply by **October 15th** each year. Travel expenses for study visits up to six weeks to and from the Republic of Ireland are awarded annually in December.

National Committee for Biochemistry Award

Awarded to a scientist who is actively engaged in biochemical research in Ireland, primarily to recognise scientific work carried out in the past decade. Nominations must be submitted by two scientists before **2nd April**. A silver medal is awarded annually and the recipient must deliver a review lecture at the Royal Irish Academy and also at a meeting of the Irish Area Section of the Biochemical Society.

Royal Irish Academy Parsons Medal for Engineering Sciences

A new award in recognition of outstanding research in engineering sciences, published in refereed international journals, by persons less than 40 years of age when the work was carried out. It will primarily recognise work carried out in Ireland in the last decade. A bronze medal is awarded. Nominations may be sought in **Autumn 1999**.

The Royal Irish Academy, Dublin.

National Commission for Microbiology Award

Awarded for scientific work carried out in Ireland in the past decade by persons less than forty years of age on the 1st January of the year the award is made. The award is biannual. Nominations must be made independently by two scientists before **1st January**. A silver medal is presented and the recipient delivers a review lecture at the Royal Irish Academy.

National Committee for Nutritional Sciences Award

Awarded in recognition of research contributions in nutritional sciences in refereed international journals. Only persons less than forty-five years of age on the 1st January of the year of the Award are eligible. Nominations must be made independently by two scientists before **1st January**. A silver medal is presented and the recipient must give a review lecture at the Royal Irish Academy.

Consultative Committee on Pharmacology and Toxicology Award of Merit

This Award of Merit recognises outstanding work carried out by a scientist actively engaged in research in Ireland within the disciplines of pharmacology and toxicology. Nominations of potential recipients by not less than two persons, who will have obtained the prior agreement of the nominee, should be submitted in **Autumn 1999**. A silver medal is awarded and the recipient will be required to give an address to the Royal Irish Academy.

National Committee for Biochemistry Science Writing Competition

Undergraduate and graduate students in third-level institutions in Ireland (both North and South) are invited to write a newspaper article suitable for the non-scientific reader, explaining the development of any new topic in biochemistry and its significance to medical, agricultural or industrial practices. The winner is awarded IR£300 and his/her article is published in *The Irish Times*. The deadline is **19th January** each year. *(A summary of the 1999 winning article will be found on page 15.)*

National Committee for Biochemistry Travel Bursary Scheme

Graduate students in biochemistry and associated areas in any third-level institution in Ireland (north and south) are invited to apply for travel bursaries. These bursaries are designed to assist research students to attend a workshop or to visit a laboratory in order to learn new techniques/information of particular benefit to their work. The deadline is **1st March**.

National Committee for Chemistry "Three Societies" Prize

This competition involves the Institute of Chemistry of Ireland, the Royal Society for Chemistry and the Society of Chemical Industry. Students at second and third levels are invited to write an essay related to chemistry and the winning entries are published. The deadline is in **January**.

Praeger Fund for Fieldwork in Natural History

Established in 1955 in memory of the famous Irish naturalist, Robert Lloyd Praeger. A number of annual grants are made available for fieldwork relevant to the natural history of Ireland. Grantees need not be based in Ireland and applications are particularly welcome from amateur natural historians. Annual deadline: **15th February**.

For information on any of the above, contact: The Royal Irish Academy, 19 Dawson Street, Dublin 2; Tel: 01-676-2570; Fax: 01-676-2346; E-mail: admin@ria.ie

Science, Technology & Innovation Awareness Programme continues through 1999

The Science, Technology and Innovation Awareness Programme, originally intended as a three-year awareness initiative, has been extended by the Minister for Science, Technology and Commerce for a further year. Minister Treacy said that, in providing funding to continue the STI Awareness Programme in 1999, he was confident that targeted activities would continue to be developed. "Widespread public information on science and technology issues is critical to our development in the technology age and, in particular, our younger people need to be well educated in science and technology, the career opportunities it provides and its importance to our economy".

The 1998 Survey of Public Attitudes to Science, Technology and Innovation undertaken by the Programme found that overall attitudes towards science and technology are generally positive and stable, and in a number of specific areas there are strong and encouraging signs.

There is a widely shared impression of increased media coverage for science, technology and innovation over the past two to three years. Regard for science and engineering as a profession is high among senior business people and civil servants: however, among students and parents there was a low level of familiarity with S&T careers and this is cause for concern.

The S&T sectors are at the core of Ireland's economic growth, and the numbers employed, particularly in the software industry, are expected to rise significantly over the next three years. It is essential that young people are made aware of these opportunities and given the chance to hone their skills. The research confirms the need to better define and demonstrate the importance of S&T in terms of role models, identifiable career opportunities and impact on the Irish economy.

With this in mind, the STI Awareness Programme has announced a number of new initiatives over 1999.

- The launch of a National Science Day for all primary schools to take place on Friday 12 November. All primary schools throughout the country will be invited to take part in hands-on science activities for the afternoon, with inter-active quizzes and fun experiments.
- The IBM/STI Science & Technology Journalism Awards were announced in August. Applications are open to all journalists who have written an article in the area of science and technology. This includes all articles printed or broadcast in the Irish media, in trade press, academic journals or newsletters. The Awards will be made in December.
  The objective of the competition is to improve the level of coherent and accessible information available to the general public through this increasingly important field of journalism.
- The publication of a National STI Poll on the general public's opinions on, and attitudes towards, science and technology issues and their impact on everyday life. The Poll will also ascertain how much parents know about career opportunities for their children in the areas of S&T, including their levels of awareness of the points required for entry to science and technology degree courses.
- The compilation of a Speakers' Directory, in conjunction with the Forfás Skills Awareness Programme, which has been sent to career guidance teachers in every secondary school in the country. The objective is to outline opportunities in the area of S&T by providing relevant role models to talk to students in their locality about their job, career path and future aspirations. Industry and academics have reacted very positively to this initiative and we are very grateful for their support.

Professor Yrjo Neuvo, Senior Vice President for Product Creation at Nokia Mobile Phones, guest lecturer at the 1998 Irish Innovation Lecture, with Mr Noel Treacy, TD, Minister for Science, Technology and Commerce, and Mr Adam Ingram, MP, Minister of State at the Northern Ireland Department of Economic Development.

In addition, several existing activities have continued:

- The National Innovation Awards, sponsored by the Awareness Programme, PricewaterhouseCoopers and *The Irish Times* are in their third year. The Awards aim to highlight our most innovative companies and the ingredients that make for successful innovation. The overall winners of the 1998 Awards, Enfer Scientific, have been extremely successful since winning the Award and, according to Riona Sayers, Operations Manager, "involvement with the National Innovation Awards has proved a very worthwhile and valuable experience for Enfer Scientific, with the company going from strength to strength both nationally and internationally".
- The 6th Irish Innovation Lecture takes place in Belfast in November. The Lecture provides an excellent opportunity for industrialists and academics from North and South to exchange ideas and experiences on how science, technology and innovation policies might evolve in the coming years.
- The third **Science Week Ireland** takes place from 7–14 November 1999. The last **Science Week Ireland** of the Millennium has been nominated an official Millennium Project and, with this in mind, we plan celebrations throughout the country during the course of the week. Details of events will be advertised locally and nationally. Last year over 150 events took place and, this year, we hope to create an even bigger and better week of fun events, activities and provide valuable public information in the important area of S&T.

The goal of the STI Awareness Programme is to show the public that S&T help to:
- Develop leading edge industry and skilled jobs
- Create exciting career options.
- Develop creativity in our children.

So, in addition to the above targeted activities, the Awareness Programme will also continue to support other S&T organisations by sponsoring events and encouraging new initiatives, and will continue in their endeavours to increase media coverage of S&T issues in the broadcast and print media.

The S&T area offers the people of Ireland the chance for continued economic prosperity, and the Awareness Programme aims to communicate that message to the general public and particularly to our young people, to encourage them to grasp the opportunities on offer.

At the announcement of the 1999 Science, Technology & Innovation Awareness Programme in the National History Museum were Mr Noel Treacy, TD, Minister for Science, Technology and Commerce, and the Z-Tron Robot.

For further information contact:
Martina McDonnell,
STI Awareness Programme,
Forfás, Wilton Park House,
Wilton Place, Dublin 2.

Irish Council for Science, Technology and Innovation

The Irish Council for Science, Technology and Innovation (ICSTI) is the body established in 1997 by the Minister for Science, Technology and Commerce, under Forfás legislation, to advise Government on the strategic direction of science, technology and innovation (STI) policy. The role of the Council encompasses all aspects of such policy, including primary, second and third level education; scientific research; technology and R&D in industry; prioritisation of State spending; and public awareness of STI innovation issues.

ICSTI has twenty-five members. The Council is chaired by Dr Edward M. Walsh, President Emeritus, University of Limerick Foundation.

The operation of the Council to date has mainly centred around the work of three Task Forces which were established in June 1997 to focus on key areas of science, technology and innovation policy. These Task Forces, whose membership was drawn from the Council, were:
- Technology Foresight, matching technology development to national needs. (Chaired by Mr Brian Sweeney, Chairman, Siemens Group Ireland.)
- Innovation Infrastructure, dealing with the development of policy aimed at removing or reducing some of the constraints in Irish society on innovation and at developing a more congenial environment for the conduct of science and technology and for the implementation of relevant research in enterprise. (Chaired by Mr Brian Trench, School of Communications, Dublin City University.)
- Public Expenditure on Science and Technology, contributing to the Council's discussions and recommendations on prioritisation and allocation of public monies for science, technology and innovation (STI). In particular, the Task Force aims to review Ireland's existing STI expenditure and to benchmark it against relevant international practice. (Chaired by Mr John Travers, Chief Executive, Forfás.)

The ICSTI Secretariat is provided by Forfás, the policy and advisory board for industrial development and science and technology in Ireland. Correspondence should be addressed to: The ICSTI Secretariat, Wilton Park House, Wilton Place, Dublin 2; Tel: +353-1-607-3186; Fax: +353-1-607-3260; E-mail: icsti@forfas.ie; URL:http://www.forfas.ie/st/sti.htm

Technology Foresight

The findings of the first ever Technology Foresight exercise undertaken in Ireland were published at the end of April. This initiative, which was undertaken by ICSTI at the request of Mr Noel Treacy, T.D., Minister for Science, Technology and Commerce, was completed within twelve months. The process sought to identify technologies that will be key to national economic development in the longer term and to make recommendations to address the opportunities and challenges associated with these technologies.

The Technology Foresight exercise, which involved a wide consultative process with scientists, industrialists, Government officials and others, achieved a high degree of consensus on the actions Ireland needs to take now to maximise future industrial competitiveness, supported by an appropriate technological infrastructure. The Irish Government is considering mechanisms for the implementation of the Report in the context of finalising Ireland's National Plan for the period 2000 to 2006.

The main recommendations of the ICSTI overview report are based on the recognition that key drivers of growth in the future will be in the areas of information and communications technologies (ICT) and biotechnology. There is a need to create, develop and attract firms which are research and technology based. The key recommendations are:
1. All Government Departments and Agencies should utilise the Foresight findings in future planning exercises.
2. Ireland should become a centre of Excellence in ICT and biotechnology niches.
3. Government policies should be more proactive in the creation of an environment conducive to technological innovation, specifically in relation to regulatory and fiscal measures.
4. The Government should establish a Technology Foresight fund of the order of IR£500 million over five years.

The objective of the Technology Foresight recommendations is to re-position the Irish economy to be widely recognised internationally as a knowledge-based economy and as an attractive location to undertake R&D. To achieve this, there is a need to develop a world-class research capability of sufficient scale in a number of strategic areas within our universities and colleges, research institutes and industry.

ICSTI Chairman, Dr Edward M. Walsh, explained that Ireland must engage with its scientific diaspora: "The additional capabilities needed to accelerate the development of world-class research capability resides in graduates of Irish institutions who have gone abroad to work in Europe and the US. Most of these people have developed specialities which would be needed to provide the initial momentum and scale for the development of strategic niche areas of ICT and biotechnology", he said.

The Technology Foresight findings are in the form of a "suite" of *nine reports*, which consists of an ICSTI overview report and *eight individual reports from expert panels*, established in the following areas:
- Chemicals and Pharmaceuticals
- Information and Communications Technologies
- Materials and Manufacturing Processes
- Health and Life Sciences
- Natural Resources (Agri-Food, Marine, Forestry)
- Energy
- Transport and Logistics
- Construction and Infrastructure

The *ninth report*, from ICSTI itself, builds on the findings of the Technology Foresight panels and develops a vision of Ireland as a knowledge-based society. Knowledge societies will exploit the enormous potential of new knowledge-intensive technologies in areas such as information and communications, biosciences, medical systems and nanotechnology. Over time, such technologies, and the industries they support, will become increasingly important in international trade.

The reports are available from the Forfás Information Office, Wilton Park House, Wilton Place, Dublin 2; Tel: +353-1-607-3134; and on the internet at http://www.forfas.ie/report/tforesight.htm

At Powerscourt House on the occasion of ICSTI's hosting of the 1999 Plenary Meeting of the Chairmen and Secretaries of the EU National Advisory Councils for Science and Technology Policy and Invited Observer from the US President's Committee of Advisors on Science and Technology.

Additional Statements on Science, Technology and Innovation Policy published by the Council within the last twelve months

Science, Technology and Innovation Culture
(Published during Science Week Ireland 1998 – November)

The statement highlights the need for scientists and technologists to increase their efforts to explain their work to the public and to become more involved in policy development. Other key points are that:

- Irish society needs to develop greater confidence in dealing with the opportunities and challenges which science and technology present in order to sustain a viable economy and develop a vibrant culture into the 21st century;
- Policy-makers, legislators and administrators from all sectors need to ensure they have access to the best available scientific and technical information;
- The public needs to understand how science is done, so that they can appreciate how scientists reach their conclusions;
- The media have an important role to play in facilitating mutual understanding between scientists and the public;
- Irish society needs to develop mechanisms for public consultation which acknowledge both the concerns of the public and the contribution of science and technology;
- The education system can play a larger part in developing the public's interest in, and engagement with, science and technology;
- Major changes in second level education are needed, specifically in content, teaching methods and assessment of science and technology;
- Professional bodies representing scientists, engineers and technologists should establish a forum which can co-ordinate efforts in the area of science and technology awareness;
- The Government should immediately commission a plan for a network of science centres, with a view to seeking EU and other funds for their establishment;
- Science and technology should be represented in the cultural activities being organised to mark the new millennium; and
- The reality that science and technology are an integral part of contemporary culture, and will play an even more important part in the culture of the next century, should be reflected in Ireland's educational and cultural policy and activities.

Innovation in Enterprises in Ireland
(Published November 1998)

In this Statement the Council emphasises the

In Forfás at the publication of the Technology Foresight Findings, Left to Right: Mr Mattie McCabe, Office of Science and Technology, Dr Killian Halpin, Head, ICSTI Secretariat, Mr John Travers, ICSTI Member and Chief Executive of Forfás, Minister Noel Treacy, T.D., Ms Helena Acheson, ICSTI Secretariat, Dr Edward M. Walsh, Chairman, ICSTI, and Mr Brian Sweeney, Chairman, Technology Foresight Task Force.

importance of enhancing innovation potential and performance at the level of the company. It recommends:

- State assistance measures to prime the technology-based innovation process and to increase interaction and collaboration;
- A tax credit for incremental research and development (R&D) expenditure should be introduced, alongside grant assistance for R&D companies and by those seeking to implement a significant shift through a company development plan based on R&D on innovation;
- A continuation of favourable tax treatment on royalties for patented inventions;
- Access to capital venture;
- Further efforts to create and promote centres of science/engineering-based competence in higher education colleges that address the needs of industry;
- A network of local and accessible "innovation officers" to assist small companies in drawing on the system of assistance and in accessing venture capital;
- Clear guidelines on the ownership of intellectual property and a "Teaching company scheme" involving graduate placements and medium-term college-industry collaboration.
- Significant support for educational programmes for life-long, technology-related learning.
- An underpinning by the State of all the above actions through promoting awareness of the need for innovation, devising strategic research initiatives and ensuring access to world-class information and communications technologies.

Priorities for State Investment in Science, Technology and Innovation in 1999
(Published November 1998)

In the Statement, the Council acknowledges the progress made on some of the actions it had recommended to Government in 1998 in relation to a number of key areas of national importance. It goes on to draw attention to a number of priorities which remain outstanding under the headings of:

- Industrial research and development
- Telecommunications infrastructure
- Industrial design
- Technology skills and expertise
- Science in schools
- Public sector research and development
- EU structural funds

Some of these areas are under consideration by Government.

Investing in Research, Technology and Innovation (RTI) in the Period 2000 to 2006
(Published January 1999)

This Statement puts forward arguments on the need to invest in science, technology and innovation as a matter of priority over the period in question if Ireland is to become an innovation-driven economy.

The Statement points out the contribution which RTI investment has made to Ireland's economic development over the past ten years through the development of indigenous firms, the creation of new technology-based firms and the attraction, retention and development of high technology foreign direct investment. It notes, however, that the level of public investment in RTI is considerably behind the EU average and has not kept pace with the growth in business spending on research. It, therefore, recommends:

- A new programme of investment in RTI should aim to invest substantially in the RTI base of the country, as a means of enhancing innovation and competitiveness in order to increase output and employment through various mechanisms;
- A new EU-supported investment programme should focus on the following four areas:
  RTI for Industry;
  RTI for Collaboration;
  RTI Infrastructure and
  Natural Resource-Based RTI.
- The promotion of sustainable development and protection of the environment should be an intrinsic part of the operation of RTI measures in the new programme.
- In deciding the balance of investment in the above areas, a strong emphasis should be placed on building up the performance, capability and skills of the business sector in terms of R&D and innovation and in promoting industry/institution collaboration.
- RTI performance in Ireland must not only be benchmarked against that of the other EU countries. It must increasingly be benchmarked against that of leading RTI investors and progressive emerging economies in other parts of the world including, for example, Israel, Taiwan, Singapore, New Zealand and Malaysia.

Knowledge as a Development Factor – Technology Foresight Ireland

Ireland has to find its position in a new international division of labour and prepare itself to deal with the changes facing society. What are the coming generations to live on and be occupied with? How can we ensure for ourselves a leading position in the accelerating development of technology? How can we predict, prevent and solve the important problems in the fields of the environment, health and food? In recent years, questions such as these have increased public awareness of the importance of research in the development of society and industry.

It is important that a democratic dialogue about the role and importance of research be developed. Last year, as a follow through to the White Paper on Science, Technology and Innovation (1996), the Government launched the Technology Foresight Ireland initiative. This initiative, managed by the Irish Council for Science, Technology and Innovation in Forfás, opened a dialogue between researchers, industry, unions, Government Departments and others about the optimum research strategy which Ireland might adopt which would underpin current economic growth and sustain this growth to the year 2015.

Technology Foresight was first pioneered in the US in the 1950s and later in Japan in the 1970s. It is now widely used in Europe, where Governments have been trying to formalise their discussion of future prospects and current choices concerning science, technology, the economy and society, in order to assist them in prioritising their investments in science, technology and knowledge.

The research effort in a small country like Ireland must necessarily be limited in relation to that of other larger industrialised economies. Therefore, it is very important that clear choices are made nationally about the areas which must be given highest priority.

Governments, enterprises and the general public all have an interest in identifying areas of strategic research and the emerging technologies likely to yield the greatest economic and social benefit. By their very nature, these new technologies require significant advances in the science base if they are to be developed successfully. Technology Foresight studies help to fuel the knowledge debate and thus contribute towards building common visions for the research community, governments, trade and industry as well as society. This helps research to become embedded and accepted in society.

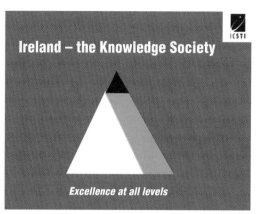

On the eve of the next millennium, technology research and development will play an increasingly central role in the way in which our societies develop. Together with education and training, it is a fundamental pillar of the "knowledge society" which we gradually see emerging – a society in which economic performance and quality of life will depend more on the production, transmission and exploitation of knowledge than on the manufacturing and exchange of material goods. In order to prepare for this future and enable this change to work effectively for the benefit of all, governments must invest, appropriately and resolutely, in research and technology and at the same time must encourage initiatives by the private sector.

Through a process of communication and consultation, the eight expert panels established to spearhead the Technology Foresight Ireland initiative reached a consensus about the actions needed to underpin and sustain economic prosperity for the eight sectors under consideration. The clear, overarching message coming from Technology Foresight Ireland is that Ireland needs to evolve rapidly to a knowledge society, capable of exploiting the enormous potential of new technologies in the pervasive areas of information & communications technology and biotechnology.

Technology Foresight Ireland concluded that the Irish economy should be repositioned, to be widely recognised internationally as a knowledge-based economy. A need for world-class research capability of sufficient scale in a number of strategic areas within our universities and colleges, research institutes and industry has been identified. The partnership of Government, industry, the higher education sector and society must combine to deliver the knowledge framework, which in future will realise:

- research and technology development (RTD) intensive and advanced technology-based indigenous and overseas companies, using high level expertise
- a vibrant, cohesive, durable and internationally recognised competitive RTD base involving industry, universities and colleges and research institutes, which provides an attractive career structure for researchers to work in Ireland
- an environment conducive to innovation
- investment in the physical and human infrastructure
- citizens well informed on scientific issues in the context of an innovation culture.

Expert Panels in 8 areas

- Chemicals & Pharmaceuticals
- Information & Communications Technologies
- Materials & Manufacturing Processes
- Energy
- Health & Life Sciences
- Natural Resources (Agri-food; Marine; Forestry)
- Transport & Logistics
- Construction & Infrastructure

Ireland has a pool of highly skilled human potential combined with an excellent education system and a good basis for developing a world class technological infrastructure. Demands made on the work force grow in step with increasing international competition. Today, a generally high level of education is not alone sufficient to ensure future prosperity. In the future, Ireland and the Irish must survive by being more skilful and quicker at research and development than our competitors – otherwise, we shall see a reduction in living standards. Thus, education, research and technology are closely linked with industrial policy. Research and its results play a vital role in modern society – the research institutions must therefore constantly meet current challenges and demands in order to retain their dynamism and value in society.

Technology Foresight Ireland strongly indicates that new roads to growth now lie elsewhere. Whilst we need to ensure a sustainable future for the manufacturing base of our economy, at the same time we must accelerate the development of a knowledge-based economy.

For further information contact:
Forfás, Wilton Park House,
Wilton Place, Dublin 2.

ENTERPRISE IRELAND — JIM CUDDY

Campus Companies Programme

The Campus Companies Programme is an initiative from Enterprise Ireland, designed to assist individuals interested in commercialising R&D on the college campus. Apart from offering valuable business development assistance to projects, a key part of the programme is the provision of financial support in the form of a CORD (Commercialisation Of Research & Development) grant. The main purpose of the CORD grant is to enable the individual to assess the commercial viability of a project. Grants of up to £30,000 are available to enable an individual undertake market research, prototype development, and the development of a business plan.

In the short period of time, Irish campus companies have achieved very promising results. There are currently 150 of them actively trading, and over 2000 quality jobs have been created in the sector in the last ten years. The recent stock market flotation of Iona Technologies, a former Trinity College campus company, highlights the potential within the Third Level sector.

Case Study: Scientific Systems Limited
Scientific Systems Limited was formed in 1994 by Dr Mike Hopkins and Ciarán Ó Móráin. The company identified a market niche in the area of plasma sensor technology - an emerging technology with applications in a wide range of sectors. However, it has specific potential in reducing costs associated with microchip manufacture. Scientific Systems has developed, and is now manufacturing, sensors for the plasma process equipment aimed at the semiconductor industry.

Dr Mike Hopkins and Ciarán Ó Móráin examine a component from a smart chip (photo by Bryan O'Brien, courtesy of The Irish Times*).*

The company's founder, Dr Mike Hopkins, was a lecturer in Physics at Dublin City University (DCU) and has only recently resigned his position to concentrate on the company's development. Mike has an international reputation in the area of plasma sensor technology and received considerable research funding during his tenure at DCU. With the assistance of the Industrial Liaison Office in DCU, Mike established Scientific Systems as a campus company. He located his fledgling company in a low cost unit within the college and, in the early stages, worked on developing the company's reputation in the area. Scientific Solutions was one of the first recipients of assistance from Enterprise Ireland's Campus Companies Programme.

Since its inception, the company has experienced high growth and has developed a solid client base which includes Siemens, IBM, Lucent Technology and Hitachi. It is envisaged that the company will employ 25 people by the end of 1999 and grow to 50+ over the next three years. It has recently attracted significant investment (£2m) from ACT Venture Capital Fund in association with Enterprise Ireland. The company hopes to open an office in the US later this year and achieve a £10m turnover by 2002.

Contact: Jim Cuddy, Enterprise Ireland;
E-mail: jim.cuddy@enterprise-ireland.com

ENTERPRISE IRELAND — BRIAN O'DONNELL

EU Fifth Framework Programme for R&D

A new Framework Programme for R&D in Europe up to 2002 has been adopted by EU Research Ministers. The research will concentrate on four main areas. These are: the Quality of Life and Management of Living Resources (including food, nutrition, health and sustainable agriculture, fisheries and forestry); Energy, Environment and Sustainable Development; User-friendly Information Society; and Competitive and Sustainable Growth. The training and mobility of researchers will also be supported, and access to major research infrastructures will be facilitated. Enterprise Ireland plays a leading role in the promotion of the Programme in Ireland.

The Programme will build on the success of previous Framework Programmes for R&D. It is anticipated that there will be considerable involvement of industrial and academic researchers from Ireland in the Programme. For industry it will assist with the harnessing of new technologies, the deepening of R&D capability, and the development of the skills and capability of technical personnel. For academic researchers it will provide support for collaborative work with industry and with researchers in other institutions.

In Cork, the National Microelectronics Research Centre, led by Prof. Gerry Wrixon, who is now President of UCC, has been a major player. It has worked on projects with all the major European electronics companies – including Philips, Siemens, Bell and SGS Thomson – in the development of state-of-the-art microelectronic devices.

Dr Daniela Zisterer performs ligand binding assays at the National Pharmaceutical Biotechnology Centre, TCD.

Small and medium-sized enterprises are particularly welcomed in EU supported R&D. A.J. Precision Components of Scarriff, Co. Clare, has worked with partners from France, Denmark, Germany, Austria and the UK in a number of projects including a two year project for the development of high speed production technology for over moulded filter membranes.

In the Biotechnology area, the Molecular Evolutionary Genetics Laboratory at TCD and the Department of Biochemistry at UCC have worked on methods and software for evolutionary analysis of genome sequence data with partners from Italy, Sweden, France and the UK.

In Framework Programme projects, financial support is usually at a rate of 50% of full costs, and projects are submitted in response to Calls for Proposals normally by consortia with participants from a minimum of two countries.

While competition for funding is quite stiff, Irish researchers are well positioned to benefit from the considerable opportunities presented by the Framework Programme.

Contact: Brian O'Donnell, Enterprise Ireland;
brian.odonnell@enterprise-ireland.com

ENTERPRISE IRELAND – INDUSTRIAL MATERIALS & ENGINEERING TECHNOLOGIES

Developing and delivering leading edge technologies for industry

ENTERPRISE IRELAND

Enterprise Ireland's programmes in Industrial Materials & Engineering Technologies (IMET) help industry to develop new and improved products and processes. Its multi-disciplinary teams of scientists and engineers provide contract R&D, consultancy and testing services to all sectors of industry. By participating in major European Technology Transfer and R&D Programmes, IMET has an extensive network of contacts within industry, research institutes and universities across Europe.

IMET provides technical consultancy, training, investigative and accredited laboratory testing services in the following areas:
- ceramics
- corrosion
- metallurgy
- organic coatings
- powder technology
- surface engineering
- non-destructive testing
- joining
- metal finishing.

Also provided is specialist consultancy, technical assessment, laboratory and site-based services to the construction, engineering and offshore oil and gas sectors. These services include:
- safety & reliability consultancy
- facade/structural engineering
- building products evaluation
- coastal & offshore engineering
- concrete technologies.

Among the technological developments in which IMET is currently involved are:

Wear resistant coatings for the textile industry

Wear resistant coatings can be applied to various components to enhance their mechanical performance and extend their lifetime. In the textile industry, significant wear occurs on needles and thread guides due to the high speed of the yarn passing over them. The Surface Engineering Group at Enterprise Ireland set out to address this problem, by applying specialised coatings, which were deposited, using magnetron sputtering, onto the thread guides. Figure 1 shows a picture of the coated (gold coloured) and uncoated thread guide. The advantage of using the wear resistant coating is clearly seen in the four fold increase in wear resistance (Figure 2). This reduced wear can lead to significant cost savings both for needle and guide replacement but, in many cases, more importantly will lead to increased process efficiency due to a decrease in equipment down-time.

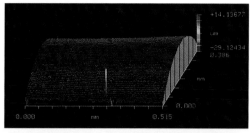
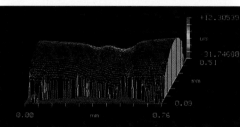

Figure 2. Comparative surface evaluation of coated thread guide (top) and uncoated thread guide (bottom) after subjecting them to wear tests. The reduction in wear exhibited by the coated guide is clearly visible.

For further information contact:
Dr Denis Dowling; Tel: +353-1-808-2403;
Fax: +353-1-808-2046; E-mail:
denis.dowling@enterprise-ireland.com

Corrosion control – the enabling technology

Computer modelling conducted by the Batelle Institute in the USA in the 1970s indicated that corrosion of metallic materials can cost an industrialised country approximately 4% of its annual GNP. It was also estimated that between 30% and 40% of this figure was avoidable through the use of an appropriate corrosion prevention technology. A recent survey reported in the journal *Materials Performance* confirmed that these figures still apply. There have been considerable advances in our understanding both of corrosion mechanisms and control of the problem, enabling many industries to function more efficiently.

Owners and operators of industrial plants and facilities can avoid many unnecessary and sometimes extremely costly corrosion problems. Enterprise Ireland can advise on the use of protective coatings (inorganic and/or organic), utilisation of corrosion-resistant alloys (such as adoption of the correct grade of stainless steel), application of inhibitors to the environment, and employment of cathodic protection technology (the imposition of an electrical potential to a structure) which are all sound corrosion mitigation strategies. In addition, good design and careful materials selection are also extremely important practices.

If corrosion damage does occur, then Enterprise Ireland can conduct examinations of affected industrial components to ascertain the mode of failure. Once a proper understanding of the failure mechanism is obtained, then this information can help prevent future costly mistakes being repeated. Examples include investigations of metals in contact with potable (drinking) and waste waters, which may sometimes be relatively corrosive. Indeed virtually any metal/fluid combinations which lead to such damage can be investigated. Laboratory-based studies, conducted preferably before materials selection is undertaken, can also assist the correct choice of alloy material for a given material/environment combination. This approach can prove very cost effective by either preventing or at least minimising

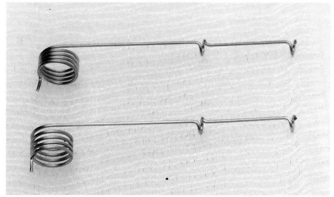

Figure 1. Coated (gold coloured) and uncoated thread guide components.

Figure 3. Corrosion of opened cast iron pipe.

unnecessary corrosion damage in service. Sometimes highly expensive alloy materials need not be used in service, provided the correct procedures to maintain alternative materials are put in place as a result of a thorough study.

A variety of techniques are available at Enterprise Ireland, including microscopic (both optical and electronoptical) and surface analytical methods, to assist such investigations. As corrosion often occurs as a result of electrochemical processes, many methods are electrochemical in nature. All of these are important tools in the understanding of corrosion problems.

For further information contact:
Dr Tony Betts; Tel: +353-1-808-2541;
Fax: +353-1-808-2046;
E-mail: tony.betts@enterprise-ireland.com

DG XI Commission's study on the potential for reducing volatile organic compound (VOC) emissions associated with the decorative paints and varnishes sector

The Council of Ministers adopted the Solvent Directive in March 1999, following ten years of discussion and consultation. This covers the use of organic solvents in static industrial plants, and seeks to limit environmental pollution associated with them, but does not cover the solvents found in products used by professional and DIY users for architectural and decorative applications.

However, several EU Member states (e.g. Germany, The Netherlands, Denmark, Austria, United Kingdom), have implemented, or plan to introduce, measures limiting the concentration of solvents in certain decorative products. Furthermore, the European Parliament has requested the Commission to submit provisions concerning the marketing of products containing solvents intended for commercial and private uses, in order to supplement the Community legislative framework dealing with the prevention of tropospheric ozone. The Commission services are currently developing a strategy on this issue, which could possibly include regulatory measures for decorative paints and varnishes.

In order to facilitate this process, the EC has commissioned a study of this sector from a consortium consisting of The Coatings Technology Services Group of Enterprise Ireland, The University of Amsterdam's Consultancy and Research Centre on Chemistry Work and Environment (Chemiewinkel), and its Institute for Environmental Management (WIMM).

The objectives of the study are:
- To present a synthesis of the relevant elements characterising the decorative paints and varnishes sector. This will cover both supply and demand sides, and an analysis of the emission limiting measures implemented or in preparation in the Member States, OECD and accession countries, with a view to establishing their cost and effectiveness.
- To explore, at a technical level, the possibilities for limiting the concentration of organic solvents in products used in this sector.

Conventional and low solvent substitute products will be classified in terms of their functional characteristics, and compared in terms of performance, technical requirements, health and environmental impact, and price. Obstacles to market penetration by more environmentally friendly products will also be analysed with a view to identifying measures which might promote wider dissemination. A proposal for reference products for each category, containing as little solvent as possible, will be given with a view to determining the corresponding potential for reducing VOC emissions.
- To analyse the environmental and economic impact of such regulatory measures.

Consideration will be given to the costs and the benefits of setting up a Community legislative framework aiming to limit solvent usage in the sector to the levels identified in the study. The impact on the following will be evaluated:
- the price of the products
- the structure of the sectors concerned, and in particular the impact on small and medium sized enterprises.
- innovation in the sectors concerned
- public authorities (follow-up and control of legislation).

Figure 4. Aerial image of Dunmore East, Co. Waterford.

For further information contact:
Philip Thornton; Tel: +353-1-808-2515;
Fax: +353-1-808-2046; E-mail:
philip.thornton@enterprise-ireland.com

The National Coastline Survey - over 7000 kilometres of digital images of the Irish coast for public use

An innovative partnership between a young Irish company and the Offshore & Coastal Engineering Unit of Enterprise Ireland, backed by the Marine Institute, has led to the development of a rapid-response digital camera system for capturing digital photographs from the air. Compass Informatics is charged with carrying out a survey of the Irish coastline using this system. To date the response from end-users to the resultant images has been extremely positive. Coastal managers, recreational divers, aquaculture operators, golf course managers, and boating organisations are among the diverse range of people who are keen to use the image resource.

Completion of the national survey is expected by Autumn 1999. The end result will be 7000 or more images covering a coastal strip 450 meters wide. These images will be provided on a series of CDs with their own viewing software.

The camera system is run from a light aircraft, which allows the survey to be carried out cost-effectively and flexibly so the it can be used as a rapid-response information gathering system. The system can also operate under clouds – which is of clear advantage given Irish weather. In this way it is more adaptable than standard aerial and satellite imaging systems.

A selection of uses include:
- marine & coastal tourism development
- harbour and marina development planning
- dune and erosion management in recreation areas, including golf courses and tourist resorts
- identifying areas suitable for aquaculture, seaweed extraction and other resource industries
- County Development Plans & Environmental Impact Assessments
- use within Geographical Information Systems (GIS).

Consistent with its brief to utilise new technologies for the sustainable development of marine resources, the Marine Institute is promoting the Survey. The project, in which the Ordnance Survey are also partners, is ERDF funded with support also from the Department of the Marine and Natural Resources.

For more details, including sample images, contact: Brendan Dollard,
Enterprise Ireland: Tel: +353-1-808-2279;
Fax: +353-1-836-8139; E-mail:
brendan.dollard@enterprise-ireland.com;
Gearoid O'Riain, Compass Informatics;
Tel: +353-1-670-5761; Fax: +353-1-670-3037;
E-mail: mail@compass.ie

IMET contact:
Dr Tom Kelly, Enterprise Ireland,
Glasnevin, Dublin 9;
Tel: +353-1-808-2642; Fax +353-1-808-2046;
E-mail: Tom.Kelly@enterprise-ireland.com

BioResearch Ireland

OVERVIEW, BY DIRECTOR
– Jim Ryan

BioResearch Ireland is the national agency for commercialising biotechnology. It was established in 1988 as a partnership between Government and the universities and now employs 165 staff within five university-based centres and its management group. The main function of BRI is to support Irish economic development by commercialising biotechnology-related research.

> BRI's mission is to work with the research community to enhance quality of life and economic well being through the application of biotechnology.

The core of BioResearch Ireland's expertise lies in its technology management capabilities, and it has a track record of successfully transferring technologies to industry from academia. While investing resources in bioscience projects, BRI also acts as a strategic business partner. BRI brings its experience in marketing, business development, patenting and licensing of technologies into the equation, thereby ensuring the optimum return for the researcher.

BRI liaises with university staff to identify research projects which have a clear commercial potential. Once an opportunity is identified, BRI works with the relevant academic researcher to guide the project towards commercial goals. This normally involves some financial investment in the research programme, advising and paying for patenting, and ultimately seeking industrial collaborators or licensees or, if appropriate, starting a new campus company to exploit the technology. The outcome is a series of collaborations and licensing deals with companies all over the world and, more recently, a pipeline of new start up companies.

BRI's five research centres and their areas of biotechnology are as follows:

- **National Diagnostics Centre, NUI, Galway**
  Immunodiagnostics, DNA Diagnostics,
  Molecular Biology, Fish Disease
- **National Food Biotechnology Centre,
  University College, Cork**
  Probiotics, Food Ingredients,
  Lactic Acid Bacteria,
  Environmental Services,
  Functional Foods
- **National Pharmaceutical Biotechnology
  Centre, Trinity College, Dublin**
  Inflammation & Cancer, Neurobiology
  & Ageing, Vaccine R&D, Pharmaceutics
  & Pharmaceutical Chemistry
- **National Cell & Tissue Culture Centre, Dublin
  City University**
  GMP Manufacture, Oncology, Apoptosis,
  Ribozyme Technology, MAb Development
- **National Agricultural & Veterinary
  Biotechnology Centre,
  University College Dublin**
  Mushroom Research, Vaccines, Immunoassays,
  Receptor Cloning, Plant Biotechnology.

Dr Jim Ryan, Director of BioResearch Ireland.

Each centre is focused on a particular area of biotechnology, reflecting the underlying strengths of the host university. The following articles highlight projects that are ongoing at BRI centres.

*Contact: BioResearch Ireland,
Enterprise Ireland, Glasnevin, Dublin 9;
Tel: 01-837-0177; Fax: 01-837-0176;
E-mail: info@biores-irl.ie; www.biores-irl.ie*

NATIONAL DIAGNOSTICS CENTRE (NDC)

Diagnostic tests for foodborne pathogens – Tony Forde, Marian Kane & Majella Maher

Today, increasing customer demand for high quality food products means that the food industry must be able to guarantee the quality and safety of its products. In the U.S., microbial pathogens in food cause approximately 30 million cases of illness, resulting in 9000 deaths, and associated costs of between $2.9-$6.7 m dollars per annum. The foods most frequently implicated include red meat, poultry, eggs, seafood, and dairy products with Salmonella, Campylobacter, E.coli 0157:H7, and Listeria monocytogenes generally isolated as the causative agents of illness. In Ireland until recently there has been considerable variation in the investigative and reporting procedures used for identifying cases of foodborne poisoning. The Food Safety Authority, in association with other regulatory bodies, is developing a standardised system with a new national centre for communicable diseases in this country.

Conventional methods of microbiological analysis of foods are time-consuming, taking several days to weeks to generate results. The development and implementation of more rapid test methods based on immunological and DNA methods offers the potential to revolutionise quality testing of food products.

A primary requirement for the widespread acceptance of novel detection systems for food-borne pathogens is convenience and ease of

DNA diagnostics laboratory at BRI's National Diagnostics Centre.

use, and a clear indication of benefit to both the producer and the consumer. Convenient immunoassays have been developed at the NDC for the detection of food borne pathogens, Listeria and E.coli 0127:H7. These assays are applied to food homogenates after an enrichment period, and can be used to identify suspect samples which are then subjected to traditional confirmation procedures. Immunoassays have also been developed to detect antibodies to Salmonella species in meat juices (both pork and chicken), indicating Salmonella contamination of specific animal population and highlighting where control measures require evaluation.

Food borne assays developed at NDC have been combined with DNA probe hybridisation into a colormetric membrane-based detection system to offer increased specificity. The test has a turn around time of 24 hours and a throughput of up to 40 samples per test run. The system may also be adapted to simultaneously screen a sample for the presence of a range of bacterial pathogens. The assays are currently used to test samples at various stages of broiler production process for the presence of Salmonella and Campylobacter, and are being adapted for monitoring environmental samples from poultry houses.

Rapid molecular identification and typing of micro-organisms is extremely important in efforts to monitor the geographical spread of virulent or epidemic pathogens. In this area, the NDC'S focus is on standardising molecular typing methods, in particular AFLP (Amplified Fragment Length Polymorphism) for food pathogens for epidemiological studies, and in European networks to validate common guidelines for these research techniques.

NATIONAL FOOD BIOTECHNOLOGY CENTRE (NFBC)

SNaPIT™
– Single Nucleotide Polymorphism Identification Technology
– A novel process for analysis of genetic variation
Patrick Vaughan

Overview of SNaPIT™ Technology

SNaPIT™ is a technology which permits the reliable, accurate and robust detection of SNPs in DNA. The system exploits the use of highly specific DNA glycosylase enzymes to excise specific substrate bases incorporated into amplified DNA. It may be applied to both fragment size analysis (e.g. gel analysis) and solid phase/immobilised (e.g. microtitre plate) formats. More importantly, SNaPIT permits the scanning of genes for new SNPs not previously characterised (SNP discovery), and permits the high throughput genotype analysis of those SNPs in target populations (SNP genotyping). This allows a direct comparison between the genomes of different population groups, such as diseased and healthy individuals.

Some of the key features of SNaPIT-check (SNP genotyping) and SNaPIT-scan (SNP discovery) which set SNaPIT apart from other SNP discovery and genotyping techniques are:

1. Its speed, accuracy and reliability.
2. The ability to automate the platform on either gel-based or non-gel-based formats means that SNaPIT may be used for the simultaneous detection of multiple different polymorphisms, thus greatly reducing capital investment, labour costs and time.
3. A similar platform is used for both SNP discovery and genotyping.

General features of SNaPIT™

By employing uracil DNA glycosylase (UDG) and deoxyuridine tri phosphate (dUTP), SNaPIT-SCAN and SNaPIT-check can detect 10 out of the 12 possible base substitution polymorphisms (83%) – i.e. Guanine (G) to Adenine (a), G to Thymine (T), A to G, A to T, A to Cytosine (C), T to G, T to A, T to C, C to A and C to T – since they involve the gain or loss of A/T or T/A base pairs, in addition to all deletion and insertion mutations. In reality, >95% of all single base substitutions are detectable with SNaPIT, since most polymorphisms and disease causing mutations are C to T (or G to A) transitions. This is due in part to the high incidence of deamination which occurs at cytosine and 5-methyl cytosine residues in DNA to yield uracil and thymine residues respectively, which increases the incidence of C to T (or G to A) base substitutions in DNA.

The remaining two single base substitutions, i.e. G to C and C to G substitutions, which occur at a very low frequency, can be detected by first converting the C residues to U residues by bisulfite treatment, thereby making them susceptible to excision by uracil DNA glycosylase. They can also be detected by screening for single stranded conformational variations of the SNaPIT-SCAN cleavage products when separated on a non-denaturing polyacrylamide gel. This technique has the advantage that all mutations can be detected with just one glycosylase. Alternatively, the use of an additional glycosylase/modified precursor nucleotide which substitutes for G or C can be used for this purpose, in this way allowing the detection of 100% of mutations. SNaPIT readily detects all other types of mutations, namely insertion and deletion mutations.

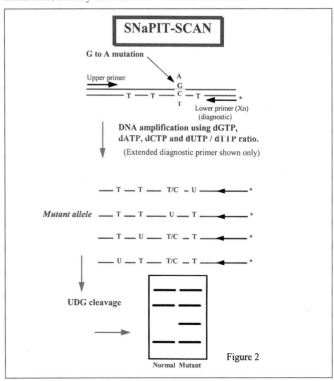

Figure 1

Figure 2

NATIONAL PHARMACEUTICAL BIOTECHNOLOGY CENTRE (NPBC)

Tim Foster & Margaret Woods

The National Pharmaceutical Biotechnology Centre, BioResearch Ireland's centre on the campus of Trinity College, Dublin, comprises 35 staff in 10 different departments working on projects supervised by 15 Principal Scientists. The four major areas of research are Neurobiology & Nutrition, Inflammation & Cancer, Pharmacy, and Vaccine Development.

In the Vaccine Development area, NPBC scientists, led by Professor Tim Foster of the Microbiology Department in the Moyne Institute of Preventive Medicine, have been working for several years on the discovery and characterization of cell surface protein adhesins from *Staphylococcus aureus* and coagulase-negative staphylococci. One application of the research is the prevention and treatment of Staphylococcal hospital-acquired infections which account for 40% of all deaths from such diseases and which add considerably to the length of time a patient spends in hospital and hence the cost to health services. The problem is compounded by the emergence of strains of *S.aureus* that are resistant to treatment by all available antibiotics. Prof. Foster and his group have discovered several novel adhesins, cell surface proteins which allow bacteria to colonise the host and initiate infection by binding to proteins present in plasma or in the extracellular matrix, or by sticking to deposits on indwelling medical devices. Two fibrinogen binding proteins have been characterized which are members of a larger family of structurally-related surface proteins with a very unusual domain consisting of serine-aspartate dipeptide repeats.

Professor Tim Foster, NPBC Vaccine Group.

In collaboration with scientists in Gothenburg, Sweden, and at Inhibitex Inc., Alpharetta, Georgia, USA, it has been shown that immunization of animals with a single protein antigen corresponding to part of one of the surface adhesins will protect against subsequent challenge with *S.aureus*. Recently BRI has signed an exclusive licensing agreement with Inhibitex Inc. (*www.inhibitex.com*), who will develop vaccines for staphylococcal infections of hospital patients. Initially, rare human blood donors with high antibody levels against particular *S.aureus* antigens will provide plasma which will be pooled, and the immunoglobulin (IgG) purified and then administered to at-risk patients. Later, volunteers will be immunized with antigens to provide IgG with much higher levels of specific antibody. Inhibitex will then expand its vaccine portfolio to include the other major causes of hospital acquired sepsis, *Staphylococcus epidermidis* and *Enterococcus faecalis*. The former project will also be in collaboration with BRI.

Another project is to develop a vaccine to prevent bovine mastitis, a disease of dairy cattle of great economic importance world-wide.

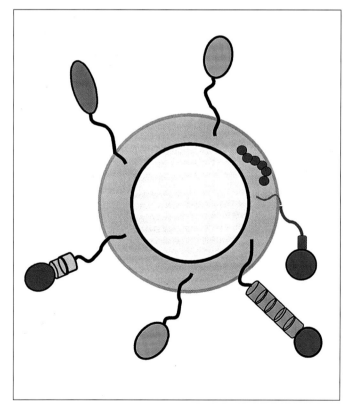

The figure shows that the surface of the S.aureus bacterial cell is decorated with a number of different proteins that interact with components of the host extracellular matrix. The binding domains are usually located in the circular segments furthest from the cell. The proteins have flexible (dark lines) and sometimes additional rigid (cylinders) stalks that project the binding domains further away from the surface. Candidate antigens for vaccines are the circular binding domains which have been cloned and expressed in recombinant form.

NATIONAL CELL AND TISSUE CULTURE CENTRE (NCTCC)

Functional Genomics – Martin Clynes

Important industrial applications of new Biotechnology discoveries are expected over the next 5-10 years in the areas of functional genomics (discovering the functions of newly-discovered genes in health and disease) and cell/tissue engineering (using differentiated or genetically engineered cells to replace damaged or malfunctioning organs). The National Cell & Tissue Culture Centre at Dublin City University is developing research programmes designed to support the growth of industry in these areas in Ireland.

A team led by Dr Carmel Daly is developing ribozyme technology. Ribozymes are RNA molecules which (surprisingly, since until recently we thought all enzymes were proteins) are enzymes which can target and catalyse the breakdown of specific mRNA molecules coding for particular proteins. As well as their potential application in Gene Therapy of diseases such as cancer and viral diseases, ribozymes are ideal tools to find out the function of novel genes and their protein products. Using cells which express the target protein, we can delete expression of this protein specifically, by targeting its mRNA with a specific ribozyme: we can then examine how the phenotype has changed as a result of deleting this protein. The NCTCC programme on ribozymes has benefited from close collaboration with one of the world's experts in ribozyme technology and its application to cancer research, Dr Kevin Scanlon in California.

The NCTCC's monoclonal antibody team, led by Dr Elizabeth Moran, can make specific antibodies to any novel gene product, based on sequence information. These antibodies can then be used to examine expression of the gene, at the protein level, in different tissues and in different diseases.

Other research groups in the Centre are developing human Cell Culture Systems for studying apoptosis and epithelial cell

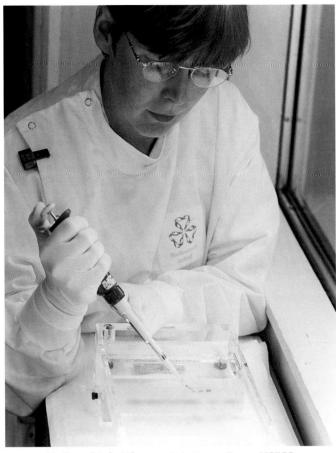

Dr Carmel Daly, Ribozyme & Antisense Group, NCTCC.

differentiation, which will be important resources for investigating the biological, and potential therapeutic/pharmaceutical role of newly-identified human genes.

This multidisciplinary approach to functional genomics represents an unusually broad and flexible resource which will support development of enterprise in this area in Ireland – it has already attracted substantial contract research from U.S. and European Biotechnology Companies.

NATIONAL AGRICULTURAL AND VETERINARY BIOTECHNOLOGY CENTRE (NAVBC)

Biocontrol – the future of horticulture and the food industry

– Elizabeth Morris & Owen Doyle

With increasing consumer demand for improved food safety standards, EU directives, along with regular national testing, ensure that maximum residue levels (MRLs) in fruit and vegetable produce are not exceeded, and that Irish food produce remains safe. However, the pests that we wish to control have begun to develop levels of resistance to our diminishing list of crop protection products. This sets a new challenge to both the food producer and the scientist. How do we control pests without pesticides? A "green" solution – the use of biological control agents – is one possibility.

Biocontrol agents, particularly for pest control in the horticultural sector, have shown great potential. The introduction of natural insect predators to control troublesome pests has been used effectively for greenhouse food crops. Pathogenic bacterial sprays (*Bacillus thuringiensis*) are already used to control cabbage white butterflies, while parasitic wasps (*Aphidius ervi, Aphidius colemani, A. matricariae*) are used to effectively control greenhouse aphids. Other cultural controls, such as insect traps and modifications to environmental conditions, have also been used to effect pest control.

The Irish Mushroom Industry

In the ever expanding Irish Mushroom Industry, commercial mushroom crops are constantly exposed to infestation by pests, which include flies, midges, mites and eelworms. The most commonly identified mushroom pest nation-wide is the Sciarid fly, which can damage up to 40% of the total mushroom crop. As with other horticultural crops, the increasing restrictions on the use of crop protection products, along with the development of resistant pest populations, has highlighted the need for alternative pest control measures.

At the Mushroom Research Group (NAVBC, UCD), we have developed a novel delivery system for a currently used biological control agent (pathogenic nematodes) to target Sciarid fly populations. It is envisaged that our "product" will significantly increase the effectivity of biocontrol agents used in the industry. Our product is introduced at the earliest possible stage of production, which allows the biocontrol agent to attack fly larvae more quickly and effectively. The novel delivery system encourages greater persistence of the biocontrol agent throughout cropping, thus eliminating the need for further applications, reducing costs and removing the need for pesticides from the system.

Biocontrol agents have no health or safety risks associated with their use. They do not leave residues in food crops, and do not encourage the development of resistance in pest populations, and are therefore an environmentally friendly way to control disease in food crops.

The Irish mushroom industry - probably the most successful in the horticultural sector.

R&D – the future of the healthcare sector

Medical device & diagnostic research in Ireland

It has been recognised by the Industrial Development Authority (IDA)[1] and the Irish Medical Device Association (IMDA)[2] that a dual focus is required to support R&D in the healthcare sector. This focus requires:
- a continuing examination of Bio Materials/Engineering R&D activities within Irish third level institutions
- an understanding of the continuing needs of the medical device and diagnostic industry.

Materials Ireland has conducted a review on the nature of current Bio Materials/Engineering R&D activities and R&D needs, both in Irish third level institutions and in companies operating in the medical device and diagnostic sector.

This sector employs 12,000 people directly, with as many employed in sub-supply companies. Exports grew from US $ 0.5 Billion in 1990 to US $ 1.4 Billion in 1996. Of the 75 companies in this sector, almost half currently undertake R&D, but most of the projected R&D effort is concentrated on a small number of companies. The continued growth of this sector is dependent on increased R&D spend in a broader range of companies. Given that the global growth of the medical device market is estimated at 7% per annum, there would appear to be significant potential for increased R&D investment in this sector in Ireland. This review was undertaken in response to these needs, in order to support and further develop R&D linkages between third level researchers and the industry in this sector.

In this study, industry expressed a need for a method of identifying research expertise in third level institutes and how to initiate contact. In addition, there is a desire by many third level researchers to realise the commercial potential of their research.

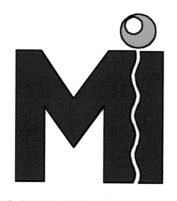

Materials Ireland

Materials Ireland: a co-ordinated approach for Industry

Materials Ireland is a national commercial R&D organisation, dedicated to increasing the competitiveness of Irish industry. This organisation is committed to providing industry with a single point of contact for research services and general support facilities, in Universities, the Polymer Development Centre (see next page), and Enterprise Ireland. Materials Ireland has been requested to co-ordinate the development of a formal R&D network between the industry and the expertise available within the third level institutes.

Structures will be developed within Materials Ireland to ensure that the relevant third level research expertise will be made available to the healthcare industry. This network structure is intended to facilitate contact between industry and the third level colleges. It is not intended to act as a buffer or barrier between individual third level researchers and the medical device and diagnostic sector, or indeed to interfere with any existing relationships.

In the last ten years, there has been a significant growth in the number of healthcare companies undertaking R&D in Ireland. Almost half of these medical device and diagnostic companies currently have some level of R&D activity.

The third level institutions are increasing their commitment to advancing Bio Materials/Engineering R&D. Proof of this is:
- the development of related undergraduate and postgraduate courses
- the organisation of key BioEngineering conferences and fora
- the establishment of a number of dedicated research centres.

The current range of third level R&D activities identified include:
- Biomaterials
- Biomechanics
- Medical Electronics/Instrumentation
- Laser Technology
- Sensor Technology
- Biomedicine.

Traditionally, the relationships between third level researchers, industry and clinicians were common in the field of Orthopaedic Biomechanics. Recently, an increasing number of partnerships have been established within cardiovascular research. However, this review has highlighted that there is significant opportunity for increased co-ordination between the Irish healthcare sector and the research abilities available within third level institutes and also with clinicians.

Industry has indicated that support for collaborative R&D is urgently required. The establishment of a formal network between industry, third level researchers and clinicians would act as a catalyst for successful development of new active Biomaterial devices and the further understanding of existing materials and products. A priority would be the establishment of a database of third level R&D capability that would facilitate the co-ordination of this formal network.

Materials Ireland has been identified as the most appropriate structure, nationally, for the co-ordination of a formal network. This proactive service will aim to satisfy a number of R&D issues identified from this review, and also highlighted recently by the IDA[1] and the IMDA[2].

For further details contact:
Patricia Chawke, Materials Ireland;
Tel: (061) 202 893; Fax: (061) 202 967;
E-mail: chawkep@mat-irl.ie;
Web: http://www.mat-irl.ie;
Joe Healy, Materials Ireland;
Tel: (01) 808-2381; Fax: (01) 808-2470;
E-mail: healyj@mat-irl.ie;
Web: http://www.mat-irl.ie

References:
1. IDA Forum, Athlone, February 1999.
2. IMDA Report, December 1998.

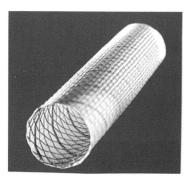
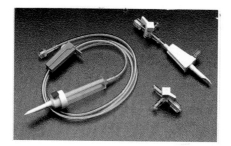

Some medical devices.

THE POLYMER DEVELOPMENT CENTRE — MIKE MORONEY

The Centre of Polymer Expertise

The Polymer Development Centre, founded in 1992, is an ILAB accredited laboratory and is one of the networks of Materials Ireland Research Centres.

The Centre is located in Athlone in an 18,000 square foot purpose built facility, which contains a clean room, processing hall, analytical laboratories and conference facilities. The Centre is fully equipped to provide the polymer industry with a wide range of technical services.

These services include:

- **Material Identification, Selection and Specification**
- **Product & Process Optimisation & Development**
  - Injection Moulding
  - Thermoforming
  - Extrusion
  - Rotational Moulding
  - Blow Moulding
  - Compounding
- **Materials and Product Testing**
  - Physical
  - Chemical
  - Thermal
  - Mechanical
- **Accredited Third Party Certification**
- **Polymer Consultancy and Training**
- **Research and Development.**

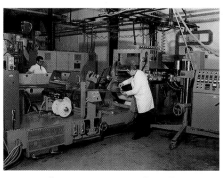

Polymer processing at the Polymer Development Centre, Athlone.

All work is undertaken by a team of multi-disciplined highly-skilled technologists. All analytical and developmental projects carried out at the Polymer Development Centre are confidential to the client with many of the standard tests available being ILAB accredited, which guarantees universal acceptance of results.

Niche areas, on which the Centre has focused its expertise include: manufacture of prototype products, validation of new devices and processes, and the development of speciality materials for the healthcare industry.

POLYMER DEVELOPMENT CENTRE

Within the automotive sector, the Centre has developed resins with specific damping characteristics, improved UV stability in material for outdoor applications, and light weight polymer composites for metal replacement. In the area of packaging, the Centre is involved in projects with various European partners in the areas of anti-counterfeiting devices and packaging, recycling of packaging materials, and the design and manufacture of novel packaging concepts.

Contact: Customer Service, Veronica Guinan, Polymer Development Centre, Athlone Business & Technology Park, Athlone, Co. Westmeath; Tel: +353-902-73088; Fax: +353-902-73090; E-mail: pdc@mat-irl.ie; www.mat-irl.ie

Materials Ireland — ILAB Irish Accreditation of Laboratories — I.S. EN 45001 Reg.No. 0067

TRINITY COLLEGE DUBLIN — WERNER BLAU

Materials Ireland Polymer Research Centre, Physics Department, TCD

The **Materials Ireland Polymer Research Centre** is based in the Physics Department of Trinity College (*Centre Director:* Professor Werner Blau; *Centre Manager:* Ms Una Moran) and acts as a support service to industry by providing both the facilities and necessary expertise to carry out research and development programmes in advanced materials. **Materials Ireland** is a national commercial research and development organisation focusing on industrial materials technologies, particularly on the processing of industrial materials *(see previous page)*.

The main expertise of the **Polymer Research Centre at Trinity College Dublin** lies in the science and technology of polymers for specialised high-tech applications. This includes polymers for the chemical and healthcare industry, biopolymers and novel polymeric materials for electrical, electronic and optical applications. We adopt an interdisciplinary approach to problems and provide cost-effective, practical and efficient solutions.

The **Polymer Research Centre** offers up-to-date expertise in most areas of polymer science and technology and affords industry access to a unique range of state-of-the-art equipment. In addition to our own personnel, the unit can also utilise the research potential of highly trained post-graduate researchers in Trinity College. The active participation by our researchers in a wide range of national, international and European Union projects guarantees that we remain at the forefront of polymer research and development.

Expertise is available at the **Polymer Research Centre** in the following areas:

SYNTHESIS OF NOVEL POLYMERS AND SPECIALITY CHEMICALS
  Hydrogels
  Fullerenes
  Reactive Polymers
  Biopolymers
  Drug Delivery Polymers
  Polymerisation Reactions
  Molecular Imprinting

ADHESIVES AND BONDING
  Recommendation of Optimum Adhesives
  Surface Wettability Analysis
  Curing Systems
  Blend Modification
  Troubleshooting

PRODUCT DEVELOPMENT
  Material Specification
  Metal Replacement
  Development of Chemical Purification Methods
  Speciality Polymer Devices
  Development of Test Protocols
  Market Studies for Niche Markets

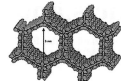

Left: Computer model of inorganic polymer MCM-41.

ENVIRONMENTAL ISSUES
  Pollution Ecology
  Advanced Materials from Renewable Resources
  Waste Disposal
  Recycling
  Biodegradable Polymers
  Non-Solvent Processes

SPECIALISED PROCESSING TECHNOLOGY
  Laser Polymerisation
  Pulsed Laser Deposition
  Excimer Laser Photoablation and Printing
  Spin Coating
  Langmuir-Blodgett Deposition

POLYMER CHARACTERISATION
  Physico-chemical Characterisation
  Molecular Analysis
  Microstructure
  Optical Characterisation
  Structure Property Relationships
  Surface Properties
  Identification of Contaminants

Contact: Materials Ireland Polymer Research Centre, Department of Physics, Trinity College, Dublin 2; Tel: +353-1-608-2404; Fax: +353-1-679-8039; E-mail: wblau@tcd.ie; http://www.tcd.ie/materials_ireland/

Research and Development in Bord na Móna Horticulture Limited

Before 1991 research in Bord na Móna was carried out on a company wide basis. However changes in the business structure of BNM in 1991 led to the formation of the Horticulture Division (now Bord na Móna Horticulture Limited). This division was charged with the development of BNM's horticultural peat business.

Concurrent with divisionalisation in 1991, it was decided to set up a group dedicated to R&D work for the Horticulture Division. Thus, in

work has been the development of a growing medium based on composted bark. This work was initiated by the desire of one major retail customer to offer a peat-free alternative to the general public. It was a step in the dark for the group, and involved finding ways to stabilize the bark as regards nitrogen retention, determining the correct particle size distribution to get an optimum balance between aeration and water holding capacity, and fertilizer inputs (on the basis of

Bord na Móna Horticulture Limited growing trials in progress at Teagasc's Kinsealy Research Centre.

Automated rapid analytical equipment used for determining plant nutrients in water extracts and acid digests of foliage samples.

the same year, Dr Munoo Prasad was appointed Chief Horticultural Scientist. Dr Prasad had previously worked for the New Zealand Ministry of Agriculture and Fisheries as a Group Leader at Levin Research Centre, and had a strong background in the type of R&D the Horticulture Division wished to pursue. Currently the group is also comprised of a Scientist (Dr Paul Simmons), Analyst (Pauline Geoghegan) and Technician (Colman Hynes). The group carries out R&D on all aspects of horticultural growing media (chemical, physical and biological) and works closely with Teagasc Kinsealy, Dublin Institute of Technology and BioResearch Ireland. International links have been formed with other similar groups, and projects are currently running with leading Dutch research stations.

One aspect of the group's work, which has had major impact, is on the relative stability of Irish and Northern European (*i.e.* Baltic) peats. Breakdown of peat in a pot over a period of months or years can impair plant growth. Thus it is important that the peat in which the plant is growing does not slump in the pot. Work carried out in co-operation with Dutch scientists, and using sophisticated techniques such as Fourier Transform Infra Red Spectrometry, has shown that Irish peat is generally more stable, due to its biochemical components, than many other peats available in Europe. These findings have become widely recognised in Europe and a number of independent articles have now appeared in the trade journals supporting these ideas.

One other successful area of the group's

growing trials). Storage life of the final product also had to be estimated. Because of the complex nature of the material, there are still many interesting areas to explore with bark. This research has been a major success story for Bord na Móna Horticulture Limited and growing media based on composted bark is now an important component of the product range.

A major innovation to the group's work came in 1995 with the development of our technical support facilities for professional customers. This support has included:

- Grower trials to optimize fertilizer and lime inputs for the vast range of plants grown in Irish nurseries
- Growing media analysis for determination of available nutrients in the compost
- Nutrient reserve analysis in controlled release fertilizers
- Foliage analysis which can determine the nutritional history of the plant
- Irrigation water analysis for determination of water quality.

These types of services have created closer bonds between Bord na Móna Horticulture Limited and their customers, and have allowed them to develop a better performing range of products.

Currently the group's work involves several unique projects which will lead to the launch of the next generation of growing media over the next few years.

Analyst Pauline Geoghegan prepares a water extract from a compost sample.

Contact: paul.simmons@bnm.ie

Satellite Imagery brought down to earth in Ireland's forests

Imagine a time when you can sit at your PC, type in the name of your local area, and up comes a high resolution aerial photography or even a satellite image of that area. Further browsing demonstrates that all the forests of Ireland can be displayed in 20 different classes. Analysis of the system produces a map of your County indicating the location of all the mature Oak forests. More detailed analysis indicates the best location for a new paper manufacturing plant. The year is 1999 and the system, Ireland's Forest Inventory and Planning System (FIPS), the largest Geographic Information System (GIS) on the island, has just gone live.

FIPS GIS user interface.

Introduction
On 23 June last, the Forest Service, now based in Johnstown Castle in Wexford, took possession of a GIS which had been under development for two years by a Coillte led consortium, including the National Remote Sensing Centre (NRSC), UK, and the Joint Research Centre (JRC), Italy. The project involved the interpretation and verification of satellite imagery followed by photo-interpretation. The consortium selected Landsat Thematic Mapper satellite imagery, which was radiometrically and geometrically corrected and subsequently classified using Silvics software developed by the JRC. Together with the high spatial resolution of the digital ortho-photography and other datasets available to the Forest Service, the Coillte consortium were able to enhance the classification results to provide 20 broad species and development classes.

A National Inventory – Why?
A number of factors have highlighted the need for a new inventory. The rapid expansion of afforestation in recent years has seen large scale landscape change in the countryside. Forest policy set out in 1996 has the objective of establishing a sustainably managed critical mass forest estate by 2030: this will be achieved through the establishment of 20,000 ha (50,000 acres) per annum of new forests. The need for up-to-date accurate data on the national forest estate has therefore become increasingly important in the context of this ambitious planting programme.

Current FIPS GIS functionality

FUNCTIONALITY	DETAILS
Planting Grant Administration	FIPS will be used to manage the various planting grant schemes administered by the Forest Service with the ability to digitise and record application details.
Felling Licence Administration	The felling licence system administered by the Forest Service will be facilitated and will aid in the updating of the inventory.
Forest Protection	The extent of any disease outbreaks or pest attacks can be digitised and proximity analysis can identify neighbouring forests at risk.
Environmental Applications	To facilitate the evaluation of the impact of afforestation or clearfelling on the environment, FIPS has the necessary data and tools for proximity and threshold analysis.
Field data	Summary field data can be stored and retrieved including the calculation of summary statistics.
Stratified random Plot Generation	The stratified random points can be selected automatically for field survey (or ultimately forecast production).
Image Processing	ERDAS Imagine™ and Silvics will provide the Forest Service with advanced image processing facilities.
Monitoring and Control	Optical and Radar methodologies using ERDAS Imagine™ and Silvics will permit the Forest Service to monitor and control felling and planting through the use of basic change detection processes.
Data Capture, Display and Retrieval	Heads up digitising will be the primary means of data capture, a display menu of all the available ancillary data will allow the user to use one or more of the many data sets whilst digitising or performing one of the other operations.
Querying Facilities	SQL statements complied in MicroStation™ as well as Seagate's Crystal Reports™ will be used to query the Oracle™ database.
Reporting and Map Production	FIPS will have the means to produce a variety of maps from both MicroStation™ and ERDAS Imagine™.

Additional FIPS Functionality
The needs of the Forest Service were closely examined, and it became clear that other GIS based information could be usefully incorporated into FIPS. Currently there are four other projects under development which when completed will add to the system. These projects are as follows:

1. A forest soils classification project with the aim of creating a forest soils productivity map.
2. The field sampling of forests in order to generate volume production forecasts.
3. A landscape project to classify landscapes into different classes depending on their sensitivity to forest development.
4. A forest grant and premium administration system. This system will manage and control the various planting schemes and will automatically update the inventory.

Summary
The first component of FIPS has now produced a forest classification: the methodology adopted used a combination of neural network and maximum likelihood classifiers to achieve the best possible classification from satellite imagery alone. Panchromatic aerial photographs were then used to enhance the spatial accuracy and where necessary the classification accuracy. The resulting inventory database includes all forest areas over 0.2 ha, in a multi-users environment that incorporates remote sensing, GIS and MIS. It also benefits from an enviable array of ancillary data, including satellite imagery, digital ortho-photography, OS raster maps, DEM, roads, rivers, lakes and archaeological sites. Four additional projects to further enhance the functionality of FIPS are also under development.

The assistance of the EU in supporting FIPS is gratefully acknowledged.

The Coillte Project Centre in Newtownmountkennedy.

Contact: The Department of Marine and Natural Resources, Johnstown Castle, Wexford; Coillte Teoranta, Newtownmountkennedy, Co. Wicklow.

DEPARTMENT OF AGRICULTURE FOR NORTHERN IRELAND & THE QUEEN'S UNIVERSITY OF BELFAST

Scientific advances in food quality and safety
Food Science at Newforge Lane, Belfast

Innovation and scientific expertise are essential for industrial development, and the food industry is no exception. The Food Science Division of the Department of Agriculture for Northern Ireland provides a source of specialist information, advice and solutions to problems, research of international standing and local relevance, as well as University graduates in food science and technology. This last is possible because the Food Science Division is also the Department of Food Science at Queen's University, an arrangement which benefits both research and teaching. Undergraduate and postgraduate students not only contribute to the research but also learn from the wide range of scientific activities undertaken.

Research

Food safety concerns us all as the number of reported food poisoning incidences continues to increase. Consequently the bacteria responsible for these outbreaks, including *Salmonella*, *Campylobacter*, *Escherichia coli* 0157 and *Listeria monocytogenes*, are gaining notoriety! The food industry has two pressing needs with regard to confirming the presence of these foodborne pathogens: speed and sensitivity. An approach based on measuring the effect of bacteria on the impedance of the growth medium (known as RABIT - Rapid Automated Bacterial Impedance Technique) has been developed for use with *Salmonella*. Impedance microbiology detects microbial growth as it happens, rather than retrospectively as is the case with conventional microbiology. This procedure allows rapid sample screening.

Other microbiological research areas include:

- the use of irradiation or high pressure in food preservation;
- surface biofilms in food processing environments as a reservoir for pathogens;
- the distribution of the bacterium *Escherichia coli* 0157 in the food chain and how it can be controlled; and
- the occurrence in unpasteurised milk of an organism, *Mycobacterium paratuberculosis*, which may be associated with Crohn's disease.

The quality of food and food products is also a topic of great interest. Research teams are studying the microbiology and chemistry of food quality.

Not all bacteria are harmful and some, like those used to produce cheese and yoghurt, are essential in food production. Investigations into how bacteria, or the enzymes they produce, can be used in food production offer novel strategies for product development.

Meat is a major product of Northern Ireland and several studies are investigating how production factors affect meat quality. Low-fat meat products often possess a less acceptable flavour than their more fattening traditional products. This is due to the role of fat as a solvent for flavour, and preliminary studies have shown that other ingredients may be able to replace this function. Other studies are examining how the welfare of animals, marinading and other factors can affect the quality of meat.

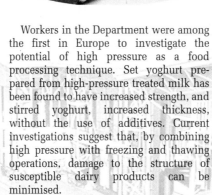

Workers in the Department were among the first in Europe to investigate the potential of high pressure as a food processing technique. Set yoghurt prepared from high-pressure treated milk has been found to have increased strength, and stirred yoghurt, increased thickness, without the use of additives. Current investigations suggest that, by combining high pressure with freezing and thawing operations, damage to the structure of susceptible dairy products can be minimised.

Other work is exploring how the modification of the lipid composition of milk can improve the nutritional and functional quality of dairy products, and the role of dietary antioxidants in a healthy diet.

Consumers demand that irradiated foods must be labelled as such, so it is essential that we can identify them. Food Science Division has been at the forefront of the development of techniques for irradiation detection (see *The Irish Scientist 1997 Yearbook*). The effects of irradiation on the nutritional and eating quality of foods have also been investigated.

Food analysis and advice

Food Science Division conducts a large number of chemical and microbiological analyses to ensure food quality and safety. Laboratories are accredited to perform most tests to the highest standards (including the United Kingdom Accreditation Service and Good Laboratory Practice).

Staff of the Food Science Division are frequently called upon to advise the Northern Ireland food industry on issues including microbial contamination, off-flavours, sensory analyses and training, processing problems and product development. Difficulties are often solved by a short telephone conversation, but sometimes require a visit or the commissioning of work to investigate the problem further. Local (and international) food industries are also involved in collaborative research projects designed to improve understanding of the food they produce.

Teaching

As members of the Department of Food Science at Queen's University, staff are responsible for the teaching of both undergraduate and postgraduate degrees. These include BSc degrees in Food Science, Food Technology and Microbiology. Students, with a primary degree in a suitable science subject, can gain a Diploma or MSc in Food Science. A new MSc in Food Safety Management, conducted part time in the evenings, has also proved very popular in its first year. All these degrees provide the scientific training and knowledge to allow students to make a positive contribution to local or national industry and, in some cases, to research. The strong research base makes the Food Science Division an ideal location for postgraduate research leading to MPhil or PhD degrees.

Thus, the wide-ranging activities of the Food Science Division enable it to contribute to the supply of qualified individuals for the food industry, and to provide scientific advice and research at the local, national and international levels.

Contact: Prof. J. Pearce, DANI Food Science Division, Newforge Lane, Belfast BT9 5PX; Tel. +44 1232 255349; E-mail: Jack.Pearce@dani.gov.uk

Silvopasture - a new land use option

In silvopastoral systems, grazing stock and trees are combined on the same land base. Research by the Applied Plant Science Division of the Department of Agriculture for Northern Ireland (DANI) and Queen's University, Belfast, has shown such a system, where sheep graze managed grassland between wide-spaced, protected broadleaved trees, to be a viable land use option delivering significant output, welfare and environmental benefits.

On their Agroforestry Unit at Loughgall, Co. Armagh, protected ash have been planted at 400 trees ha^{-1} in 1989 into managed ryegrass pasture (160 kg Nha^{-1}) and grazed with sheep from March to November. The system is being compared with agricultural and woodland controls.

It has been found that:

Trees can be successfully grown in an intensively managed ryegrass pasture in combination with sheep. In the silvopasture, trees now average 8.3m tall (9m in woodland) and show good form if carefully pruned.

Swards can be managed as easily in silvopasture as in open grassland and yields have not been reduced by the trees. Leaf phenology of ash (late in leaf, early drop of edible leaves) complements pasture growth.

Sheep output and individual animal performance are not reduced by 10 year old trees.

Welfare – Sheep spend more time in the shade and shelter of trees than in the open in bright, sunny and wet weather.

Biodiversity is greater in silvopasture than the other systems. Some diversity in grassland *flora* is occurring near the base of the trees; there are more *carabid beetle, spider, bird* (particularly in winter) and *young earthworm* species in silvopasture than in agricultural or woodland controls.

Silvopasture.

Economic predictions for the system are favourable. The current downturn in meat product prices and increased interest in agri-environment support payments towards enhanced biodiversity are likely to enhance the attractiveness of silvopasture.

The site is part of a UK national network of trials and a comprehensive review of the **sustainability** of the system has proved favourable.

Farmer attitudes are positive – they see silvopasture as a highly flexible land use option delivering environmental benefits.

On this basis, Greenmount College has taken demonstration systems onto a network of commercial farms and then to the wider farming public.

The DANI unit at Loughgall offers a modern and well equipped farming and scientific resource for collaborative research, particularly into the more fundamental environmental issues arising from what is proving a viable future land use option.

Contact: Dr Jim McAdam;
Fax: 028-90668372;
E-mail: jim.mcadam@dani.gov.uk

Agri-environmental indicators

Support for farming is being increasingly directed away from conventional crops and livestock towards agri-environment and direct income support measures. As a consequence, the impact of agriculture and agricultural policies on the environment are now major issues in the EU. In order to quantify these impacts, it is necessary to develop agri-environmental indicators. The purpose of such indicators is to:

- provide information to policy makers and society in general on the state of the agri-environment and on trends in environmental conditions;
- assist policy makers in understanding the linkage between agriculture and effects of agricultural policy on the environment;
- contribute to the evaluation of the effectiveness of policy measures on environmental sustainability.

For example, the EU is currently demanding information on the effectiveness of the Rural Environment Protection Scheme (REPS). The demand for information on agri-environmental relationships is growing not only from policy makers such as the EU or the national government but also from farmers and other users of land, including recreational users and, more importantly, by society in general.

The environmental performance of commonly-used agricultural systems must be quantified. Without this knowledge, it will be impossible to (i) develop a measurable definition for what is meant by "sustainable agricultural development"; and (ii) identify factors of concern, corrective actions that might be taken, and the potential costs of implementing the necessary changes in practice. Restrictions on agricultural production in response to environmental constraints can only be applied in an inefficient, trial-and-error fashion (through adoption of the restrictive precautionary principle) in the absence of this information.

Right: Unsustainabe grazing management.

Teagasc, at its Soils & Environment Research Centre, Johnstown Castle, is addressing some of the issues involved as part of the research programme on Sustainable Farming Systems and has identified the following priority areas where it is necessary to establish indicators:

- Nitrogen balance
- Pesticide use
- Water quality
- Countryside stewardship
- Phosphorus balance
- Greenhouse gases
- Soil quality
  Physical degradation
  Biological degradation
  Chemical degradation
  Lime status
- Biodiversity

Future land-use systems will have to reconcile and achieve multiple objectives. Whilst maintaining profitability in market goods will continue to be a core objective, there will be an increasing recognition of the non-agricultural production value of land, and also of the impact of farming on the surrounding environment. A new land-use ethic is rapidly developing, embracing agricultural, forestry, landscape, ecological, environmental and social elements, with the concept of sustainability being at the core of future land-use policy development. Clearly, land-use research in its broader context of embracing all the elements of the rural economy as listed above will require a new focus, and will include the quantification of the parameters of sustainable farming systems.

Contact: Dr John Lee, Teagasc Environmental Research Centre,
Johnstown Castle, Wexford; Tel: 053-42888; Fax: 053-42002;
E-mail: jlee@johnstown.teagasc.ie

Food research – the key to competitiveness

The central aim of government policy for the Irish food industry is to maximise its value to the economy. Under CAP, price support systems favoured the production of certain dairy commodities but, in future, with falling price supports, greater diversity into value-added consumer products and ingredients for targeted customer needs will be required.

The interests of the producer are also best served by diversification as a buffer against failing price supports. The ability of the dairy and food ingredients industry to adapt, and the capacity to confront demands of safety, functionality, flavour and nutritional quality that premium markets require will determine future competitiveness and growth. Hence sustained, and indeed increased investment in both public and in-company R&D is essential to meeting these challenges.

On behalf of the Government, Teagasc through its two food divisions – the Dairy Products Research Centre at Moorepark, and the National Food Centre in Dublin – are committed to technology transfer to the Irish Food Industry through a comprehensive programme of strategic and applied research, consultancy and training, backed up by world class laboratory and pilot process plant facilities.

At the Dairy Products Research Centre, research on cheese food ingredients and the quality and safety of dairy foods represents most of the public research programme. This provides an essential base of scientific expertise which feeds into the food industry and strives to bridge the gap between Irish food companies and global leaders.

The primary aims of the Centre are to maintain scientific excellence in strategic areas of research, and to actively manage the innovation process to drive the growth and competitiveness of the Irish dairy food and ingredients sectors. The latter is achieved by monitoring the progress of research projects, identifying outputs with commercial potential, and quickly progressing from pre to full commercial exploitation with industry client or partners. This process typically involves securing intellectual property, applications and pilot plant validation, followed by industry trials and test marketing *(see Figure 1)*.

A unique asset of the Centre is its world class pilot processing plant, operated by the Teagasc subsidiary Moorepark Technology Limited (MTL) which is now in its sixth year of operation. This facility, jointly owned by Irish dairy processors, is widely used by industrial and research technologists in supporting the technology transfer process, providing a vital link in the scale-up and commercial exploitation of scientific knowledge.

Moorepark Technology Limited – world class pilot processing plant, bridging the gap between research and commercial development.

Research projects are formulated in consultation with industry and are managed to ensure that they progress to commercial implementation. Following expansion of the programme from 1994, with the assistance of EU Structural Funds, many projects have reached maturity and are now feeding into the technology transfer process. The following is a brief review of the main highlights of that Programme.

CHEESE: A GROWTH PRODUCT

Markets for cheese are immensely diverse, and growth opportunities exist in retail, industrial, food service and fast food outlets. In the move towards a free market, expansion of Ireland's cheese output is predicted. Hence continued research is required at a pre-competitive level to support the attainment of new targets of consistency, variety, texture and flavour. Moorepark research in cheese has a number of major achievements in recent years resulting in major commercial initiatives.

Low-fat cheese
Dietary trends have favoured the growth of low-fat products throughout the world and will continue to drive demand for such products.

However, low fat Cheddar on the market lacks the flavour and mouth-feel of the full fat product. The research at Moorepark is aimed at overcoming this deficiency through a detailed investigation of the technology of manufacture and an understanding of the key microbiological and biochemical factors impacting on flavour. The result – a *Moorepark Process* – has been successfully developed (patent pending) which produces a high quality product with taste properties close to those of full-fat Cheddar. This is acknowledged by industry to provide a product superior to any on the market, which should give an edge to Irish manufacturers in further evolving this market segment.

Pizza cheese
The technology for manufacture of Mozzarella cheese and other pizza cheese varieties is new to the Irish dairy industry, and there has been much dependence on bought-in technology in entering this growth product area. Strict quality specifications for shredability, meltability, flowability, chewiness, stretchability, flavour and colour are demanded by customers, and the ability of manufacturers to customise their products according to these criteria will determine future competitiveness. Cost, and the ability to produce a consistent product year-round, are essential.

Hence the Dairy Products Research Centre has established a strong base of research expertise in Pizza (Mozzarella) cheese which is underpinning product differentiation by Irish manufacturers.

A successful blueprint for all year round production of quality Mozzarella, based on the *Moorepark Milk Production System* and a *Standardised Cheesemaking Process*, has been developed. Key features of the *Moorepark Milk Production System* include the maintenance of cows on a satisfactory plane of nutrition, especially in late lactation, and drying off cows when the yield drops to 9kgs (2 gals) per day.

New cheese varieties
Dairy Products Research Centre research on new cheese varieties is stimulating increasing interest in industry in cheese diversification. In line with national policy to broaden the product base of the cheese industry, technological research is carried out on new or modified cheese types. This follows earlier research which led to the development of a successful new cheese now marketed commercially under the brand name *Dubliner*.

The objectives of the current programme are to present further options to industry for market assessment, to establish a data base of technology for certain *continental* varieties, and to overcome quality problems associated with a seasonal milk supply.

Figure 1. Moorepark Model for Technology Transfer.

Research on new cheese varieties presents options to industry for market assessment.

The availability of the pilot plant facilities of Moorepark Technology Limited for small scale manufacture permits interested companies to commission the manufacture of product for initial market tests without risking major investment in plant, while availing of the manufacturing expertise at the Centre.

To-date several prototype cheeses have been developed for initial evaluation and market tests by cheese manufacturers. This research has provided a model for diversification which should ease the path and limit the risk for Irish companies embarking on the development of new varieties.

Key elements of the Moorepark Model:
- *Development of a pre-commercial knowledge and skills base for new cheese varieties.*
- *Assisting commercialising partner to establish a market presence using MTL facilities as an initial manufacturing base.*
- *Backup for industrial scale-up in manufacturing plants.*

Improving cheese flavour

Ireland produces 80,000 tonnes of Cheddar annually – most of which is exported. Quality and consistency are vital in a fiercely competitive international market. Starter bacterial cultures, and indeed other bacteria, play a critical role in flavour development.

Moorepark Technology Limited – Flexible cheese-making facility for small scale manufacturing and test marketing.

Moorepark, in collaboration with the faculty of Food Science and Technology, University College Cork (UCC), have acquired an international reputation as world leaders in starter culture research. Through this research, new culture strains for flavour enhancement and improved phage (bacterial virus) resistance have been successfully developed and validated by pilot and commercial scale cheese production.

Probiotic cheese

The demand for foods for specific health benefits (called *functional foods*) is growing rapidly. Cultured dairy products such as bio-yoghurts are an ideal platform for functional food development because of the possibility of incorporating beneficial microbes (or probiotics) in the fermentation process.

Again in collaboration with UCC, microbial strains of known probiotic properties are being developed and evaluated for their propagation and survival in cheese up to the point of consumption. To date, probiotic bacteria have been shown to survive well in Cheddar cheese during the ripening and maturation period – often over one year. This represents a major commercial opportunity to add value to our most important cheese product and, not surprisingly, is stimulating significant industrial interest.

FOOD INGREDIENTS – ADDING VALUE

Innovation partnerships between the manufacturers and users of food ingredients are an increasingly important feature of the global food industry. Through expert knowledge of formulated foods, the manufacturer of dairy-based ingredients can attain greater customisation of his products and move up the value chain in market supply.

The Dairy Products Research Centre has a vibrant research programme in this area and has contributed to food ingredients innovation in the Irish food sector in several areas.

Functional and nutritional proteins

The functional and nutritional value of milk proteins for specific market segments can be enhanced by the emerging protein technologies. Moorepark has contributed to a number of successful innovations which are bringing Irish companies to the forefront of developments in this area, and several products from the programme are now being commercialised. Those include a protein derivative which is rich in the amino acid glutamine and which is suitable for athletes as an aid to recovery from exertion, a flavour ingredient from yeast, and a novel whey protein product enriched with the component beta-lactoglobulin. A new process for production of total milk proteinate, an ingredient which combines both casein and whey protein in a functional form, has been patented, and a special form of milk protein for humanised infant formula has been developed with nutritional properties that are superior to currently used protein products.

The development of such specialised products is essential to enable Irish companies to extract maximum value from all milk components. Apart from innovative process technologies, such developments require a range of research skills in biotechnology, biological analysis and nutritional science, which are not normally

Confocal Laser Microscope is a powerful tool to study food microstructure.

available in-house to Irish companies, and without which access to these specialised developments would not be possible.

New ingredient powders

Innovative technology has been developed for ingredient powders for use as functional or flavouring ingredients in baked products and snack foods. Access to such technologies is important for Irish companies wishing to establish more specialised ingredient outlets.

Through research carried out on the world class evaporation-drying facilities of Moorepark Technology Limited, a database was compiled on the complex relationship between process conditions and the properties of milk-based powders, which has provided a unique capability for powder specialisation. This was complemented by research on the technology of micro-encapsulation which permits the manufacture of high-fat, and sometimes highly unstable ingredients in powder form, and the entrapment of flavour agents for delivery to a food product. Such pilot-scale processing facilities play a key role at the process design and validation step in the course of technology transfer from the laboratory to the factory.

A natural ingredient (in powder form) with a smoked bacon flavour, developed at the

A natural ingredient with smoky bacon flavour won a prestigious international award at Food Ingredients Europe in Frankfurt.

Centre in collaboration with two Irish companies, won a prestigious award at the Food Ingredients Europe Exhibition held in Frankfurt in November last.

Flavour ingredients for snack foods, high free-fat powder for chocolate manufacture, and high-fat powders for baked products are other examples of successful products from this research.

QUALITY: A CUSTOMER IMPERATIVE

The excellent safety record of dairy foods cannot be taken for granted. New microbial risks are emerging from changing market specifications and by diversification into more sensitive and vulnerable product areas such as soft cheeses or savoury dairy products. The positive nutritional aspects of dairy products and, in particular, the increasing interest in health promoting products (functional foods) can be the stimulus for a new growth phase in consumer markets. However, fundamental research back-up is necessary to exploit these new market opportunities and to support product claims. Such research is well beyond the scope of most Irish food companies.

Hence the Dairy Products Research Centre plays a vital role, not only in exploiting and promoting the positive attributes of dairy foods, but also in protecting the excellent record of the Irish Dairy Industry in quality and safety.

The Centre has had a number of major successes in this area.

New pathogen inhibitor

A new approach to suppressing the growth of pathogenic organisms in certain susceptible dairy products has been developed. This is based on the discovery and patenting by Moorepark and UCC scientists of a natural inhibitory agent [lacticin 3147] produced by a lactic acid producing bacteria, which kills a number of dangerous pathogens.

The lacticin-producing organism has been successfully used to prevent listeria growth in soft cheeses, thereby enhancing consumer safety. It also has a role in accelerated ripening of hard/semi-hard cheeses since it suppresses the growth of spoilage organisms and thereby permits a higher ripening temperature and accelerated flavour development. The same inhibitor has been found to suppress mastitic organisms (*streptococci* and *staphylococci*) in dairy cows, and has been licensed to an Irish veterinary products manufacturer for use in teat-seals for mastitis prevention during the dry period. This breakthrough has attracted major international interest, placing Irish food scientists at the forefront of food molecular biology and, while the genetic sequencing of this natural inhibitor was in itself a major scientific achievement, its full potential in food quality and safety, as well as in animal and possibly human health, have yet to be exploited.

Butter – at last the good news

Despite the bad press for butter (and indeed other fats from ruminant animals), some good news.

Research in the USA in particular has shown that a fatty acid in ruminant animal fats, called *conjugated lineoleic acid* (CLA), protects against cancer as well as offering a number of health benefits.

Research on CIA at Moorepark has shown that milk (and beef) produced from grass contains significantly higher levels of CLA than that produced on concentrates. In addition, CLA may be further boosted by supplementing cows on pasture with rapeseeds and soybeans. This not only boosts CLA levels, but also produces a softer spreadable butter, high in the acclaimed health promoting mono-unsaturated fat. Butter – the new health food?

Premium quality lactic butter

Lactic butter, one of our major export successes as a branded product in recent years, has been prone to flavour inconsistencies due to variability in diacetyl (the main flavour compound) production during manufacture.

However, research at the Dairy Products Research Centre has made a major breakthrough in optimising the flavour of lactic butter and, in conjunction with manufacturers, industrial trials proved highly successful in enhancing and standardising diacetyl production, and hence butter flavour.

This new technology has now been successfully adopted by Irish manufacturers, giving them a new competitive edge.

TECHNOLOGICAL SERVICES TO INDUSTRY

Research at the Dairy Products Research Centre is supplemented by a range of services, including dissemination and technology transfer, which have the common purpose of boosting the technical base of Irish food companies. Apart from a substantial programme of contract research, the Centre organises regular major conferences which keep industry updated on forefront developments internationally. Access to sophisticated analytical services as well as to specialised high technology equipment is also provided.

The Centre also provides manufacturing loss measurement and control services to dairy plants and, in general, maintains a number of research and service activities in process efficiency.

Traditionally, the Dairy Products Research Centre has played a central role in developing and evaluating analytical instruments for dairy laboratories, and it continues to support these laboratories through the regular provision of Milk Standards for instrument calibration for milk testing. The Centre also runs an *Inter-laboratory Proficiency Testing Programme* for dairy laboratories which provides independent validation of the competence of the participating laboratories. A service to test and approve dairy equipment cleaning products is also provided by the Centre.

In recent years, a range of Certified Training Courses for dairy plant operatives have been developed in collaboration with Fás. The MTL pilot process plant provides an ideal, cost-effective facility for this training programme.

In the recent past, the Centre has collaborated with farmers' organisations and dairy companies in setting up an agreed, planned split sampling appeal scheme for dairy farmers. Moorepark's contribution to the new appeal scheme is as technical co-ordinator and provides a split-sample testing service in its Milk Standards Reference Laboratory.

Contact: The Dairy Products Research Centre, Moorepark, Fermoy, Co. Cork; Tel: 025-42222; Fax: 025-42340; E-mail: reception@moorepark.teagasc.ie

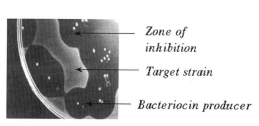

Lacticin 3147, a natural inhibitor produced by a lactic acid bacteria, kills a number of dangerous pathogens.

Grass-fed beef: a natural health food!

The Teagasc Research Programme on profitable production of high quality beef is carried out at Grange Research Centre, Dunsany, Co. Meath in collaboration with The National Food Centre, Dunsinea, Co. Dublin, other Teagasc Research Centres and Academic Institutions. One objective of this programme is to consistently produce beef with sensory characteristics such as tenderness, juiciness and flavour that are requested by consumers in the many countries into which Irish beef is exported. Another objective is to enhance the perception of beef as a wholesome, healthy food. Consumption of red meat continues to decline worldwide due in part to changes in lifestyle and the increased choice of foods available to consumers. A major influence on the fall in consumption is the concern of health-conscious consumers with dietary fat, which is fuelled by medical advice to restrict the consumption of calories from fat and, in particular, saturated fat.

Beef is often considered a fatty food but, while beef fat contains a higher proportion of saturated fatty acids (SFA) than plant foods, it also contains an array of fat molecules which have beneficial effects on human health. The medical profession now recognises that some mono-unsaturated fatty acids, i.e. fatty acids with one double bond in the chain of carbon atoms, are protective against heart disease; that longer chain polyunsaturated fatty acids (PUFA), in particular those with the first double bond at the omega-3 position, such as those found in oily fish, are anti-thrombogenic; and that conjugated linoleic acid (CLA), a geometric isomer of linoleic acid (an 18 carbon fatty acid with two double bonds), is protective against cancer, obesity and heart disease.

Beef from grass-fed cattle has a "healthier" profile of fatty acids than beef from grain-fed cattle.

We examined the potential of different diets, in particular diets based on grass, to increase the concentrations of the "healthy" fatty acids in beef. Steers were offered five diets for 85 days before slaughter. All animals were offered sufficient feed to ensure a similar growth rate for each diet. The composition of the diets were (g/kg total diet dry matter):

- 617 grass silage
- 900 concentrate + 100 straw
- 510 grazed grass
- 770 grazed grass
- 1000 grazed grass

The remainder of the forage-based diets was the same concentrate as that offered in the concentrate-based diet. After slaughter, samples of the *m. longissimus* muscle (the striploin) were collected and the profile of fatty acids measured.

The intramuscular or marbling fat concentration averaged 41 g/kg. Lean beef, as produced in this experiment, should therefore be considered a low fat food, especially when compared to the fat concentration presented for beef in older tables of food composition (>100 g/kg). There was little difference between the grass silage and concentrate-based diets for those variables measured. However, as the quantity of grass in the diet was increased, there was a decrease in SFA concentration, an increase in the omega-3 PUFA concentration, without an effect on the omega-6 PUFA concentration, and an increase in the conjugated linoleic acid concentration. The means for the extreme treatments are illustrated in the *Figure*.

Some medical authorities suggest that consumers should restrict calorie intake from fat to 30% of total calories, to increase the ratio of PUFA to SFA, and to decrease the ratio of omega-6 to omega-3 PUFA in the diet. Together with the measured increase in conjugated linoleic acid, our data indicates that, by virtue of the Irish climate facilitating the efficient growth of grass as an animal feed, Irish beef producers have a natural advantage in producing beef that is more compatible with current medical advice on diet composition.

Our current research seeks to identify the mechanisms to allow us to design feeding practices which will optimise the concentrations of "healthy" fatty acids in meat produced not only from grass but from other feedstuffs commonly used in Ireland. Information from these studies will allow Irish farmers to produce beef with pre-defined characteristics, thereby improving the competitiveness of Irish beef in the lucrative markets of Europe and beyond.

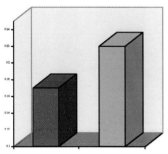
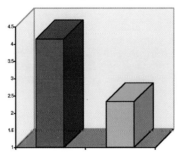
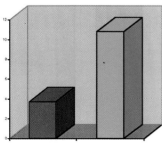
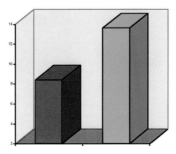

*Selected fatty acids in beef from cattle fed **grass** or **concentrates** before slaughter.*

Contact: amoloney@grange.teagasc.ie

Distribution pattern of supplementary concentrates for finishing cattle

Introduction
In Ireland, cattle finished for slaughter in spring are normally fed grass silage *ad libitum* plus four to seven kilograms of concentrates per head daily, or a total concentrate allowance of 600 to 1000 kg per animal. Until recently, this concentrate allowance was fed at a flat rate per day; but now, this may have to change. This is because cattle are sometimes retained for longer than necessary for adequate finish to qualify for premia which are age/date based. In the absence of a change in feeding regimen, feed costs would be excessively high and carcasses would be overfat. The weight gain response to a particular concentrate input is influenced by the potential for compensatory growth. This is where animals, which previously had a lower growth rate, grow faster than animals which previously had a higher growth rate. The objectives of this study were:
1. to compare the growth and carcass responses to a fixed concentrate input fed at a flat rate daily or fed *ad libitum* after various intervals during which the animals received silage only, and
2. to quantify the compensatory growth response to concentrate supplementation following various periods on silage only.

Cattle finishing indoors on silage and concentrates.

Experimental
Fifty-six Charolais x Friesian steers, 19 months of age and 568 kg initial liveweight were used in 4 treatments:
1. Silage *ad libitum* plus a flat rate (6.35 kg) of supplementary concentrates daily for 126 days.
2. Concentrates *ad libitum* for 83 days.
3. Silage only for 35 days, then concentrates *ad libitum* to 126 days.
4. Silage only for 70 days, then concentrates *ad libitum* to 149 days.

The same total concentrate allowance (800 kg per animal) was fed in all treatments.

Results
Actual concentrate intakes for Treatments 1, 2, 3 and 4 were 788, 815, 798 and 817 kg respectively, with corresponding total dry matter intakes of 1148, 818, 1119 and 1277 kg. Daily carcass gains were lowest for the group fed concentrates at a flat rate *(Figure 1)*. Otherwise, daily carcass gains declined as the length of the feeding period increased. The daily liveweight gain data *(Figure 1 also)* for the three treatments fed concentrates *ad libitum* show the magnitude of the compensatory growth response following delays in concentrate introduction. When the concentrates were introduced immediately at the start of the finishing period, and fed *ad libitum* until the target allowance had been consumed (Treatment 2), mean daily gain was 991 g. By delaying concentrate introduction for 35 days (Treatment 3), mean daily gain thereafter was 1040 g, and a further delay to 70 days further increased daily gain to 1282 g.

Figure 2 shows that delaying concentrate introduction for 35 and 70 days, compared with immediate introduction, increased carcass gain by 19 and 27 kg, respectively. It would be expected that Treatments 1 and 3, which were fed the same concentrate allowance over the same length of finishing period, would have similar total carcass gains. The 10 kg greater carcass gain in Treatment 3 may have been due to a digestive upset suffered by the Treatment 1 animals at the start of feeding. There was no difference between the three treatments fed concentrates *ad libitum* in the efficiency of utilisation of feed energy for carcass production, but these were 14% more efficient than the group fed concentrates at a flat rate.

Implications
Since all other treatments resulted in better animal performance than the standard flat method, the data question if this traditional approach is indeed the correct one. Separating silage and concentrate feeding could have practical management advantages, and allows for the feeding of concentrates at a high level in the period immediately before slaughter, something which should be beneficial to both meat colour and tenderness. In terms of predicting the performance of animals fed concentrates *ad libitum*, a baseline of about 1 kg/day can be assumed for animals with no compensatory growth potential. This increases by about 30 g/day for each week that animals are on silage only before the introduction of concentrates *ad libitum*.

Contact: Dr Gerry Keane, Teagasc Grange Research Centre, Dunsany, Co. Meath; Tel: 046-25214; Fax: 046-26154; E-mail: gkeane@grange.teagasc.ie

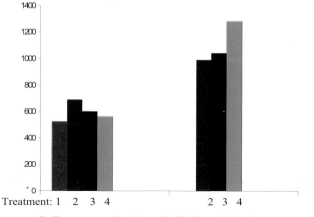
Figure 1. Carcass and liveweight gains (g/day).

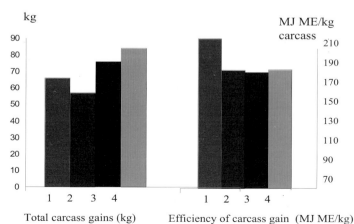
Figure 2. Carcass gains and efficiency for the total experiment.

Enhancing the texture and sensory quality of meat products – use of high pressure treatment

Although many in the food industry perceive high pressure treatment as a relatively new technology, early investigations examined the ability of high pressures to increase the shelf-life of milk treated at ambient temperatures. Since then, the effect of high pressure on micro-organisms, proteins and, more recently, in food processing has been studied. Work at The National Food Centre investigated the use of high pressure (150 MegaPascals - MPa) to enhance the functionality and acceptability of frankfurters with reduced salt and phosphate concentrations, and breakfast sausages with reduced phosphate, in an effort to address consumer demands for high quality products with fewer added ingredients.

Sodium chloride (salt) and phosphate are essential ingredients in meat processing, where they act by solubilising the myofibrillar proteins that contribute to water-holding capacity, thus reducing cook loss. These ingredients also enhance fat binding in meat products, entrapping other ingredients to form a uniform and cohesive mass. Salt also decreases water activity, which significantly extends the shelf-life of the product. However, consumption of excess sodium chloride has become a major issue in the food industry because of the relationship between sodium and hyper-tension in humans. According to the US Department of Agriculture, phosphates are safe when added within the permitted concentration of 0.5%, but such additives generate adverse reactions from the consumer.

Research Officer Clodagh Crehan sets the operating parameters on the High Hydrostatic Pressure Unit.

Our research has examined high pressure to compensate for reduction in salt and/or phosphate in frankfurters and sausages. High pressure is obtained using a hydraulic pump and is applied to the products via a pressure-transferring medium. This results in uniform pressure transmission throughout the product, independent of product shape and size. Being a non-thermal process, product flavour and nutritional qualities are maintained.

Our results indicated that frankfurters treated with high pressure (150MPa) had lower cooking losses than non-pressure treated products. Emulsion stability and texture of pressurised products were as acceptable as control frankfurters. Cook yield was also enhanced when phosphate concentration was reduced to 0.1%, and texture was improved in reduced-salt frankfurters after 300MPa pressure. Pressure treated (150MPa) sausages had similar stability and overall acceptability as non-treated products and texture profile analysis was enhanced when the phosphate was reduced to 0.25% after the pressure treatment. Overall flavour intensity of breakfast sausages with 0.25% phosphate was not affected by the pressure treatment.

Although high pressure treatment is a batch process, these results are positive for the processed meat industry as a novel route to the production of additive-reduced meat products. Further investigations into minimal processing in the Meat Technology Department are concentrating on the use of high intensity ultrasound as a rapid, energy efficient technique to cook pressurised meat products.

This work is supported by the European Regional Development Fund.

Contact: Declan J. Troy: The National Food Centre, Dunsinea, Castleknock, Dublin 15; Tel: 01-805-9500; Fax: 01-805-9550; E-mail: d.troy@nfc.teagasc.ie

Extracting residues from food with carbon dioxide

Checking food for residues of chemicals such as pesticides and veterinary drugs is important to ensure the safety of the food supply to the consumer. The conventional methods for residue analysis use a range of organic solvents to extract residues from food samples and subsequently to remove interfering substances before residue determination.

Carbon dioxide (CO_2), an inert, low cost, non-toxic gas, may be used as an efficient alternative to organic solvents for residue extraction from food samples. CO_2, like other substances, can occur in different phases such as a solid, a liquid or a gas depending on the conditions of temperature and pressure applied. In addition, at a particular combination of temperature and pressure conditions, CO_2 can be in the supercritical fluid phase. As a supercritical fluid (SF), CO_2 has physical properties which are intermediate between those of a liquid and a gas. The solvating power of SF-CO_2 is more like that of liquid CO_2 while its diffusivity and viscosity are more like those of CO_2 gas. These properties of SF-CO_2 make it highly suitable for residue extraction, involving penetration of the food sample and dissolving of the residues to achieve an efficient extraction.

To use SF-CO_2, an instrument which can control pressure and temperature to those suitable for supercritical conditions (31°C, 73.8 bar pressure) is required. The food sample is dispersed on hydromatrix, sand or sodium sulphate and packed into a stainless-steel cylinder. SF-CO_2 is pumped through the sample to extract the residues. After extraction, the SF-CO_2 is depressurised and the residues are trapped on liquid or solid traps for determination, usually by chromatography or immunoassay.

SF-CO_2 has been applied to the analysis of residues of pesticides and veterinary drugs in food samples. At The National Food Centre, this technology was applied to the analysis of antibiotics and beta-agonists in animal tissue samples. The beta-agonists, such as clenbuterol, may be used illegally as growth-promoting agents in beef production. They pose particular challenges for the use of SF-CO_2 because they are relatively polar compounds and are not easily extracted due to competition between the water in the sample and SF-CO_2. A procedure of drying the tissue samples before extraction was developed to facilitate the release of the beta-agonist residues into the SF-CO_2. The extraction efficiency by this method was found to be as good as those based on extraction using organic solvents.

Mandy O'Keeffe loading a sample into the SF-CO_2 equipment.

Research on the use of SF-CO_2 is continuing with its application to other important veterinary drug residues in meat, such as the anti-parasitic agents ivermectin and benzimidazoles.

This work was part-funded by the US-Ireland Co-operation Programme in Agricultural Science and Technology. The Eastern Regional Research Centre, United States Department of Agriculture, Philadelphia, and the Department of Analytical Chemistry, University College Cork, have participated in this project.

Contact: Michael O'Keeffe, The National Food Centre, Dunsinea, Castleknock, Dublin 15; Tel: 01-805-9500; Fax: 01-805-9550; E-mail: m.okeeffe@nfc.teagasc.ie

Eating quality of deep-water fish species

The decline in traditional fish stocks such as cod has led to an interest in alternative fish species. At The National Food Centre, twenty-three unusual fish species have been assessed for their eating quality. The fish were caught by personnel from the Fisheries Research Centre in two deep-water (about 1000m deep) surveys of the eastern slopes on the Rockall Trough, north-west of Ireland.

The different fish species were presented as fillets to twelve taste-panellists. They tasted four unusual species and cod in each panel session and scored them according to acceptability. The mean scores showed that six species (orange roughy, black scabbard, morid cod, greater argentine, snake mackerel and Portuguese dogfish) were preferred to cod. The taste panellists commented that ling and cod were particularly bland, while some of the dogfish shark species had a sweet flavour.

Processing can change or disguise the fish texture, flavour or appearance, and has the potential to add value to those species which are less desirable as fillets. Nuggets were made from minced flesh, shaped and coated with batter and breadcrumbs, then deep-fried. Most species were more acceptable as nuggets than as fillets, and nuggets from ten species were preferred to cod nuggets. Taste panel scores for fish-cakes made from mashed potato and flaked cooked fish showed that eight species were preferred to cod fish-cakes.

Orange roughy – the most acceptable fish as fillets and nuggets.

The composition (protein, water, salt and fat contents) of each fish species was analysed, and lead, cadmium and mercury levels were determined for six species. There was concern that the fish might have accumulated high heavy metal concentrations because some species are long-lived (orange roughy of over 80 years old have been caught!). Fortunately, however, the levels in the fish were much lower than the maximum European limits set for fin-fish in 1992.

Overall, the results show that there is considerable potential for exploiting many deep-water fish. Our current work is directed at reducing damage to the fish during freezing and thawing, while sensory tests to compare fresh and frozen fish are planned.

This work is funded under the Marine Research Measure of the Operational Programme for Fisheries, administered by the Marine Institute and part-financed by the European Regional Development Fund.

Contact: Dr Martine Brennan,
The National Food Centre, Dunsinea,
Castleknock, Dublin 15;
Tel: 01-805-9500; Fax: 01-805-9550;
E-mail: m.brennan@nfc.teagasc.ie

The Veterinary Laboratory Service*

The objectives of The Veterinary Laboratory Service are to implement Department of Agriculture and Food policy in respect of animal health and welfare and veterinary public health. This is done through:

- providing the state veterinary service with technical support and diagnostic capability in respect of statutory and regulatory animal disease eradication and control programmes.
- assisting in the development and implementation of plans for the prevention and control of such diseases.
- providing an efficient laboratory diagnostic service to the livestock industry through practising veterinary surgeons.
- providing a consultancy service when required, to complement laboratory diagnosis and further assist disease investigations.
- monitoring and collating data on diseases at national level.
- providing expert advice and education on disease diagnosis and control to veterinary practitioners, public authorities, private organisations, agriculturists and stock owners.
- identifying diseases with zoonotic implications and collaborating with the public health specialists and the Food Safety Authority.
- research and development.

The Central Veterinary Research Laboratory (CVRL) comprises three divisions; Bacteriology/Parasitology, Virology and Pathology. In addition to providing a routine diagnostic service, each division operates a number of specific surveillance programmes for endemic, emerging and exotic diseases. The CVRL is also collaborating in a number of research projects with other research institutions, both nationally and internationally.

The CVRL is recognised by the EU as the Reference Laboratory in Ireland for a number of diseases – including salmonellosis, newcastle disease, avian influenza, classical swine fever, and Aujeszky's disease.

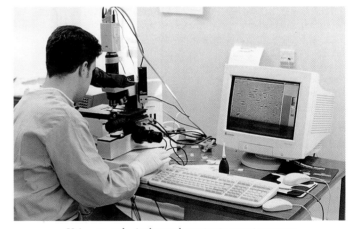

Using stereological morphometry to count neurons in a BSE-affected bovine brain.

In this role, the CVRL validates the competency of private laboratories testing product/samples collected for disease surveillance or trade purposes as required under various EU directives or national legislation.

*The Veterinary Laboratory Service comprises the Central Veterinary Research Laboratory (CVRL) and the Dublin Regional Veterinary Laboratory at Abbotstown in Dublin; the Brucellosis Laboratory, Cork; and five "stand-alone", multi-disciplinary Regional Veterinary Laboratories located in Athlone, Cork, Kilkenny, Limerick and Sligo.
Contact: Mr John Ferris, Director;
Tel: +353 1-607-2869; Fax: +353 1-821-3010.

Forestry research in the assessment of tree health from nursery to field

The importance in economic, environmental and cultural terms of forestry in the Irish landscape has received considerable attention over the past two decades. Political interest in, and motivation for, the extension of the forest resource has been boosted by various European and international protocols.

Plantation forestry can be managed on a sustainable basis, and the principles of sustainable forestry management are being implemented and incorporated into current forestry practice by Coillte, the state forestry body. It is still, however, of paramount importance to ensure the highest potential economic return and successful establishment of forests.

One of the most crucial periods in the forestry cycle is that of the nursery. The growth and health of tree seedlings both in the nursery, during periods of cold storage, and at the time of planting, has been shown to be crucial for successful establishment and subsequent health of forest plantations.

One of the projects underway in the Department of Botany, UCD, funded by COFORD, is, for the first time, looking at the potential application of novel ecophysiological techniques in nursery tree quality assessment. To date the work has been concentrating on the development of quality assessment protocols for Douglas fir *(Pseudotsuga menziesii)*. In association with Coillte's Ballintemple Nursery, and the forestry research group at UCD (Dr Conor O'Reilly), the over-winter frost hardiness development and cold storage tolerance of Douglas fir has been monitored with a view to developing more rapid tests of plant physiological status.

Chlorophyll fluorescence equipment (right hand side) and infra-red gas exchange equipment (left hand side) used in the assessment of plant quality for Douglas fir seedlings.

Conifer trees are lifted over winter when they are "hardy" and tolerant of handling and then either planted directly at the forest site or stored at low temperature, to maintain plant dormancy, until spring. The use of cold storage, even in climatically mild climates such as Ireland, allows greater flexibility in the management of nursery tree crops, and thus offers significant potential improvements in out-planting success. In European climates, over-winter dormancy in Douglas fir exhibits substantial variation in response to climatic variation at the nursery site, which can confound development of general lifting and cold storage guidelines.

Current measures of plant quality include morphology (i.e. root-shoot ratio) and physiology (e.g. root electrolyte leakage REL, visual assessment of shoot frost hardiness, and root growth potential RGP). These tests can, however, be quite time consuming, with visual shoot hardiness assessment taking 14 days and RGP 21 days. Therefore rapid screening methodologies are required which provide the nursery manager with "instant" assessment of the plant's physiological status.

Chlorophyll fluorescence provides an ideal technological solution, as it is a rapid, non-destructive ecophysiological tool that allows measurements to be made *in situ* and can detect reductions in plant performance before any visible signs are evident. This technique provides an estimate of the efficiency of the photosynthetic tissues. Various parameters relating to photosynthetic performance can be measured under ambient conditions or after dark adaptation of the tissues.

Preliminary investigations using this methodology with Douglas fir are already showing some excellent results, providing direct practical applications that promise to reduce the time periods involved for plant quality assessments. This work should, therefore, assist managers in refining their lifting, storage and planting times to ensure successful establishment of nursery-grown material.

To date, we have found that estimates of chlorophyll fluorescence have shown significant correlation with freeze induced damage to needles obtained by the more common visual assessment method. This technique, therefore, appears to allow more rapid estimates of shoot hardiness (within 24 hours of freeze testing), than by visual assessment methods (which require 14 days). We have also been monitoring plant vitality in the field in relation to time of lifting and temperature during cold storage using gas exchange technology.

Another area of expertise in our research group is centred on the root zone, and the potentially beneficial effects of mycorrhizal (root-fungus) associations for the tree (this work is being carried out by Ms Suzanne Monaghan). Low temperature induced plant water deficits before or during cold storage and after planting can have a deleterious effect on tree establishment. Fundamental studies of ectomycorrhizal colonisation of roots have implicated a role in the regulation of plant water balance, but the practical applications of any influence of ectomycorrhizas upon drought resistance have received little attention. Seedling trees, inoculated with known mycorrhizal fungi, are being studied to obtain practical information on the influence of direct inoculation upon drought resistance both during cold storage and field establishment.

It is intended that other coniferous species will be tested over the next season as well as continuation of our work on Douglas fir. It is important to obtain data from several seasons as climatic differences will influence plant tolerance, and hence the ideal period for lifting, storing and planting of stock will vary.

For further information on this program or on plant quality assessment services, please contact Dr Mike Perks or Dr Derek Mitchell; Tel: 01-706-2251/2253; Fax: 01-706-1153; E-mail: Michael.Perks@ucd.ie

Extending the salmon's range

Concrete fish ladders are now a standard "tool" for the fishery manager internationally. They are usually built at waterfalls to extend the area of catchments available to migratory fishes like salmon and trout, thereby increasing their production. Over the period 1995 to 1999 a major salmonid habitat enhancement programme has been undertaken in Ireland. Most of the budget (£11.8m) spent to date in the salmonid area has been used to repair damage to river beds and banks caused by land management practices in the past. This programme has been funded by the EU Tourism Angling Measure.

The baseline surveys of catchments, carried out initially to allow one to design and cost these schemes, illustrated the presence of many waterfalls, most of which were at relatively high altitudes in the catchments concerned. The cost/benefit of building fish ladders at these locations was prohibitive because of the limited extent of salmonid spawning and nursery water upstream of the individual falls in question. However, a few high falls were located at relatively low altitudes and had extensive lengths of potentially valuable salmonid waters upstream. Our colleagues in the Engineering Division of the Department of the Marine and Natural Resources kindly provided us with costings for the construction of concrete fish ladders at a number of sites. Costs for individual "ladders" were generally in the range of £60,000 to £260,000. While

1. An impassable falls for salmon and trout.

considering the cost/benefit of implementing these programmes, a colleague in the Office of Public Works, Mr John Gilmore, suggested that we might try a novel technique to achieve our objective in this area and reduce costs. He suggested that we use a "rock-breaker" to "cut" a series of steps through a falls, thereby providing fish passage. A rock breaker is simply a hydraulic tracked machine fitted with a large "jackhammer" instead of a bucket.

Initial trials with this tool were very encouraging. This option has since been used to "stepout" five waterfalls which were impassable to salmonids, and ease passage over a falls at a sixth location which salmon could only ascend in some years.

The end product (2) and (3) is not as symmetrical as a concrete fish pass – the individual shape and length of the pools is often determined by the natural folds and fissures in the rock.

Individual pools need to have a certain form, with the deepest area being beneath the falls and the "tail" of each pool sloping up to the next fall (3). This ensures that fish have the shortest possible jump from one pool to the next. It also helps to ensure that most cobble, rubble and flotsam moving through the falls in a flood event will wash through the pools.

The largest falls "stepped out" in this fashion to-date is on the lower reaches of the Clydagh River in the Moy Catchment. This "opened up" an additional 30km of salmonid spawning and nursery channel. The estimated cost of a concrete fish pass at Clydagh Falls was in excess of £260,000. The "rock breaking operation" cost £25,000! Large numbers of adult salmon and lake trout from Loughs Conn and Cullin were observed ascending this falls as soon as works were complete.

Contact: Dr Martin Grady,
Senior Research Officer,
Central Fisheries Board,
Mobhi Boreen, Glasnevin, Dublin 9;
Tel: 01-837-9206; Fax: 01-836-0060;
E-mail: info@cbf.ie

2. The falls has been "stepped out".

3. The ideal shaped pools in a "stepped" waterfall.

Algal control in waterways using barley straw

Dense growths of filamentous algae pose serious problems for amenity and recreational exploitation in many Irish watercourses. Excessive algal growth impedes boat traffic, obstructs angling, clogs sluices and lock chambers, in addition to creating unsightly and malodorous masses. The growth of large algal populations can also cause serious diurnal fluctuations in dissolved oxygen levels and result in fish kills. Likewise, the death and decay of a large algal biomass can deoxygenate the water, killing fish and other aquatic fauna. Floating mats of algae can reduce the level of incident light that reaches submerged plants, thereby restricting growth and reducing overall productivity.

Control of filamentous algae in waterways using mechanical methods (cutting, raking, harvesting) has been largely unsuccessful. This is because large numbers of plant fragments remain and vegetation regrowth is rapid. The use of algicides in weed control trials conducted by the Central Fisheries Board on the Royal and Grand Canals provided moderate to good results but, because these herbicides are not selective, they also killed ecologically important submerged plant species.

There is a considerable body of research that demonstrates the antialgal properties of rotted barley straw, properties bestowed without having any discernible adverse impact on higher plants, invertebrate fauna or fish. A primary requirement for the successful use of barley straw is the maintenance of aerobic conditions. Unstable, short-lived algal inhibitors are released during the aerobic decomposition of the straw. These are highly selective against planktonic and filamentous algae and are algistatic rather than algicidal. There is strong evidence that these algal inhibitors are derived from oxidised polyphenolics released from solubilised lignin, although the precise nature or mode of action of the inhibitors remains unknown.

The effects of more than 100 barley straw treatments in the UK and Ireland were assessed, and results reveal that algal control was achieved, to at least some extent, in all types of water body, but was better in smaller watercourses (<5ha).

Detailed trials conducted on an algal infested section of the Royal Canal between 1990 and 1993 demonstrated the effectiveness of barley straw in inhibiting nuisance algal growth. Mattresses of loosened straw, retained in garden netting, were anchored along the banks in the trial section. These were spaced at 50m intervals and alternated from bank to bank. A quantity of straw to provide a dose of 10g per cubic metre was applied. The straw was replaced at roughly six month intervals. A contiguous, untreated control section was monitored for comparative purposes.

Algal growth in the section broadly followed a cyclical pattern, with peak biomass between July and September and low production in February and March. In the treated section, however, algal biomass decreased from the time the straw was first introduced. Thereafter, as long as rotted straw was present, no filamentous algae were recorded. The absence of algae in this section between August 1991 and Spring/Summer 1993 permitted the recolonisation of higher plants, which are commonly less troublesome and more ecologically useful in fishery waters than algae. Further trials have produced similar results, and barley straw is now routinely used for algal control in many aquatic situations.

An algae infested section of the Royal Canal near Abbeyshrule.

Mattress of barley straw being placed in the Royal Canal at Mullingar.

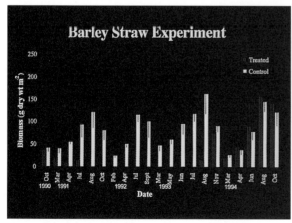
Effect of successive barley straw treatments on the biomass of filamentous algae in the Royal Canal.

Contact:
Dr Joseph Caffrey,
Senior Research Officer,
Central Fisheries Board,
Mobhi Boreen, Glasnevin, Dublin 9;
Fax: 01-836-0060.

Water quality in Ireland 1995-1997

The EPA has recently published a review *Water Quality in Ireland 1995-1997*. This covers some 13,000 km of river and stream channel, 120 lakes and 23 estuarine and coastal waters. In addition, a national assessment of the quality of groundwaters, based on sampling at some 200 representative locations, is presented, as well as information on the condition of the main canals. The assessments of water quality in the rivers and lakes set out in the report constitute the formal basis for the implementation of the recently adopted Regulations on Phosphorus Standards.

A further deterioration of the quality of river quality was recorded in 1995-1997, the proportion of unpolluted channel decreasing to 67 per cent from 71 per cent in the 1991-1994 period. This was due to an increase in the length of channel classified as slightly or moderately polluted, where the main effect is considered to be eutrophication (excessive enrichment from nearby soil resulting in the growth of algae and bacteria, which use up the oxygen needed by aquatic fauna). While the extent of serious pollution remains low

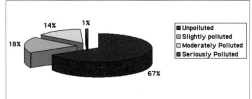

Water quality in rivers 1995-1997 (13,200 km).

(~1 per cent), there was a significant increase in this category in 1995-1997, reversing a downward trend which had been evident since the earliest national overview. Most of the slight and moderate pollution was attributed to agricultural activities, but municipal wastes discharges were the main cause of the serious pollution.

The position in lakes and tidal waters was more satisfactory than that in rivers. Of the 120 lakes surveyed, 97 were assessed as unpolluted, and there was little change in the overall position since 1991-1994. The incidence of pollution in the estuaries and coastal waters was mostly localised and intermittent: the generally satisfactory condition of bathing waters and the low levels of potentially toxic pollutants in commercial fish and shellfish provide further evidence of the limited impact of waste discharges in these tidal waters.

While there was no indication of widespread pollution of aquifers, a substantial number of samples showed evidence of organic pollution and nitrate contamination at specific locations. Of particular concern is the detection of faecal coliforms in over 30 per cent of samples subjected to bacteriological examination: this emphasises the need to disinfect all groundwaters before use.

It is concluded that the reduction of phosphorus inputs to control eutrophication in the freshwaters remains the chief task facing the water pollution control agencies. In view of this situation, new management initiatives at catchment level, along with the statutory requirements under the Phosphorus Standards Regulations, are welcome and should contribute significantly to the achievement of good quality in rivers and lakes.

Innovative Web site

The EPA is committed to promoting environmental awareness in Ireland through the provision of public access to objective, reliable and up-to-date environmental information. In support of this commitment, it has recently developed an innovative Web site which it is hoped will heighten awareness of environmental issues, encourage public participation in the Agency's activities, and act as a valuable information resource on the environment.

Much of the information produced by the Agency is by its nature technical in content. Given the diversity of its audience, however, the web site has been designed to ensure that the required level of technical detail is available to industry and consultants while, at the same time, the information is presented in a manner which can be readily understood by non-experts. The site therefore contains various sections which reflect the broad and diverse audience it is seeking to reach.

General information about the EPA, its powers and functions are provided, along with answers to environmental questions which are frequently asked of the Agency. Other sections include details of Research and Development projects being co-ordinated by the EPA, Publications, Newsletter articles on topical environmental issues, and Education. Two sections which merit particular mention are:

Licensing – This provides information on the licensing of industrial complexes (Integrated Pollution Control licensing), licensing of waste management facilities (e.g. landfills), and the regulation of Genetically Modified Organisms. Users can download draft and final licences for any company or waste management facility licensed by the Agency. In addition, the environmental guidance notes for the various sectors of industry are available for downloading, along with the EPA's annual report on licensing compliance and control.

Interactive Maps – One of the more innovative parts of the EPA web site is the presentation of river water quality and air quality data for Ireland using interactive national and local maps and dynamic access to databases. The use of the maps allows detailed technical information to be presented to the general public in a user friendly, easily understood manner.

The river maps show several thousand monitoring stations around the country which are colour-coded according to their level of pollution. Each of these monitoring stations can be selected, and the trend of water quality (often back to 1971) can be examined. For those who require further technical information, results of chemical analyses of the river water are also given in many cases.

The air quality maps function in a similar manner, and allow users to compare the results of air quality monitoring around Ireland with the specified standards and thresholds set at National and European level.

The key to these applications is that they allow detailed technical information to be presented in a simple pictorial manner to the general public through user-friendly interfaces which require little further programming to incorporate additional new data.

The EPA Web site.

Water quality contact: Dr Paul Toner,
Environmental Protection Agency,
Tel: (01) 285 2122; Fax: (01) 285 1766;
E-mail: p.toner@epa.ie;

Web site contact: Ms Yvonne Doris,
Environmental Protection Agency,
Tel: (053) 60600; Fax: (053) 60699;
E-mail: y.doris@epa.ie;

The Prevention of *E. coli* O157:H7 Infection – A Shared Responsibility

Since its establishment in January this year, the Food Safety Authority of Ireland (FSAI) has already published a number of scientific reports prepared by its Scientific Sub-committees. One in particular has received widespread attention from both media and the food industry.

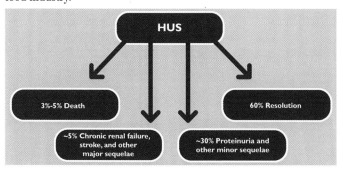

Range of Haemolytic Uraemia Syndrome (HUS) symptoms.

Verocytotoxin-producing *Escherichia coli* (VTEC) of which *E. coli* O157:H7 is a member, has emerged in the last decade as a global public health concern. In Ireland the number of reported cases of *E. coli* O157 has risen dramatically from 8 in 1996, to 76 last year. The incidence rate in 1998 was 2.1 per 100,000. This is high by comparison with other European countries. The prevention and control of VTEC infections is a priority, and as a result the Authority convened an expert working group to assist its Microbiology Sub-Committee in the development of a programme for the control of VTEC in Ireland. The result of this was a report entitled *The Prevention of* E. coli *O157:H7 Infection – A Shared Responsibility*.

Escherichia coli (E. coli) is the name given to a large family of bacteria commonly found in the gut of humans and animals. While the majority of *E. coli* are harmless, some types can cause illness. The *E. coli* O157:H7 strain causes serious illness in humans ranging from diarrhoea to kidney failure, and even death.

Human infection with VTEC has been increasing since the early 1980s and has been reported from over 30 countries on six continents. Some of the reported increase is undoubtedly due to improved surveillance, since some laboratories now examine all stool samples for VTEC, while others test only bloody stool samples.

Livestock are the reservoir for most VTEC, with cattle being the principal source of *E. coli* O157:H7. A recent study in the UK showed that VTEC was isolated from 752 (15.7%) of 4800 cattle. There is limited information on the prevalence of VTEC in the animal population in Ireland. VTEC has been isolated from cattle and is excreted in their faeces. VTEC is also present in the intestines of other animals including sheep, goats, deer, horses, dogs and cats. Seagulls, pigeons and geese are also known to carry the organism.

The Authority's report highlights the need for a shared approach and responsibility throughout the food chain to control the incidence of this bacteria. The Authority is keen to communicate the potential for serious disease and notes the number of large-scale outbreaks worldwide. These include a large beef burger related outbreak in the USA; a Japanese school lunch outbreak involving over 9,000 children in 1996 and an outbreak in Scotland in 1996, which resulted in 501 cases and 21 deaths in elderly people.

It would be naïve to think that a major outbreak of *E. coli* O157 couldn't occur in Ireland. At present there is no way to eradicate the germ in livestock so control measures must be taken from "farm to fork" to reduce the risk of spreading the infection and prevent people from becoming ill. The FSAI's detailed report provides comprehensive recommendations for every section of the food chain.

Farmers, Industry, Caterers Have a Responsibility

The report recommends that campaigns should be undertaken to raise farmers' awareness of the serious nature of the illness and their role in its control. In addition, it suggests that advice be made available to farmers on the best animal husbandry practices required to produce clean livestock; only cattle meeting the standards of cleanliness of categories 1, 2 and 3 of the Clean Livestock Policy of the Department of Agriculture and Food should be supplied to the abattoir, and fruit and vegetables should be produced under a food safety management system based on HACCP (hazard analysis and critical control point).

Also amongst the many recommendations are specific steps that should be undertaken by meat processors including abattoirs and retailers. Recommendations include a clean carcass initiative in abattoirs to ensure that faecal contamination from the hides and skins of animals is not passed onto the meat; batch numbers to be placed on minced meat and minced meat products for traceability to source in the event of an outbreak; all high risk meat products to be labelled with clear cooking instructions; and training of staff in food hygiene.

The Authority has also called on Government to make *E. coli* O157:H7 a notifiable infection immediately it is identified by both laboratories and doctors.

Codes of Best Practice/Effective Labelling

The report calls for the food processing industry, hotels, restaurants and everyone in the commercial catering business and the retail sector to fulfil their statutory obligation to implement a food safety management system based on HACCP and adopt the recognised Codes of Best Practice for their sector. They must also ensure staff are adequately trained.

Even with the very best practices, raw meat may still contain some *E. coli* O157 and the report calls for clear labelling to alert consumers and to provide instructions on handling and cooking.

Cases of E. coli *O157:H7 infection per 100,000 population in England and Wales, Northern Ireland and Republic of Ireland 1996-1998.*

Consumers – the Last Line of Defence

The report states that consumers, and those preparing and cooking food in domestic and commercial kitchens, must assume that raw meat may contain germs and pay particular attention to a number of crucial areas. Raw meat needs to be refrigerated to prevent any *E. coli* O157 present from multiplying. Transferring *E. coli* O157 from raw meat directly or via hands or utensils to other foods that will not be cooked before eating must be avoided by proper hygiene practices; and high risk meats, such as mince and burgers, should be thoroughly cooked until the juices run clear.

Copies of *The Prevention of* E. coli *O157:H7 Infection – A Shared Responsibility* are available from the FSAI on (01) 817 1300 or through Government Publications – cost £5.00.

The report and six leaflets aimed at farmers, abattoirs, food processors, retailers/caterers, consumers and carers of vulnerable groups are also accessible through the Authority's website: www.fsai.ie

THE HEALTH RESEARCH BOARD

HRB Research Project Grants 1999

Grantholder	Committee	Institution	Project
Dr Dermot Cox	Cardiovascular Diseases	Department of Clinical Pharmacology Royal College of Surgeons in Ireland 123 St Stephen's Green, Dublin 2	NGR – A novel fibrinogen binding site for aIIbB3
Dr Ann McGinty	Cardiovascular Diseases	Department of Medicine & Therapeutics Mater Hospital/UCD 41 Eccles Street, Dublin 7	An investigation into the potential immuno-modulatory role of cyclooxygenase II
Dr Stuart J. Bund	Cardiovascular Diseases	Department of Human Anatomy & Physiology University College Dublin Earlsfort Terrace, Dublin 2	Vascular structure and myogenic tone as the combined background for resistance artery contraction in hypertension
Professor David Coleman	Dental Sciences	Dept. of Oral Surgery, Oral Medicine & Oral Pathology, School of Dental Science & Dublin Dental Hospital Trinity College, Dublin 2	An investigation of the mechanisms of azole antifungal drug resistance in Candidia Dubliniensis
Professor Keith Tipton and Dr O'Sullivan	Dental Sciences	Department of Biochemistry Trinity College Dublin 2	Semicarbazide senstive amine oxidases in murine and human dental pulp and the effects of oxidative stress on murine odontoblasts
Dr Lynda Fenelon and Dr O'Farrelly	Dental Sciences	Education & Research Centre, St. Vincent's University Hospital, Elm Park, Dublin 4	Streptococcus Mutans adherence to hydroxyapatite beads and its prevention
Dr Alan Baird	Gastroenterology	Department of Pharmacology University College Dublin, Belfield, Dublin 4	Nuclear Factor kB – rational target for inflammatory bowel disease therapy?
Dr James F.X. Jones	Gastroenterology	Department of Human Anatomy & Physiology University College Dublin Earlsfort Terrace, Dublin 2	Central control of the crural diaphragm
Professor Fergus Shanahan et al	Gastroenterology	Department of Medicine NUI, University College Cork Cork University Hospital, Cork	Multiple gene array analysis of the inflammation-dysplasia-cancer sequence in inflammatory bowel disease
Dr David Croke and Dr Mayne	Genetics	Department of Biochemistry Royal College of Surgeons in Ireland 123 St Stephen's Green, Dublin 2	Development of a haplotyping system for the human Galactose-1-phosphate Uridyltransferase locus and mutation-haplotype correlations in transferase-deficient glactosaemia
Dr Gwyneth J. Farrar et al	Genetics	Smurfit Institute of Genetics Genetics Department, Trinity College, Dublin 2	Mutation-independent therapeutic approaches for dominantly inherited brittle bone disorders
Dr Finbarr O'Sullivan	Health Services Research	Statistical Laboratory Department of Statistics University College, Cork	Statistical segmentation methods for small area estimation problems in epidemiology and public health
Dr Sheelagh O'Brien and Dr McMahon	Health Services Research	Occupational Health Department, St Vincent's University Hospital, Elm Park, Dublin 4	Prevention of needlestick injuries – hospital intervention trial
Dr Geraldine Moane	Health Services Research	Department of Psychology University College Dublin Belfield, Dublin 4	The meausrement of treatment effects on denial in a sample of sex offenders using the ICD (Inventory of Cognitive Distortion)
Dr Aideen Long	Immunology & Pathology	Department of Biochemistry Royal College of Surgeons in Ireland 123 St Stephen's Green, Dublin 2	Interaction of PKC0 and the cytoskeleton in signalling through T cell surface molecules
Dr Derek Doherty	Immunology & Pathology	Education & Research Centre and Liver Unit St Vincent's Hospital, Elm Park, Dublin 4	Characterisation of CD1-restricted T cells in the human liver
Dr John J. O'Leary and Ms. Ring	Immunology & Pathology	Department of Pathology Coombe Women's Hospital Dublin 8	Cdc 6, Mcm5, qFish telomerase/telomeric repeat assay and HPV status as prognostic markers in cervical screening
Dr Luke O'Neill	Immunology & Pathology	Department of Biochemistry Trinity College Dublin 2	Cloning and characterisation of A46R and A52R, two novel viral members of the interleukin-1 / Toll receptor family
Dr Philip Newsholme and Dr Nolan	Metabolism & Endocrinology	Department of Biochemistry University College Dublin Belfield, Dublin 4	Determination of the role of complement-fixing anutoantibodies and activated complement in the B-cell dysfunctional and destructive phase of Type-1 diabetes mellitus
Dr Christopher Thompson	Metabolism & Endocrinology	Department of Endocrinology Beaumont Hospital, Beaumont Road, Dublin 9	Metabolic factors controlling Aquaporin-2 expression

Grantholder	Committee	Institution	Project
Dr Angus Bell	Microbiology & Virology	Department of Microbiology Moyne Institute, Trinity College, Dublin 2	FK506-binding proteins (FKBPs) – roles in malarial pathogenesis and antimalarial drug action
Professor Gregory J. Atkins	Microbiology & Virology	Department of Microbiology, Moyne Institute of Preventive Medicine, Trinity College, Dublin 2	Destruction of pathogenic cells using Semliki forest virus vector
Dr Wim Meijer	Microbiology & Virology	Department of Industrial Microbiology National University of Ireland, University College Dublin, Belfield, Dublin 4	Fuelling intracellular pathogenic bacteria: The role of lactate metabolism of Rhodococcus Equi in growth and survival within macrophages
Professor Cyril J. Smyth	Microbiology & Virology	Department of Microbiology, Moyne Institute of Preventive Medicine, Trinity College, Dublin 2	Genetic diversity of Helicobacter Pylori isolates through microevolution In Vivo
Dr Mary McCaffrey	Molecular & Cell Biology	Department of Biochemistry, National University of Ireland, Cork, Lee Maltings, Prospect Row, Cork	Molecular characterisation of a Rab4interacting gap-like clone
Dr Seamus Martin	Molecular & Cell Biology	Molecular Cell Biology Laboratory, National University of Ireland, Maynooth Maynooth, Co. Kildare	Molecular regulation of apoptosis in man: Identification and characterization of components of the TRAIL 'death receptor' signalling pathway
Dr Finian Martin	Molecular & Cell Biology	Department of Pharmacology University College Dublin, Belfield, Dublin 4	Characterisation of the role of specific MAP kinases in modulating mammary epithelial cell survival
Dr Catherine Godson	Molecular & Cell Biology	Centre for Molecular Inflammation & Vascular Research, Department of Medicine & Therapeutics Mater Misericordiae Hospital/UCD 41 Eccles Street, Dublin 7	Lipoxin mediated gene expression in macrophages
Dr Ronald W.G. Watson and Dr O'Neill	Molecular & Cell Biology	Department of Surgery, UCD/Mater Misericordiae Hospital, 47 Eccles Street, Dublin 7	The role of the inhibitors of apoptosis (IAP) family of proteins in delayed neutrophil apoptosis
Dr John O'Connor	Neurosciences & Mental Health	Department of Human Anatomy & Physiology University College Dublin, Earlsfort Terrace Dublin 2	The role of Jun N-terminal kinase in long-term potentiation in the dentate gyrus of the hippocampus
Dr Marina A. Lynch	Neurosciences & Mental Health	Physiology Department Trinity College, Dublin 2	Analysis of the mechanisms underlying the inhibitory effect of whole body irradiation on long-term potentiation: A role for IL-1B
Dr Louise Gallagher and Dr Gill	Neurosciences & Mental Health	Department of Psychiatry, Trinity Centre for Health Sciences, St James's Hospital James's Street, Dublin 8	The genetics of autism: candidate gene and linkage disequilibrium screen
Professor Brian Lawlor	Neurosciences & Mental Health	Department of Psychiatry for the Elderly St James's Hospital, Dublin 8	Factors determining the nondetection of dementia in the community dwelling elderly
Dr Kieran McDermott and Dr Sullivan	Neurosciences & Mental Health	Department of Anatomy University College, Cork	The role of GDF-5 and GDNF neurotrophins and glial-neuronal interaxtions in models of Parkinson's disease
Dr Niamh Moran and Dr Ryan	Oncology & Haematology	Department of Clinical Pharmacology Royal College of Surgeons in Ireland 123 St Stephen's Green, Dublin 2	Identification and characterization of disulphide isomerization in the platelet itegrin, GpIIb/IIa
Dr Margaret Worrall	Oncology & Haematology	Department of Biochemistry University College Dublin, Belfield, Dublin 4	The role of collagen in maspin tumour suppressor function
Dr David Croke and Dr Long	Oncology & Haematology	Department of Biochemistry Royal College of Surgeons in Ireland 123 St Stephen's Green, Dublin 2	Characterisation of gene transcripts associated with colon cancer metastasis
Dr Patrick Prendergast and Dr Blayney	Other	Department of Otolaryngology Mater Hospital, Dublin 7	Computer simulation for improved biofunctionality of implants for middle-ear surgery
Professor Hannah McGee et al.	Public Health & General Practice	Health Services Research Centre, Department of Psychology, Royal College of Surgeons in Ireland, Mercer Street, Dublin 2	Health status and health care access of homeless women and their children in Dublin hostels and temporary rented accommodation
Dr Anthony Staines and Dr Crown	Public Health & General Practice	University College Dublin Earlsfort Terrace, Dublin 2	The aetiology of lymphomas – A European case-control study
Dr Patrick C. Brennan	Public Health & General Practice	UCD School of Diagnostic Imaging Herbert Avenue, Dublin 4	A national investigation into patient radiation doses and the setting of reference dose levels for contrast based, routine diagnostic examinations
Professor John Morrison	Reproductive Medicine	Department of Obstetrics & Gynaecology, NUI, Clinical Science Institute, University College Hospital Galway, Newcastle Road, Galway	Chorionic COX-2 expression and role of COX-2 inhibitors in treatment of human preterm labour
Dr Paula Gallaher and Dr O'Neill	Respiratory Disease	Department of Respiratory Medicine Beaumont Hospital, Beaumont Road Dublin 9	Community acquired pneumonia – regulation of cytokine expression and investigation of gene polymorphisms associate with disease severity
Dr Charles Gallagher	Respiratory Disease	Department of Respiratory Medicine St Vincent's Hospital, Elm Park, Dublin 4	Exercise limitation in cystic fibrosis

NORTH EASTERN HEALTH BOARD & HEALTH RESEARCH BOARD • TOM O'CONNELL

Prevalence of Hepatitis B virus in the Republic of Ireland

Hepatitis B virus (HBV) is one of the world's most common and serious infectious diseases. It is estimated that about two billion people who are alive today have at some time been infected with HBV. About 350 million people are chronic carriers of HBV - i.e. over 5% of the world's population. This represents a large reservoir of infection. Persons infected with HBV are at risk of developing liver cirrhosis and liver carcinoma. HBV accounts for most of the one to two million deaths estimated to occur each year from all forms of viral hepatitis.

There is no good information on the prevalence of HBV infection in the general population of the Republic of Ireland. The seroprevalence of Hepatitis B surface antigen (HBsAg) in the Irish blood donor population is of the order of 0.025%. However, these are a self selected "healthy" population.

The aim of this study is to use postal oral fluid testing for hepatitis B anti-core antibody (anti-HBc) to determine the level of HBV exposure in the population of the Republic of Ireland. This information can be used to contribute to the debate on whether universal HBV vaccination should be introduced in this country, as advocated by the World Health Organisation. An additional aim is to study the feasibility of postal oral fluid collection for epidemiological purposes.

The sample collection phase took place between November 1998 and January 1999. A total of 962 households were written to.

Dr Tom O'Connell demonstrates the use of the swab - it is rubbed firmly along the gum until the sponge is wet, then placed in a tube, capped, and posted back.

Households received an initial letter outlining the purpose of the study. This was followed by a swabs letter containing six swabs for collection of oral fluid, along with easy-to-follow instructions. The respondents were asked to collect oral fluid from their other household members, and to mark their age and sex on the plastic swab holders. These swabs were then posted to the Virus Reference Laboratory in Dublin.

A total of 1770 specimens have been received, which represents a household response rate of 58.9%. The age and sex profile of respondents closely match that of the Irish general population. Specimens are now being tested for anti-HBc combined IgG/IgM, using the Immune Capture Enzyme Immuno-Assay (ICE) technique for oral fluid. The positive cases will be statistically analysed to produce a prevalence figure for the general population of the Republic of Ireland.

The project is being led by Dr Tom O'Connell (North Eastern Health Board) and Dr Lelia Thornton (Eastern Health Board). The co-workers are Dr Jeff Connell (UCD), Ms Grainne McCormack (UCD), Dr Anthony Staines (UCD), Dr Darina O'Flanagan (National Disease Surveillance Unit) and Seamus Dooley (UCD). The project is being part-sponsored by the Health Research Board.

Contact: Dr Tom O'Connell, Department of Public Health, North Eastern Health Board, Railway Street, Navan, Co. Meath; Tel: 046-71872; Fax: 046-72325.

ST PATRICK'S & ST JAMES'S HOSPITALS, DUBLIN 8 & HEALTH RESEARCH BOARD • BRIAN LAWLOR, AISLING DENIHAN, MICHAEL KIRBY, IRENE BRUCE & DAVIS COAKLEY

Depression in the community dwelling elderly: no longer the "common cold" of psychiatry

Depression is common in the community dwelling elderly, affecting at least 10 to 15% of people over the age of 65. Most cases of depression in this age group are mild to moderate in degree and are associated with chronic physical illness, bereavement and other negative life events. From a clinical point of view, patients with late-life depression in the community commonly present with anxiety symptoms masked by physical complaints, and it can be difficult to detect such cases. Even when depression is detected, family practitioners often believe that the symptoms are self-limiting and do not warrant intervention.

We have been studying depression in the community dwelling elderly over the last three years and have examined its impact on patient functioning and quality of life. We have confirmed that depression is common in this age group and that the most important risk factor is **chronic** physical ill health. However, contrary to popular belief, this type of depression is not self-limiting and persists over time when untreated, with 70% of subjects either dead of suffering case-level mental disorder at follow-up three years from base-line.

Furthermore, depression is significantly associated with decreased functioning and poorer quality of life. Because elderly patients with this form of depression present primarily with a mixture of anxiety and depressive symptoms, the anxiety aspect of the illness is often detected and treated symptomatically with benzodiazepines. While this type of medication gives some symptomatic relief, it does not improve the underlying depression. In addition, we have found that caring for an elderly person with depression contributes significantly to psychological ill-health amongst carers.

Follow-up studies over a three year period indicate that those patients who are untreated remain symptomatic and continue to be depressed. However those who have received suitable anti-depressant treatment appear to improve. We have also found that anxious depressives, when compared to depressives who are not anxious but are just as depressed, are prescribed more benzodiazepines and antidepressants compared to the silent cases. This would indicate that GPs are detecting cases because of their obvious anxiety, but misdiagnosing them as anxiety rather than depression and treating them inappropriately.

In conclusion, our studies confirm that depression in the community dwelling elderly is not a benign condition in that it persists over time, impacts significantly on the quality of life and function of the individual, and is associated with excess mortality. We have also found that anxiety symptoms catch the attention of the doctor and the family, and that this aspect of the depression gets treated symptomatically but inappropriately with benzodiazepines.

Recommendations from our studies are that this type of depression should be detected as early as possible and treated with anti-depressants, because treatment seems to improve outcome. A second recommendation is that anxiety symptoms or features should alert the clinician to the possibility that there may be underlying depression and, if depression is present, this should be treated with antidepressants rather than symptomatically with benzodiazepines. Our studies in the community dwelling elderly are continuing to explore the cause and the effect of depression in this age group. We expect that the findings of our ongoing studies will help improve the detection and treatment rate of this type of depression among the community dwelling elderly.

This study has been supported by The Health Research Board.

*Contact:
Professor Brian Lawlor,
St Patrick's Hospital & St James's Hospital,
James's Street, Dublin 8.*

The Forensic Science Laboratory

Dr James Donovan, Director.

The Forensic Laboratory was established in 1975 to provide a scientific analytical service in crime investigation to An Garda Siochana. This service has been slightly extended to the Military Police and to investigate sections of the Department of Agriculture.

Over the years, the service provided to the Gardaí has been extended by provision of new techniques and by the purchase of new equipment.

The Laboratory is divided into three sections: **Biology, Chemistry and Drugs**.

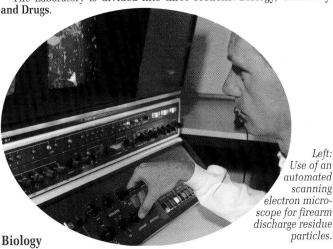

Left: Use of an automated scanning electron microscope for firearm discharge residue particles.

Biology

Biology deals with the scientific evidence arising from murders and sex offences. In these areas, the use of DNA profiling is of particular significance, either to demonstrate someone's guilt or to eliminate them as suspects. The lab is now providing DNA analysis by the Polymerase Chain Reaction method.

Chemistry

In the Chemistry section, shoeprints left at scenes of crime are powerful evidence, frequently allowing absolute identification. A computer package known as SICAR allows the identification of up to seven thousand different shoeprints.

Kits are provided to the Gardaí to swab hands suspected of having fired a gun. The finding of special particles of lead, antimony and barium on the swabs indicates that the hands very recently fired a gun, and an automated scanning electron microscope with an x-ray analyser is used for this purpose.

Other work in Chemistry involves explosive analysis, examination of hit and run debris, and reading of tachograph charts for accident investigation.

Drugs

The Drugs section is the busiest, with a huge throughput of controlled substances – involving cannabis resin, heroin, cocaine, the various ecstasys and amphetamines, etc. The size of exhibits varies from a fraction of a gram to dozens of kilograms.

Loading a gel onto a sequencer for PCR analysis of DNA

Contact: Dr James Donovan, Tel: 01-671-3813 or 01-677-1156, Ext. 2910; Fax: 01-679-4667.

Genetically modified (GM) foods

Consumer groups rank public concern with food as follows:
1. Safety – will the food make my family ill?
2. Health – is this food good for me?
3. Quality – will I enjoy it?
4. Price – is it good value?

The development of genetic engineering over the past twenty years has enabled the transfer of genes carrying desired traits, such as disease resistance, from one organism to another. These developments, in the provision of new crops, have to a large extent neglected consumer concerns or benefits, and now genetically modified foods are viewed with varying degrees of alarm and scepticism by the general public.

The genetic make-up of living organisms is determined by the DNA (deoxyribonucleic acid) composition. Altering the sequence of the bases, adenine [A], thymine [T], guanine [G] and cytosine [C], by the insertion of non native DNA, will give a genetically modified product. These modified foods would not occur under natural conditions, either by cross-breeding or through natural recombination.

In gene technology, promoter (DNA from a cauliflower virus) and terminator (DNA from a plant bacterium) sequences are used as regulation sites ("On" and "Off" switches) for newly inserted genes. They mark the beginning and end of an altered DNA genome and flank the introduced genes.

The detection of these promoter and terminator sequences is a

Farm fresh food.

powerful generic approach to GM food detection. In the State Laboratory, we can identify these promoter or terminator sequences in maize and soya using techniques such as the polymerase chain reaction (PCR). In this procedure, an enzyme duplicates the target sequence from the promoter or terminator, using short segments of synthetic DNA, called primers. The amplified DNA, from the promoter or terminator, can then be visualised and confirmed.

There is no requirement in the US to inform the consumer about the presence of genetically modified crops. In the US in 1998, it is estimated that, of all soya produced, about 30% was genetically modified. Also, modified maize (disease resistant), rape seed, tomato (slow ripening) and potato are available. These can be found in numerous consumer products such as corn flakes, cakes, biscuits, chocolate, popcorn, ketchup, tomato puree, potato salad, chips and soy sauce.

In the EU, the inclusion of GM soya and maize must be labelled under EU Regulation 1139/98. The challenge now facing the State Laboratory is to show compliance with this regulation at a national level.

Contact: Dermot Hayes, State Laboratory, Dublin 15;
E-mail: dhayes@statelab.ie

The Irish National Accreditation Board

What is Accreditation?

Accreditation may be defined as the formal recognition of the competence of a body to conduct a specific activity such as testing or certification.

With the extensive range of products and services available, it is difficult to choose which product or service meets the customer's quality requirements. To facilitate the buyer, the supplier can have the service or product assessed objectively by an organisation such as a certification body or test laboratory, as appropriate to the product or service, and have a certificate of conformance or test report issued accordingly.

The reliability and usefulness of any such report or certificate depends critically on the competence of the organisation or individual conducting the test.

The accreditation of the organisation for a defined scope is the mechanism which provides that confidence.

Benefits of Accreditation

The National Accreditation Board (NAB) through its membership of international multilateral agreements with EA (European co-operation for Accreditation) and ILAC (International Laboratory Accreditation Cooperation) ensures that Irish accreditation status is recognised internationally.

The main benefit of accreditation is that it plays a key role in guaranteeing the access of Irish products and services to international markets and greatly reduces technical barriers to trade.

Who Regulates Accreditation in Ireland?

NAB is the National Body with responsibility for accreditation in accordance with the harmonised EN 45000 series of European Standards and the relevant International Organisations for Standard-isation (ISO) Standards and Guides.

NAB was established in 1985 to accredit calibration and testing laboratories. It has since expanded its responsibilities to also accredit organisations involved in the certification of:
- quality and environmental management systems
- products and personnel
- inspection bodies
- attestors

Who can be accredited?

Any organisation or individual can be accredited to carry out a defined conformity assessment activity.

NAB FUNCTIONS

Laboratory Accreditation (ILAB)

Laboratory accreditation granted by the National Accreditation Board is commonly referred to as "ILAB" accreditation. The following documents are used in accreditation: EN 45001, ISO Guide 25, ILAB Regulation and ILAB General Criteria of Competence.

Who is Eligible?

The ILAB scheme is voluntary and open to any laboratory that performs objective testing or calibration. Quality of work and objectivity are the major elements considered when establishing suitability of a laboratory. The system operates in all areas where objective testing or calibration is carried out – such as chemical, mechanical, electrical, medical, biological, metrological, and non-destructive testing fields.

The NAB accredits all types of laboratories both private and public, whether providing a service to outside customers or solely to their parent companies.

Accreditation of Certified Bodies

Standards used in accreditation are EN 45011, EN 45012, EN 45013.

Who is Eligible?

The NAB accredits certification bodies operating product certification, quality system certification, and certification of personnel.

Certification bodies for products issue product certificates or licences to manufacturers, who then become entitled to display a mark of conformity on their products or issue certificates indicating their product's conformity with specific requirements.

Quality system certification may be issued by certification bodies to suppliers whose quality systems meet the requirements of a particular quality system standard, such as the internationally agreed ISO 3001/2/3.

Certification bodies for personnel confirm the competence of a named individual to perform specified services or duties by issuing a certificate of competence.

Accreditation of Inspection Bodies

Standard used in accreditation is EN 45004.

Who is Eligible?

NAB provides an accreditation service for organisations requiring accreditation to EN 45004. This standard covers the functions of bodies whose work may include the examination of materials, products, installations, plant processes, work procedures or services and the determination of their conformity with requirements, and the subsequent reporting of results of these activities. The requirement for the independence of inspection bodies may vary according to legislation and market needs.

Accreditation of Attestors and/or Attestation Bodies

Standard used in accreditation is EN 45503.

Continued on next page

RADIOLOGICAL PROTECTION INSTITUTE OF IRELAND & UNIVERSITY COLLEGE CORK   STEPHANIE LONG (RPII), SIMON BERROW & EMER ROGAN (UCC)

The effect of Sellafield discharges on harbour porpoises

It is clearly important to monitor radioactive contamination in the marine environment and its transfer through the food chain. A joint study was carried out by the Radiological Protection Institute of Ireland and the Department of Zoology and Animal Ecology, UCC, to assess radionuclide contamination in cetaceans (porpoises, dolphins and whales) from Irish and British coastal waters. Cetaceans are top-level predators and thus are vulnerable to contaminants. The study focused on harbour porpoises (Phocoena phocoena), which are probably the most abundant of the Irish Sea cetaceans.

A wide range of radionuclides is present in the marine environment. The majority of these are naturally occurring radionuclides such as potassium-40 (^{40}K), uranium and polonium-210 (^{210}Po). Man-made sources include nuclear weapons testing, the Chernobyl accident and discharges from nuclear fuel reprocessing plants. In the Irish Sea, discharges from the British Nuclear Fuels' reprocessing plant at Sellafield are the main source of artificial radioactivity. Of the radionuclides discharged from Sellafield, caesium-137 (^{137}Cs) is the most important in terms of the dose to both man and marine organisms.

Stephanie Long measures radioactivity in porpoise at the RPII's radioanalytical laboratory.

Muscle samples from twenty-five porpoises stranded on the coasts of Britain and Ireland or by-caught in fishing nets in Irish coastal waters were analysed for artificial ^{137}Cs and naturally occurring ^{40}K. The average levels of ^{137}Cs measured were some ten times higher in porpoises from the Irish Sea (17.8 Becquerels per kilogram - Bq.kg^{-1}) compared to those from the Atlantic (1.9 Bq.kg^{-1}), Celtic (1.7 Bq.kg^{-1}) and North Seas (2.4 Bq.kg^{-1}). These results are consistent with monitoring studies which show that levels of ^{137}Cs are higher in marine life from the Irish Sea than from the Atlantic, Celtic and North Seas. Average levels of ^{40}K (94.7 Bq.kg^{-1}) in porpoises were much greater than those of ^{137}Cs and did not vary between the different sampling locations.

The average level of ^{137}Cs measured in fish landed at Irish Sea ports during the same period was about 3.0 Bq.kg^{-1}. Porpoises consume a range of fish species and so it is clear that the levels of ^{137}Cs are increasing with trophic level (the different feeding levels within an ecosystem). The dose to porpoises from radioactivity in the marine environment was calculated and found to be insignificant. Thus, despite the elevated levels of ^{137}Cs in Irish Sea porpoises relative to those from the Atlantic, Celtic and North Seas, the resulting radiation dose is unlikely to have had a detrimental effect on their health.

Contact: Stephanie Long; Tel: 01-269-7766; E-mail: steph@rpii.ie

---

THE IRISH NATIONAL ACCREDITATION BOARD   MARIE O'MAHONY

Continued

Who is Eligible?

Attestation is a service unique to utilities, and provides for independent, suitably qualified persons or bodies (accredited Attestors and Attestation Bodies) to inspect the procurement procedures of contracting utilities in the water, energy, transport or telecommunications sectors (utilities) subject to national rules implementing EU Council Directives 93/38/EEC and 92/13/EEC in order to verify whether the utility operates its procurement systems in compliance with the EC rules on Public Procurement so that interested parties are given a fair chance to secure the award of contracts.

Accreditation of EMAS Certification Bodies and EMAS verifiers

Standards used in accreditation: In the EMS field NAB accredits against EN 45012 and the requirements of EAC Guide 5. EMAS verifiers are accredited against EAC Guide 5 and the EMAS Regulations. Sites registered meet the requirement of (EEC) No. 1836/93.

Who is Eligible?

NAB accredits Certification Bodies to certify to Environmental Management Systems (EMS) standards (such as the EN ISO 14000 series of standards).

NAB accredits environmental verifiers (Certification Bodies or individual verifiers) who meet the EC's Eco-Management and Audit Scheme (EMAS) requirements.

NAB also acts as the competent body for the registration of sites participating in the EMAS scheme.

GOOD LABORATORY PRACTICE

NAB is the national monitoring authority for the inspection and verification of Good Laboratory Practice (GLP) under S.I. No. 4 of 1991 European Communities (Good Laboratory Practice) Regulations.

These regulations were issued to give effect to European Directives requiring tests on chemical products to be carried out in accordance with the criteria set out in the OECD document *The OECD Principles of Good Laboratory Practice*.

They ensure that test results are of a high quality, mutually comparable, and that resources are not squandered by repetition of tests.

Contact: nab@nab.ie

The seabed project

Raymond Keary — Geological Survey of Ireland

The recent announcement that the Government has decided to invest IR£21 million in a programme of seafloor mapping is unprecedented. It is the largest ever investment by the State in Earth Science. It may well be the largest investment in a single scientific project.

The programme, which will be controlled by the Geological Survey of Ireland, is in fact a many-faceted operation. Over the past few years, personnel of the Geological Survey have developed the idea, and enlisted the support of the great majority of those involved in Marine Science in the country.

The core of the programme is coverage of the entire Irish seabed area, an area some ten times the size of the island, with multi-beam sonar.

Three technical developments over the past decade have made this possible. First, the development of multi-beam sonar, a technique which measures the water depths over a swathe of seabed. The width of the swathe is some three times the water depth.

Second, the multi-beam technique has been made possible by the number crunching capacity of modern computers. Without the third development, both the above would be of limited use. This is the high accuracy navigation made possible with the use of the Global Positioning System.

The multi-beam system will enable the construction of a reliable base map over the whole sea floor area. This will be of use to engineering, fishing and mineral exploration interests, as well as providing a reliable basis, for the first time, to the scientific workers interested in all aspects of seabed biology, chemistry and physics.

In addition to the seafloor mapping, it is planned to collect a number of other data-sets, including gravity, magnetics, sea surface temperature and salinity.

Certain areas will be selected, on the basis of the preliminary data, for more detailed examination. These will yield material for benthic biologists and microbiologists, as well as geological and geophysical material.

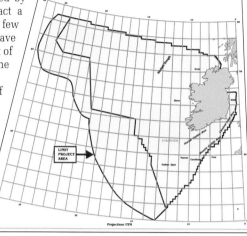

The Irish seabed area

Contact: Geological Survey of Ireland, Beggar's Bush, Dublin 4; Fax: 01-668-1782.

Protecting our geological heritage

Matthew Parkes & John Morris — Geological Survey of Ireland

Heritage and conservation are now high visibility areas of public concern - yet little or no attention has been paid to the physical foundation upon which all landscape, human and cultural heritage is based - the rocks beneath our feet. In Ireland, this imbalance is now beginning to be redressed through the Irish Geological Heritage Programme (IGH), which started in 1998. This is operated by the Geological Survey of Ireland (GSI), in partnership with Dúchas, the Heritage Service, who will administer designation and management of selected sites as Natural Heritage Areas (NHA).

The Wildlife (Amendment) Bill, long awaited by conservation groups, is expected to go through the legislative process in 1999 and will give legal status to NHAs. It will also offer protection against specific activities which may damage the particular interest of each site.

The IGH site selection process is based upon geological themes, the first two completed being (1) Karst and (2) Precambrian to Devonian Palaeontology. The final selection of the most important sites will be a consensus judgement of GSI, the theme contractors and an Expert Panel, ensuring all sites are defensible against objections. The Karst theme will protect some key areas, of the Burren for example, that fall outside existing designated areas or National Parks.

One particular site from the Palaeontology theme stands out as a special case, and has been developed as a flagship geological heritage project for the GSI. On Valentia

A view over the trackway.

Island, in Co. Kerry, a fossil trackway was discovered in 1993. It has enormous potential for interest and education through the communication of simple geological concepts to a wide audience. In conjunction with the Valentia Heritage Society, GSI drew up a plan, adopted by Dúchas, for the site to be purchased by the State. This has been completed, with safe public access being a prime reason. Visitors will be able to get very close to see the trackway, but not walk over it and destroy it in the process. Interpretation will be available on site as signboards and a cast for touching, as well as a leaflet guide.

The site is of international importance, being well dated at older than 385 million years; probably second oldest in the world. About 200 prints represent the passage of a tetrapod, a primitive four-legged vertebrate, across the soft sediment of a large river floodplain in Devonian times. It is a key record of the important evolutionary step of vertebrates leaving aquatic environments and breathing air on land. This site has the scope to fire people's understanding and appreciation of geology that ranks with the best the world has to offer.

Contact: Geological Survey of Ireland, Beggar's Bush, Dublin 4; E-mail: parkesma@tec.irlgov.ie

Part of the trackway (Swiss Army knife for scale).

GEOLOGICAL SURVEY OF NORTHERN IRELAND AND GEOLOGICAL SURVEY OF IRELAND — PATRICK J. MC KEEVER & ENDA GALLAGHER

Landscapes from stone

Starting in June 1997, the two Geological Surveys embarked on a major collaborative effort to publicise the great variety of geological attractions in Ireland's twelve northern counties. Operating under the brand name, **Landscapes From Stone**, the aims of this programme are twofold. Firstly, the Surveys will develop and publish informative literature for the non-specialist on the region's rocks and scenery. By including elements of archaeology, mythology, botany and zoology, as well as geology, it is hoped that the visitor to the region will better appreciate, enjoy and understand the region's natural attractions. Additionally, the Surveys are working to attract geological field-excursion groups back to the area. Although some universities have continued to visit the region for many years, the present numbers are only a fraction of those attracted to the area during the 1960s. In conjunction with tour operators, the Surveys are working to provide visiting groups with competitive travel and accommodation rates. Both aims have obvious economic benefits on the ground.

Funded by the EU Special Support Programme for Peace and Reconciliation, the EU INTERREG Programme, and by contributions from local government councils within the region, the project has, to date, seen the publications of the book, *A Story Through Time* and the accompanying *Landscapes From Stone* map. Both these publications highlight geological attractions across the twelve counties. This year, work is on-going to publish a series of geological walking and touring guides for areas as diverse as Lough Neagh, the Sperrin Mountains, Donegal, and the Sligo-Leitrim uplands.

Following a geological familiarisation trip for UK university geology departments last autumn *(see photograph)*, and attendances by the Landscapes From Stone team at several international geological conferences, field parties are now returning to the region. This year, several British and North American groups are due in this summer and

Field-trip co-ordinators from major UK geology departments on a geological familiarisation trip with staff from GSNI and GSI at Horn Head, Co. Donegal.

several others are actively planning trips for next year. By the end of 2000, it is hoped that geological field parties will once again be a common feature in the northern counties, and that an increasing number of people will explore and enjoy the geological wonders of the region.

For further information contact: Dr P.J. Mc Keever, GSNI,
20 College Gardens, Belfast, Co. Antrim; Tel: (0)1232-666595;
E-mail: p.mckeever@bgs.ac.uk;
or Mr E. Gallagher, GSI, Beggars Bush, Haddington Road, Dublin 4;
Tel: 01-604-1381; E-mail: gallaghe@tec.irlgov.ie

MARINE INSTITUTE — PAUL CONNOLLY

Into deeper waters

Traditional commercial fishing activities in the waters of the north Atlantic have focused on species living at depths up to 300m (e.g. cod, haddock). However, over the past ten years, the exploitation and marketing of fish species living in depths up to 1,200m has expanded. The deep water fishery to the west of Ireland and Scotland is now well established and provides important revenue for the French, Spanish and Scottish fleets which take the majority of the catches (circa 20,000t). The main species taken are roundnose grenadier, black scabbard and deep water shark. Little is known about the biology of deep water species and, with no management measures in place (e.g. Quotas), there are serious concerns about future survival of the stocks.

In view of the prohibitive cost of deep sea research, and to maximise the benefits from future work in the north-east Atlantic and Mediterranean, the EU awarded a three year contract to a consortium led by Dr John Gordon of the Scottish Association for Marine Science (SAMS), Oban Scotland, with partners from the UK, Ireland, Iceland, Norway, France, Spain, Portugal, Italy and Greece. The work focused on the developing

Otolith from a black scabbard showing the rings used to age the fish. An otolith is a tiny bone in the head which is responsible for balance. It is laid down in layers as the fish gets older.

deep-water fisheries, and specifically on understanding their interaction with and impact on a fragile environment. The Marine Institute were partners in the project, and Dr Paul Connolly, Dr Ciaran Kelly, Maurice Clarke and Colm Lordon of the Institute's Abbottstown laboratory undertook the analyses of Irish deep water survey data and to examine the life history of several species of deep water fish. A deep water database was developed, and data from Irish deep water surveys carried out from 1992 to 1998 was used to map the distribution of fish and shark species off the west coast of Ireland. Age estimation work carried out on grenadiers show the species to be long lived (up to 60 years), slow to mature (at 15 years of age) and to have a low fecundity.

The final project report is due to be submitted to the EU in late July 1999. The data provided by the project partners will be used to formulate appropriate management measures for the deep water fisheries of the North East Atlantic.

The Marine Institute is currently involved in a new project proposal with the UK, France, Norway, Faroes, and Canada aimed at identifying the "stock identity" of grenadier and shark using genetic and computer image analyses techniques. This project will continue the Marine Institute's international collaborative work on the biology of deep water commercial fish, and take Irish deep water fisheries research into the new Millennium.

Contact: Dr Paul Connolly,
Marine Institute Laboratories,
Abbottstown, Castleknock, Dublin 15.

Into the Age of the Ocean . . .

We live in exciting times – moving through the Age of Aquarius and into the new Millennium. Time perhaps, for Ireland as a nation to review its whole outlook on its greatest natural resource – the sea. Under the United Nations Convention on the Law of the Sea, Ireland can lay claim to 900,000 square kilometres of territory beneath the waves – territory that has been, until recently, not only underwater, but also undeveloped and undiscovered.

It is the job of the Marine Institute, which was set up in 1991 to undertake, promote and assist marine research of all kinds, to make sure that this territory is not only underwater, but also understood.

The research vessel Celtic Voyager deploys an ICAMS buoy off the South-West coast.

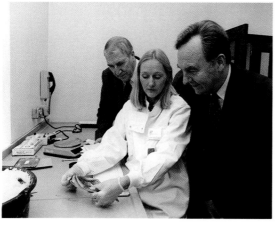

Liz Barnwell demonstrates how to age fish to Minister for the Marine and Natural Resources, Dr Michael Woods, T.D. and Dr Sean P. Crowley, Chairman of the Marine Institute, at the Abbottstown Laboratory.

Abbottstown

In February 1998 the Minister for the Marine and Natural Resources, Dr Michael Woods, turned the sod on a £1.25 million extension to the Marine Institute's laboratories at Abbottstown, Co. Dublin, which was completed by the end of the year. The new extension contains specialist facilities for fish stock assessment – including a micro-tagging facility, wet labs, cold room, electronic laboratory, fish ageing unit and dry lab – and environmental monitoring – including a plankton laboratory. The new facility has also allowed the "repatriation" of the Marine Institute's Fish Health Unit onto the Abbottstown complex, into new laboratories containing facilities for histology, virology, sample preparation, bacteriology and post-mortem. This puts the Abbottstown laboratory into a lead position to provide RTDI services to both government and industry relating to a whole range of scientific and technical issues. The Marine Institute has also initiated an on-board observer at sea programme (EU funded) to monitor discarding by the Irish fishing fleet. The on-board observers are known as FATs (Fleet Assessment Technicians) and are based in five fishing ports around the country. Between 1993 and 1998, the FATs have sampled 175 fishing trips (852 days at sea), 2,878 hauls and have collected 443,839 individual fish records. The FATs also play an important role in liasing with fishermen for the Marine Institute.

Galway

Meanwhile in Galway, last year the Marine Institute laid the foundations on its brand new Technical Support Base on the Parkmore Industrial Estate. This campus of three buildings includes a joint venture company – Marine Technical Development Services Limited – between the Marine Institute and Marine Technology Limited whose core role will be to develop and service Irish marine technological industries, including the Marine Institute's own research vessel Celtic Voyager. Other buildings "on campus" in Galway include a specialist marine equipment store, and a large engineering facility for the storage and maintenance of sea-going scientific equipment, as well as a purpose built office/laboratory facility that will act as home to a number of West coast marine service facilities, including Single Bay Management for the aquaculture industry, research vessel services, and the Institute's Marine Technology Program. The proximity of the base to the internationally acclaimed facilities at the Martin Ryan Marine Science Institute at NUI Galway, greatly upgrades Ireland's marine science capability in the west.

Newport

This year the Salmon Research Agency of Ireland at Newport, Co. Mayo – which was originally set up as a joint venture between Arthur Guinness and the then Minister for Fisheries in 1955 – was incorporated into the Salmon Management Services Division of the Marine Institute, and is currently undergoing a £500,000 upgrade. This unique facility, with its access to both the wild fisheries and aquaculture systems, is a vital addition to the Institute and will play a crucial role in the development of Ireland's Atlantic salmon and sea trout resources for both marine food and recreation.

Celtic Voyager

Out at sea, the research vessel Celtic Voyager has enjoyed a busy year on a wide variety of duties, including a major project on the deployment of real-time data buoys as part of the ICAMS – the Integrated Coastal Analysis and Monitoring System – which represents the first ever deployment of data buoys in Irish coastal waters with the capacity to "nowcast" ocean conditions (i.e. to measure the ocean and report back their findings as they actually happen), information which could be of great importance in understanding such phenomena as Red Tide outbreaks, pollution incidents, fish migrations, harbour design and even the weather. This data will be used by a whole range of people including: Bord Iascaigh Mhara, the Southern Regional Fisheries Board, Met Eireann, Cork County Council, the Department of the Marine and Natural Resources, and the Kenmare Bay Aquaculture Association, as well as being beamed direct to the Space Applications Institute, at the EU Joint Research Centre in Italy.

In deeper water, results of the survey project using GLORIA (Geological Long Range Inclined Asdic) has born fruit with the discovery of a number new submarine canyons and extraordinary carbonate mounds, each up to 300 metres high. These deep water coral reefs, consisting as they do of layer upon layer of slowly grown coral deposits, form a historic record of the climate beneath the sea, dating back up to half a million years.

Other notable events during the year include the completion of the first ever comprehensive assessment of the Irish marine environment through the Quality Status Report team based in Shannon. This 388 page report is Ireland's contribution to a forthcoming assessment of the North-East Atlantic by the OSPAR (Oslo-Paris) Convention, and concluded that, while Ireland's coastal waters are no longer pristine, the state of the marine environment is healthy overall.

Ireland has some 900,000 square kilometres of territory under the sea – territory that, up until recently, has remained undeveloped, undiscovered and underwater. As we pass into the new Millennium, the Marine Institute is working to ensure that this vast resource is understood, protected and developed as never before.

Contact: The Marine Institute,
80 Harcourt Street, Dublin 2;
Tel: 01-478-0333; Fax: 01-478-4988;
more information is available on our home page at: http://ww.marine.ie

Ultraviolet radiation and the skin

Ultraviolet (UV) radiation is part of the terrestrial solar spectrum. It is invisible, but produces biological effects which are dose and wavelength dependent. UVB (290-315 nm) is mainly responsible for sunburn. Although UVA (315-400 nm) is 1000-fold weaker photon for photon than UVB, its greater prevalence in sunlight, and the fact that it varies less with time of day and season, makes it very significant biologically, contributing to the effects of UVB and being largely incriminated in ageing of the skin.

UV radiation is a complete carcinogen, having the ability to initiate and promote skin cancer. Ninety per cent of skin cancers are produced by sun exposure, and are therefore preventable. Ultraviolet radiation is also immunosuppressive – thus cells which normally recognise antigenic cancer cells are ablated by low doses of UV, and immunosuppressive cytokines are produced. A consequence of UV-induced immunosuppression is that the skin fails to recognise induced cancer cells and permits clonal expansion, enabling progression of skin cancers.

Occasionally individuals deliberately chose to expose themselves to UVA on a repeated basis, and may therefore accumulate large doses of artificial UVA by using sunbeds. It is now known that those who accumulate such large doses increase the risk of skin cancer. Skin cancers have been found in such individuals in areas which would not normally be sun exposed: this, together with an increased risk of malignant melanoma in sunbed users, make us aware that UVA is harmful to the skin and in sufficient dose is carcinogenic in man.

There are three main types of skin cancer, basal cell carcinoma arising from the basal layer of the epidermis, squamous cell carcinoma arising from the keratinising cells of the epidermis, and malignant melanoma arising from congenital naevi or occurring as a consequence of genetic mutation within a melanoma prone family group, or from sun-exposure. Repeated sunburn, particularly in childhood, is a further risk factor. All those with fair skin types are at risk of all types of skin cancer.

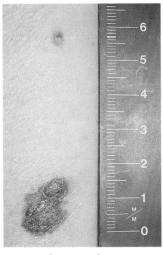

Malignant melanoma on the lower leg.

There is a genetic basis to skin cancer susceptibility which extends far beyond the observable phenotype. Thus those who carry the genetic "UV-susceptible" trait fail to recognise antigen applied to the skin at the same time as the UV exposure occurs. The UV-resistant individuals continue to recognise the antigen. Those who develop skin cancer are much more likely to be UV-susceptible than matched controls, and are also more likely to be deficient in antioxidants such as glutathione, and less able to deal with free radicals generated in skin by UV-exposure. Rates of DNA repair are less efficient in individuals developing basal cell carcinomas at an early age than expected.

We are only beginning to unravel the importance of the interaction of these various parameters but, over the next few years, as a consequence of our current EU-funded projects, we hope to gain insight into why some people are more prone to UV-induced cancers than others.

Contact: gillian@iol.ie

Susceptibility genes for Attention Deficit Hyperactivity Disorder

Attention Deficit Hyperactivity Disorder (ADHD) is one of the most prevalent syndromes affecting 2-5% of children from diverse cultural and geographical regions, with an over-representation of boys by approximately 3:1. The disorder affects children from pre-school, school age, through adolescence into adulthood, albeit with age and gender related changes in its manifestation. Symptoms of ADHD include an inability to sit still, difficulty organising tasks or activities, an inability to pay attention, and excessive risk-taking behaviour. These symptoms cause significant impairment in family and peer relationships and school performance. The disorder can cause serious additional psychopathology in adolescence and adulthood, such as drug abuse and accidents.

Large twin studies have shown monozygotic (identical) twin concordance rates consistently higher than dizygotic (non-identical) concordance rates. Adoption studies also support the genetic hypothesis. These studies suggest a strong genetic component in the aetiology of ADHD. Heritability, estimated from a study of 2,350 twins, is between 0.6-0.9. The mode of transmission is unknown, but is likely to be due to many genes each of small effect.

Dopaminergic abnormality in the brain, in particular a hypodopaminergic hypothesis, has been implicated in ADHD. Pharmacological information and recent molecular genetic studies support this hypothesis. Medications which inhibit the dopamine transporter (and thus increase dopamine availability in the synaptic cleft), such as methylphenidate and dextroamphetamine, improve behaviour in ADHD and are the mainstay of treatment. In addition, mice lacking a functional dopamine transporter gene (DAT1) display marked hyperactivity similar to that of ADHD. Genetic association studies can identify genes of small effect. Case/control designs are common, but we have used a more robust design which studies genetic variation transmitted or not from parents to affected offspring.

Using a sample of 110 Trios (father, mother and affected child) collected throughout the Dublin region with financial support from the Health Research Board, and examining genes related to the dopamine hypothesis, we have identified three genes, where certain DNA variants were preferentially transmitted to the ADHD children from their parents. These are:

1. The Dopamine 5 receptor (DRD5)
2. The Dopamine transporter gene (DAT1)
3. The Dopamine beta hydroxylase gene (DBH).

Our DAT1 findings confirm previous work by others, but the DRD5 and DBH findings are novel. We are currently undertaking the task of further localisation of the associated DNA variation. Ultimately we aim to clone the variants that increase risk for ADHD and examine their functional effect. Identification of the DNA variants that predispose to the ADHD would increase our knowledge of the biology of this disorder and facilitate the search for therapy that might reduce and control the symptoms. This in turn will be of major social and economic importance.

Contact: Dr Michael Gill,
Department of Psychiatry, Trinity Health Sciences Centre,
St James's Hospital, Dublin 8; Tel: 01-608-2241; Fax: 01-608-2241;
E-mail: mgill@mail.tcd.ie

Liver lymphocytes: new insight into liver disease

It is notoriously difficult to perform research on normal adult human liver because of its inaccessibility. The lack of appropriate research in the past led to the assumption that the liver did not have a significant immunological role. Researchers at the Research Laboratories and the National Liver Transplant Unit, St Vincent's University Hospital, are in the privileged position of being able to obtain a biopsy from the donor organ at the time of transplantation. These tiny pieces of tissue have yielded enough cells, molecules and nucleic acids to build up a unique picture of the immune system in healthy liver.

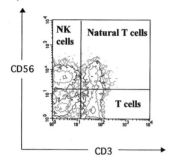

Flow cytometric profile of liver lymphocytes showing T cells, NK cells and the unique population of lymphocytes termed Natural T cells.

In particular, scientists and doctors at St Vincent's University Hospital have identified unique populations of white cells in normal adult liver, which were not previously described. These unusual cells combine the characteristics of T lymphocytes which are highly specific in their activity and Natural Killer (NK) cells which are the body's primary defence mechanism against malignancy. It appears from studies on diseased tissue that these hepatic lymphocytes have a key role in protecting the liver from malignancy and also in controlling the progress of damage caused by infections in the liver, particularly Hepatitis C.

In the first series of experiments, sections of liver tissue were stained with labelled antibodies which were specific for T lymphocytes, and these cells were clearly seen throughout the liver. Then flow cytometry (a technique used to quantify sub populations of cells on the basis of cell surface characteristics) was used to examine surface marker expression by liver T cells. This approach revealed cells which had characteristics of conventional T lymphocytes and NK cells. This technique was also adapted to show that these cells were capable of secreting large amounts of inflammatory cytokines, which may have an important role in mediating the local response to infection or malignancy. Lymphoid stem cells and molecular evidence of local T cell maturation suggest that these unusual T lymphocytes may actually be able to differentiate in the liver.

Significant changes in the relative proportions, phenotype and function of these cells in disease states suggest an important regulatory role. The ongoing aim of the Liver Research Group is to define this role.

Contact: Dr Cliona O'Farrelly,
Director, Research Laboratories,
St Vincent's University Hospital, Dublin 4;
Tel: 01-283-9444; Fax: 01-283-8123.

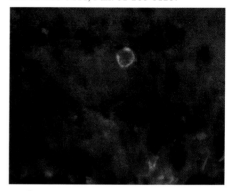

Dual immunofluorescence staining of a Natural T cell in the normal human liver tissue.

Irish BioIndustry Association

Background

The Irish BioIndustry Association (IBIA) was formally launched by the Irish Business and Employers Confederation (IBEC) in November of 1997. IBEC decided to establish such an association because it believes that biotechnology offers tremendous opportunities for Ireland in terms of jobs and wealth creation. IBIA draws its membership from all sectors of the biotechnology industry – those who directly manufacture products, those who distribute products, and those who support the industry either through services or provision of advice, finance, capital, etc.

From the outset, IBIA has recognised the vital role of research and development in supporting bio-industries, and therefore intends to work closely with research institutions such as universities and Bio-Research Ireland and proposes to establish a scientific advisory board to its association. IBIA has identified its mission and strategic objectives, which are outlined below.

Mission Statement

To ensure a platform is developed which will allow a successful biotechnology-based industry sector to flourish in Ireland, resulting in additional wealth and job creation for the country and its citizens.

Strategic Objectives

1. To co-operate with all relevant Government Departments and Authorities to promote a regulatory framework which encourages biotechnology companies to establish and prosper.
2. To effectively communicate the benefits of biotechnology to society.
3. To actively promote research and development in biotechnology and related sciences in Ireland.
4. To actively promote an educational infrastructure which will facilitate the growth of the biotechnology sector.
5. To adopt a code of practice designed to ensure relevant regulations for the safety and efficacy of products are adhered to by members of the group.
6. To actively participate in the work of the biotechnology industry in the rest of the European Union.
7. To support a business environment which comprises the key components, essential for the development of a successful biotechnology sector.

IBIA's current Chairman is Dr Michael Comer, Managing Director of Caramed. IBIA's Executive Director is Matt Moran. IBIA is a full-time member of EUROPABIO (the European Association for BioIndustries).

For more information on the Irish BioIndustry Association contact its Director, Matt Moran at the offices of IBEC, Confederation House, 84-86 Lower Baggot Street, Dublin 2; Tel: 01-605-1567; Fax: 01-638-1567; E-mail: matt.moran@ibec.ie

The Irish Pharmaceutical & Chemical Manufacturers' Federation

Matt Moran, Director IPCMF.

THE IRISH PHARMACEUTICAL & CHEMICAL MANUFACTURERS FEDERATION

CONFEDERATION HOUSE,
84/86 BAGGOT ST.,
DUBLIN 2.
TEL.: (01) 6601011
FAX: (01) 6601717

RESPONSIBLE CARE

Background

The Irish Pharmaceutical and Chemical Manufacturers' Federation (IPCMF) is a sector within the Irish Business and Employers' Confederation (IBEC). It represents the needs of the pharmaceutical and chemical manufacturing industry in Ireland. Currently representing approximately 54 companies, the IPCMF is the lead representative body for the manufacturing sector in this industry.

The Federation

The IPCMF forwards the views of the industry to a wide range of stakeholders. It achieves this through the operation of a well-defined structure comprising a number of specialist committees.

Through the Brussels office of IBEC (The Irish Business Bureau) and also through CEFIC (the European Chemical Industry Council), the IPCMF works on behalf of its members at a European level to ensure that European policy is compatible with its own objectives for the development of the sector. The IPCMF deals with a wide range of issues, including environment, industrial policy, education, health and safety, quality, human relations and training, taxation policy, communications, linkages with other industry and suppliers. The wide proportion of IPCMF members is drawn from globally based companies and the Federation maintains an international presence and working relationship with companies and agencies to ensure that the best needs of Ireland as a source for the manufacture of pharmaceutical and chemical products are maintained and protected.

Industry Background

The pharmaceutical and chemical sector as we know it today started to become established in Ireland towards the late sixties. The establishment and subsequent rapid growth of this sector was a result of a focused strategy by the Industrial Development Authority (IDA) to develop and encourage high technology knowledge-based industries in Ireland. The rapidity of the growth of the pharmachem sector is borne out when one can observe that between 1973 and 1995 exports grew by a massive 6,373% to £5 billion from a fairly modest £79 million. Between 1982 and 1995 the Balance of Payments figure increased from £110 million to £2.45 billion – an increase of some 2,200%.

Much of this growth was driven by a huge influx of global pharmaceutical and chemical companies, many of them coming from the United States. This now means that Ireland has plants operating from 16 of the top 20 pharmaceutical companies in the world. A quick run around the country reads like a *Who's Who* of the pharmachem sector, with such high profile companies as Pfizer Pharmaceuticals, Merck Sharp & Dohme, Warner Lambert, Henkel, Janssen, Bristol Myers Squibb, Schering Plough, Roche, Eli Lilly, Leo Laboratories, SmithKline Beecham and many more having plants in Ireland. The sector has always been a steady employer. The *Table*, which compares the employment growth between 1995 and 1997, reveals that employment within the industry continues to grow steadily.

The sector can be split into three broad areas as the *Table* demonstrates – basic industrial chemicals, pharmaceuticals, and then other chemicals – which is mainly fine chemicals but also includes the manufacture of man-made fibres, polymers, etc. The table indicates that the largest contribution to employment comes from the pharmaceutical sector, and this is also the higher growth part of the industry.

Central Statistics Office figures published in 1997 valued exports from the pharmaceutical and chemical sector at £8.8 billion. This is equivalent to 25% of all Ireland's total exports, and represents an increase of 54% in the preceding two-year period. This provides compelling evidence of the remarkable recent growth rate of the sector in value terms.

	1995	1997
Basic industrial chemicals (including fertilisers)	2400	2200
Pharmaceuticals	9900	11900
Chemicals, remainder (includes man-made fibres)	6300	6800
Total	18600	20900

Table: Employment within the Chemical Sector

---

*Contact: Matt Moran,
Director IPCMF, Confederation House,
84-86 Lower Baggot Street, Dublin 2;
Tel: 01-605-1584; Fax: 01-638-1584;
E-mail: matt.moran@ibec.ie*

Industrial Research & Technology Unit

Innovation and the ability to compete in a global marketplace are universally recognised as fundamental to the future prosperity of business in Northern Ireland.

Companies today are faced with an increasingly competitive environment, where the pace of change means that high technology products can, very quickly, be matched and improved by competitors from anywhere in the world.

Consumer needs are also becoming more sophisticated and today's customers are more conscious of their options and of the latest technological developments. In this environment, no company can afford to stand still. However, many companies recognise the importance of research and development investment for long term survival and growth.

In Northern Ireland, the Industrial Research and Technology Unit (IRTU) plays a central role in promoting and supporting a culture where innovation and technology development can thrive. IRTU aims to raise awareness of the importance of innovation and to complement this with a range of programmes that provide the necessary investment support for research and development.

Over the last five years, initiatives such as those outlined below, have contributed to a 31% increase in real terms in industrially relevant research and development by Northern Ireland companies, compared to a decrease for the UK as a whole.

Scientific Services – Provision of a wide range of advisory, consultancy and laboratory services across disciplines such as biology, chemistry, environmental science, metallurgy and information technology.

Compete – Support for market led product and process development. Project definition phase attracts assistance up to 50% of eligible costs to a maximum grant of £15,000. Project development phase attracts assistance up to 40% of eligible costs to a maximum grant of £250,000.

Start – Support for technology based, industrially relevant, pre-competitive R&D. Particular attention is given to projects involving technologies highlighted in the Foresight exercise. Assistance is available for up to 50% of project costs, to a maximum grant of £2 million per project.

Smart - Support for individuals and small firms to develop innovative ideas. Winners initially receive up to £45,000.

TCS – Support for collaboration between industry and universities through graduate placements.

Design – Advice and information to promote world class design standards.

For further information on any of IRTU's programmes or services, please contact Dr Deirdre Griffith, IRTU, 17 Antrim Road, Lisburn; Tel: 028 9262 3000; E-mail: info@irtu.dedni.gov.uk; or visit the IRTU website at http://www.irtu.nics.gov.uk.

IRTU CASE STUDIES

An enlightened approach to business success

John and Gerard Meenan set up FSL Electronics in 1988 to develop a remote control system that could be used in a wide range of industrial and commercial environments.

Through contact with the Department of the Environment (NI), the Company was asked to develop a system that could detect faults in street lights.

Recognising the considerable resources that would be required to complete the project, the company approached IRTU and in 1992 was successful in the Smart Awards, and proceeded to obtain further Smart funding the following year for a Street Light Monitoring system.

Having completed this project, prototypes have been sold successfully to the Electricity Supply Board and a number of companies across America.

The company has also developed an industrial remote control system for use in the mining and excavation industry, construction, building materials processing, forestry and timber processing, food and animal feed manufacturing and security shutters.

FSL's "Ultrabeam" has become recognised worldwide as a quality, state-of-the-art remote control system that can operate in some of the harshest environments. As part of its unique approach to business, FSL will provide a complete bespoke service for customers from initial consultancy and design to manufacture and commissioning.

Other products include a range of outdoor scoreboards and timing systems for cricket, GAA, hockey, rugby and soccer.

On the right wavelength

In 1982, Radiocontact Limited identified the need for technologically advanced but competitively priced broadcast, transmission and telemetry equipment. Previously this had only been available from a limited number of companies that concentrated on large scale applications at high prices.

Today the company is one of the most innovative and technologically sophisticated in the United Kingdom, designing and developing a range of unique products that have been marketed successfully on UK and world markets.

The Leaky Feeder Perimeter Movement Detection System, for example, was developed by Radiocontact Limited to detect movement within the area of a perimeter fence or generated protective field.

Radiating cables are similar to normal co-

axial cables but with reduced screening or ports in the screening which allow or control radiation leakage. The radiating/receiving (leaky feeder) cable consists of an outer screen of copper braiding separated from an inner conductor by a dielectric sheaf. The electromagnetic radiation is generated when a current flows in a circuit through the cable. This will also induce some of the external radiation that the cable is exposed to. The radiating cable therefore can be used either as a transmitting or receiving device, like a radiating or receiving antenna.

Leaky Feeder Cables are mounted on the perimeter fence or wall and connected to a radiating source to generate an electro magnetic field around the protected area.

This system will also provide volumetric detection, both portable and permanent, where a centrally mounted antenna is used to generate the electric field along a "ring" of Radiating Cable laid on the surface of the ground around the perimeter of the area to be protected. Its versatility allows it to adapt to any particular site characteristics from undulating ground to across rivers using a combination of LF cables and antennas.

Speaking about the project, Radiocontact's Managing Director John Glen said: "The scale and development costs in projects of this type are such that they could not have been undertaken within a viable timescale and without jeopardising the ongoing expansion and development of the existing core business without the assistance and grant aid that the IRTU Smart scheme offers".

From willow plant to heat and power plant

The production of energy from renewable energy sources has received considerable political focus and attention within the European Union for many years and has been the subject of extensive pre competitive research.

In 1997, the European Commission issued a draft Strategy and Action Plan for renewable Energy sources with a target to double the contribution of renewable energy sources by 2010.

IRTU's Compete programme is assisting Rural Generation Limited (RGL) to develop the World's first "commercial" biomass Combined Heat and Power (CHP) unit run on biogas produced from willow coppice chips.

An Energy Technology Support Unit Survey of the renewable energy resources in Northern Ireland revealed that the potential accessible resource of electricity generated from coppice energy crops is 470GWh/year, equivalent to the installation of a 54MW power station.

Energy crops, RGL argues, are one of the greatest potential sources of renewable energy for electricity and heat, and if energy crops, such as Short Rotation Coppice (SRC), are included in the general mix of agricultural crops there would be significant environmental and ecological benefits, alongside the development of a fully sustainable energy resource.

Rural Generation's objectives are to take fairly basic systems pioneered at Enniskillen Agricultural College and Brook Hall Estate near Londonderry and "value engineer" them into a commercially viable product. The willow coppice is grown and harvested then chipped, dried and gasified to fuel a CHP unit and provide heat both to dry the willow and provide space heating of farm buildings. Electricity is also sold to Northern Ireland's power grid at an index linked price.

The Company is also broadening its outlook to include other fuels – e.g. waste wood, wood from woodland management, and the briquetting of saw dust and MDF dust to examine the gasification possibilities, resulting in opening new potential markets for this technology.

John Gilliland, Chairman of Rural

The Prime Minister fuelling the biomass Combined Heat and Power unit with willow coppice chips.

INDUSTRIAL RESEARCH & TECHNOLOGY UNIT

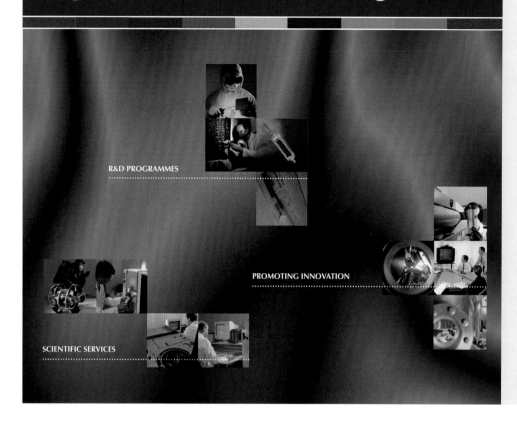

Together, we can make things better.

Now, more than ever in Northern Ireland, technology is crucial to our economic well being and innovation is the key to our success. The Industrial Research and Technology Unit can help with financial assistance, information and advice for encouraging research, technological development and technology transfer. Call IRTU today and together we can help you do whatever you do, better.

Call us on 028 9262 30000.

E-mail: info@irtu.dedni.gov.uk.

Harry Cherry

Generation Limited, says the experience has convinced him that there is considerable potential to replicate these biomass-to-energy plants on a much wider scale in the British Isles.

Innovation the key to business success in the Millennium

Millennium Products is a DTI and Design Council initiative supported by the IRTU Design Directorate in Northern Ireland to identify innovative and creative new products to mark the Millennium.

The Design Directorate has been encouraging and facilitating applications from all over the Province, with 23 products developed in Northern Ireland having been awarded Millennium Product status, a formidable total in a UK context.

Award winning innovation on the Waterfront

Valpar Industrial Limited's core business is the manufacture of insulated clusters of tubing for draught beverage dispensers. As part of its long term commitment to product innovation, the company developed new technology for jacketing tubing bundles which, with further adaptation, using recycled materials, produced Docksafe.

The winner of an IRTU Innovation Award in 1998 and a Millennium Product, this unique, energy absorbing fendering system for marina pontoons enables leisure boat owners to moor boats safely, even at night, with cushion protection against hull damage, without the need to deploy conventional fenders.

Before berthing, traditional inflatable fenders are suspended from boats and provide only localised hull protection. Pontoon edging strips offer little protection and are expensive and difficult to install.

Docksafe has unique energy absorption and deflection ("cushioning") properties, is easily installed and provides continuous protection along the length of the pontoon.

What's more, its outer coating absorbs and later releases ultra violet light, creating a highly visible boundary between pontoon edge and water.

Looking smart for the Millennium

Developed by William Clark & Sons in Maghera, Tailorluxe uses revolutionary technology to enable manufacturers to produce a garment offering a traditional, hand-tailored look and feel using normal methods of bulk production.

Another Millennium Product, Tailorluxe's secret is a revolutionary adhesive that drops rapidly in viscosity under conditions achievable in normal garment production during the pressing/steaming stage. It opens the opportunity for manufacturers to produce a garment that is loose-lined without the need and expense of basting and hand-stitching. The interlining is fused easily at 60°C to allow conventional manufacture of the jacket but, unlike any other standard interlining, Tailorluxe is designed to be released totally from the outer fabric, using steam and vacuum at the final pressing stage.

Not only does it challenge the way we think about jackets, Tailorluxe also uses natural fibres or blends and environmentally friendly coatings… so it's also good for the planet!

Such is the interest in the new development that a major international manufacturer of interlinings has purchased the marketing rights for much of Europe, although it is accepted that, because of the radical nature of the innovation, it will take some time to establish itself in the market place.

Industry Research & Development Group

Tony Gannon, Chairman, IRDG

Dick Kavanagh, Managing Director, IRDG

Anne Marie Dolan, Office Manager, IRDG

IRDG is the lobby group for the research, development and innovation orientated companies in Ireland. It is a company limited by guarantee with its own board of directors and is entirely funded by members' annual subscriptions. IRDG is firmly established as the voice of the research and development orientated companies in Ireland, is recognised to be close to its member companies and operates in a pro-active and non-bureaucratic manner. IRDG is an affiliate of the Irish Business and Employers' Confederation (IBEC) and the IRDG chairman is a member of IBEC's National Executive Council.

The criterion for membership is that a company should be a manufacturer with a commitment to research and technology development. The main objectives of the Group are to identify the needs of members, to advise and assist them on research and technology development matters and to lobby Government, Government Agencies and the EU on their behalf. In practice, the Group tries to ensure that policies and funding mechanisms are in place from which members can benefit directly.

The Group provides the following service to its members:-
- Regular newsletters
- General meetings
- Information and advice on what is happening, while it is happening and what to do about it
- Provide channels of communication and a forum for discussion for members
- Establish sub-groups in areas of interest
- Meetings and topics of special interest (e.g. Research, Technology and Innovation and the Fifth Framework Programme).

Vision Statement
Employment and growth are the two major objectives shared by all, including industry. Competitiveness is the key strategic issue facing industry and is the single most important factor in retaining existing and in creating the opportunities for new employment. The main drivers of increased competitiveness are Research, Technology and Innovation (RTI) in new and modified products and processes.

Strategic Objectives
- The national spend on RTI needs to double in real terms by 2005.
- IRDG has developed an RTI strategy for the post-1999 situation and promoted this strategy to Government and the EU. The strategy elements include: a continuation of the existing RTI Initiative, with objectives of increasing the number of first-time R&D performers, increasing the spend of existing R&D performers and increasing the number of critical mass R&D performers; a new industry-led collaboration scheme (industry - industry and industry - third level institution); and a new in-company technical skills development programme. The proposed strategy would be financed by a combination of industry funds, Structural Funds, Exchequer Funds, etc.
- IRDG will try to develop and improve in real terms the interface between industry and the Universities, Institutes of Technology, and Research Institutes through co-operation, collaboration and joint action.
- IRDG will seek to promote participation in EU Programmes.
- Produce a route map/guide for companies on the EU Framework Programmes.
- Develop linkages with other R&D activities in Ireland and elsewhere.
- Disseminate information using the internet.

Contact: Tony Gannon, Chairman;
Dick Kavanagh, Managing Director;
Anne Marie Dolan, Office Manager,
Industry Research & Development Group (IRDG), Confederation House,
84/86 Lower Baggot Street, Dublin 2;
Tel: 01-605-1608; Fax: 01-661-1095;
E-mail: irdg@iol.ie

Pan-European R&D yields new developments, new opportunities in Microprocessor Design Equipment

The high rate-of-change in microprocessor technology is a constant challenge to the semiconductor manufacturers who make these increasingly-complex devices, and to the suppliers of the development hardware and software that's essential for creating, debugging and testing a microprocessor-based product. Ashling Microsystems Limited is an Irish company that manufactures In-Circuit Emulators, the complex hardware and software instruments that designers use to replace the microprocessor in their product for debugging purposes, and to "wake-up", exercise, monitor, test and debug their electronic designs. From the beginning, Ashling's products were based on microprocessor designs using the ubiquitous and very successful 8051 8-bit microcontroller, or on one of its many derivatives. Philips Semiconductors, who licensed the 8051 design from Intel, developed a wide range of derivatives from this basic core, for applications as diverse as car radios, Smart Cards, cordless telephones and computer keyboards; and Ashling developed a close business relationship with Philips by developing appropriate Emulators or adapters for each new microcontroller. But it was clear to both Ashling and Philips that the market needed a microprocessor architecture more powerful than the 8051; and thus Philips commenced the design of the "XA", a 16-bit microprocessor architecture that retained the best features of the 8051, while executing programs 15 times faster.

Ashling's headquarters and R&D laboratory is at the National Technological Park, Limerick.

The task to be undertaken by Ashling and Philips was a challenging one: to rapidly design and produce a new range of derivatives based on the XA architecture, and to design a full range of hardware and software tools to help designs of embedded systems that use the new processor. With these objectives in mind, Ashling made contact with several European companies who had already notified their interest in microprocessor systems and design techniques to the European Commission. After a first meeting of the companies to define our R&D work programme, we began the detailed work of preparing a proposal for Domain 5 of the European Union's Fourth Framework Programme for Research and Technological Development in Information Technology. Domain 5, the Open Microprocessor systems Initiative, relates specifically to innovative design methods for embedded microprocessors, based on open standards. After correcting the shortcomings and omissions in our original proposal, in June 1995 Ashling submitted, on behalf of the project-consortium, a revised full R&D proposal entitled "STEPCAM": A System-level development Toolchain for Embedded Performance-Critical Applications on a universal Microcontroller.

By August we received news that our proposal had been accepted for funding by the Commission, and we were asked to submit a full Work-Plan that formed the basis of the contract between the Commission and all of the participating companies. Our consortium is relatively large by ESPRIT standards, consisting of:

- Ashling (Ireland), responsible for developing the In-Circuit Emulator and Source-Level Debugger for the new microprocessor, and for managing the project as a whole;
- Philips Semiconductors (Germany), responsible for developing the new XA microcontroller;
- University of Ulster (Northern Ireland), responsible for researching the current state-of-the-art and for selecting the appropriate open-standards for the project;
- Verilog (France), who developed a powerful and sophisticated CASE (Computer Aided Software Engineering) system for XA designs;
- Tasking (Netherlands), who supplied a Compiler and Linker for the new XA architecture;
- Etnoteam (Italy), who developed an RTOS (Real-Time Operating System) for the XA;
- Philips Consumer Electronics (Netherlands), who undertook the task of applying the XA processor in a major new TV design.

Work on the project started in late 1995, and the project was completed two years later. The project met its objectives: the XA architecture is now firmly established in the embedded world, with a comprehensive range of powerful development tools whose mutual compatibility is guaranteed by adherence to the STEPCAM standards, and the latest Philips high-end TV is based on a new XA derivative and was developed using the STEPCAM tools.

The results of the project have also been very beneficial for all partners in the project, including Ashling. Although the work involved in learning the details of ESPRIT program participation and in co-ordinating the project was considerable, participation in the Fourth Framework Programme has considerably expanded our product range, has enabled us to develop essential new technologies and skills in microprocessor development, and above all has allowed us to create and develop enduring technical and business relationships with other European companies who complement our position in the microprocessor-design market.

Contact: Mr Michael Healy, Managing Director, Ashling Microsystems Limited, National Technological Park, Limerick;
E-mail: michael.healy@ashling.com

BIOTRIN INTERNATIONAL LIMITED, MOUNT MERRION, DUBLIN — FIONA MANNING

The 4th RTD Framework Programme

Research and Development (R&D) is an expensive business! As a biotechnology company, the establishment and maintenance of a leading edge R&D programme is central to creating a platform from which future revenue would be generated. However, R&D costs can account for 30% of a company's revenue. The availability, therefore, of EU support for R&D activities provides companies with an alternative means to develop aspects of their R&D programme which might otherwise not be carried out. Indeed, the need to improve the scientific and technological basis for Community businesses and allow them to compete at a global level was one of the fundamental driving forces behind the introduction of EU RTD activities. With a budget of just under 12 billion ECUs, the 4th Framework Programme (FP) provided an excellent opportunity for European researchers to expand their R&D activities.

Biotrin is currently involved in two projects funded under the Standards, Measurement and Testing (SMT) and BIOTECH 2 programmes of the 4th Framework Programme. The aim of the SMT project is to develop a standardised laboratory-based system for testing toxicity in the liver and kidney.

The primary objective of the BIOTECH 2 project is to develop a laboratory-based model of the human liver and kidney and investigate the effects of cytokines on the organs. These cytokines have potential therapeutic benefits, but their effects on the liver and kidney are unknown.

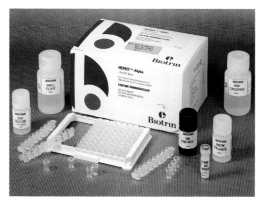
Biotrin's Hepkit™ EIA for the detection of liver damage.

Our role in both of these projects was to develop assays which could detect a protein which leaked out of certain cells in the liver and kidney when they are damaged (by drugs or cytokines in these cases). By monitoring levels of this protein, it would be possible to get a realistic and accurate picture of the amount of damage occurring in the organ.

As one of Biotrin's main product areas was in the area of biomarkers of organ and tissue damage, these projects provided an exciting opportunity for us to expand our R&D program to investigate laboratory-based model systems. Most of our products are used in a clinical environment (monitoring organs after transplant or drug regimes for possible side-effects to the kidney). Participation in these projects gave us the opportunity to investigate the usefulness of the products in a cell culture system, thereby potentially opening new markets. It also gave us the opportunity to work with leading European scientists and set up the possibility of future collaborations

In conclusion, participation in the 4th FP enabled Biotrin to develop aspects of its R&D program that may not otherwise have been accomplished, and to establish invaluable contacts with leading scientists.

Contact: fiona.manning@biotrin.ie; http:\\www.biotrin.ie

GOLDEN VALE PLC — JIM CODD

Novel Trans-European project on fish oil incorporation into functional foods

Technical difficulties, together with the need for a coherent approach, led to an innovative three-year research project on fish oil incorporation into dried functional foods. This project brought together a Trans-European group of experts in the fields of food technology, quality control and nutrition.

Golden Vale spearheaded the research project by identifying the need to develop a more in-depth and forward-looking approach to the use of fish oil as a highly nutritional and beneficial functional food ingredient. Previous nutrition research concentrated mainly on high levels of fish oil consumption over short periods. This project looked at the effects of low levels of consumption of fish oil over a longer period, a true reflection of the levels likely to be consumed.

Funding was provided by the European Commission from the "FAIR" programme of the Fourth Framework Programme of EU funded research. It was one of the biggest EU funded projects in the food sector that had Irish involvement and was the only project to date co-ordinated by the Irish Dairy Industry.

The partners involved in the project were:

Partner nº 01	Golden Vale Plc, Ireland
Partner nº 02	TEAGASC, Ireland
Partner nº 03	Deutsches Institut für Lebensmitteltechnik, Germany
Partner nº 04	Ytkemiska Institutet, Sweden
Partner nº 05	Consejo Superior de Investigation Cientificas, Spain
Partner nº 06	Trinity College Dublin, Ireland
Partner nº 07	University College Cork, Ireland
Partner nº 08	University of Ulster, Northern Ireland, United Kingdom
Partner nº 09	Danish Institute for Fisheries Research, Denmark

Project objectives were
- To determine the lowest threshold intake of n-3 polyunsaturated fatty acids (PUFA) which will exert positive nutritional effects.
- To evaluate the bioavailability of n-3 PUFA from dried ingredients.
- To develop a process to incorporate n-3 PUFA into a range of products with a good initial quality and an acceptable shelf life.

Jim Codd, Golden Vale Plc, Project Co-ordinator.

Main achievements
- Manufacture of food ingredients and food products containing fish oil that have good initial quality and stability to the development of off-flavours. The spray drying process developed is the subject of a patent application.
- The optimum level of intake of n-3 PUFA having a positive nutritional effect was established.
- Positive health benefits of fish oil consumption with implications for heart disease and cancer were observed.
- The successful completion of a complex project involving many disciplines and many countries.
- The network of Trans-European contacts set up is likely to be invaluable in future research.

For further information contact Jim Codd; Fax: +353-63-35001; E-mail: jcodd@goldenvale.com

GUINNESS RESEARCH AND DEVELOPMENT, DUBLIN, & UNIVERSITY OF ULSTER   EDWARD CUMMINGS[1], BRIAN EGGINS[1], ERIC MCADAMS[1], CLAIRE COLEMAN[2], MICHAEL CLEMENTS[2] & DAVID MADIGAN[2]

Biosensing to ensure the perfect pint

When a consumer picks up a pint of cold beer, it is generally recognised that four factors influence his or her perception of quality: colour, flavour, foam and clarity. Although this latter parameter, clarity, may be achieved through the effective removal of yeast following fermentation, the clarity of beer may also be influenced by the presence of proteins combined with secondary plant products known as flavanols. Flavanols are naturally occurring polyphenolic compounds in barley malt and hops, which remain in the finished beer to an extent which is dependent on the brewing processes used. When flavanols combine with beer proteins, colloidal particles may be formed, which are large enough to scatter light (a little like dust particles seen in a shaft of sunlight) and these may confer a hazy appearance on the beer. Although this haze is harmless, it is generally regarded as undesirable in ale and lager beers. Brewers therefore seek to control the concentration of flavanols in beer through the use of highly reproducible brewing processes.

Researchers from the University of Ulster at Jordanstown, in a CAST (Co-operative Award in Science & Technology) project, have been working with Guinness to develop biosensors for the detection of flavanols in beer. The technology is based on the use of an enzyme which reacts with flavanols to indirectly measure the concentration of these substances in beer. A naturally occurring plant enzyme (in this case polyphenol oxidase) is incorporated in a polymeric matrix, which is then used to construct an electrode which may be immersed in the beer sample to be analysed. The reaction of flavanols in the beer sample with the entrapped enzyme generates a measurable current, which is proportional to the concentration of flavanols in the sample.

The sensor developed has been successfully applied to the analysis of beer samples, producing results which agreed with more established (but less rapid) analytical methods. It has also been demonstrated that the sensor may be used in an on-line capacity,

Edward Cummings (left), demonstrates the sensor system to (L-R) David Madigan, Brian Eggins, & Michael Clements.

enabling very rapid analysis of beer during production. It is hoped that, in future, brewers will be able to control flavanol levels in beer to a greater accuracy than before as a result of the application of this new technology. The end result, we hope, is more satisfied consumers who get a perfect pint every time.

[1]University of Ulster at Jordanstown
[2]Guinness Research and Development, Dublin

Contact: Dr David Madigan, Guinness R&D, St James's Gate, Dublin 8; Tel: 01-453-6700; Fax: 01-408-4816;
E-mail: dave.madigan@guinness.com

LAKE COMMUNICATIONS, BALLINODE, SLIGO   JAMES CLARKE

Broadband urban rural based open networks

In the Fourth Framework of the European Commission's Research and Development (R&D) Programme, LAKE Communications was awarded the European Project Co-ordination of an Advanced Communications Technologies (A.C.T.S.) project entitled BOURBON – which stands for **BrO**adband **U**rban **R**ural **B**ased **O**pen **N**etworks. With headquarters in Dublin and an Advanced R&D facility located in Sligo, LAKE Communications is one of Ireland's leading suppliers of datacommunications products, networks and services.

The main goal of the technical work in BOURBON has been to specify, set up and run real life field trials of broadband technologies involving Small and Medium Sized Enterprises (SMEs). This work was carried out in trial sites and (pre)commercial heterogeneous broadband Asynchronous Transfer Mode/Internet Protocol (ATM/IP) and ISDN environments in Ireland, Finland, France, Austria, Germany, Scotland, Italy, Greece and North Holland. The BOURBON trial's broadband core networks support several access types (ISDN, xDSL, optical).

The Irish Trial is being co-ordinated by LAKE Communications in Sligo with the support of Telecom Eireann. The trial involves SMEs in the fields of Computer Aided Design (CAD)/Computer Aided

Participants from Austria, England, Finland, France, Germany, Greece, Ireland, The Netherlands and Scotland at an A.C.T.S. BOURBON consortium meeting hosted by LAKE Communications in Dublin. James Clarke is on the right of the front row.

Manufacture (CAM), Toolmaking and Engineering, running real time applications using ATM in the core network and Asymmetric Digital Subscriber Line (ADSL) in the customer premises to access the network. ADSL allows the transfer of digital data over the same copper cables as the standard telephone voice traffic.

Applications being facilitated include remote teleworking in real time with office workstations by a CAD company, Tecnocad. Without the unique combination of ATM and ADSL, this would not have been possible due to the very large nature of CAD files. Other SMEs involved in the Irish trial are STET Engineering, ELON Design and Jennings O'Donovan, whose engineers can use the speeds of ATM over ADSL to work collaboratively in real time with designers in other companies and exploit the high speed ADSL access to the Internet which the project enables.

This project clearly shows that the use of the latest broadband technologies, which the larger companies have been exploiting already for years, remarkably increases the competitiveness of Small and Medium Enterprises.

Other Irish partners participating in the project include Tellabs in Shannon, CIRCA Group Europe and Armstrong Electronics in Dublin.

Acknowledgement: The project participants wish to express their appreciation of the co-funding and support provided by the European Commission DG XIII-B during the course of this project.

For further information, please contact:
Mr James Clarke, Project Manager,
LAKE Communications,
Business Innovation Centre, Ballinode, Sligo;
E-mail: jclarke@lake.ie; WWW: www.lake.ie

Card technology for a smarter, faster world

Smart Cards are a part of everyday life in areas such as banking, transportation, access and mobile communications. In Ireland we know them in the form of Telecom Eireann phone cards and in miniature as SIM cards in mobile phones. However, many other applications are known world-wide and many future opportunities exist.

Since the concept was developed in the mid 1970s, the technology has advanced and the market has grown rapidly in size, geography and application areas. The French were the first to adopt the homegrown technology and, in the late 1970s, Smart Cards were first used there as prepaid telephone cards and banking cards. Although still centred in Europe, the technology has moved steadily to Asia and the US. In Ennis, Co. Clare, Visa Cash Smart Cards and terminals are being used in an "electronic purse" trial organised by Allied Irish Bank and Bank of Ireland. Smart cards are capable of carrying out more complex transactions than the magnetic strip option that prevails in the rest of Ireland, and are the best enablers for future banking and communication requirements.

What exactly are they?

Smart Cards are credit card-sized plastic cards into which a die (small integrated circuit (IC) – typical dimensions 2mm x 2mm) has been embedded. The die *(see illustration of module assembly)* is carried on an active site seen as a small gold square found at one end of the card. The die, depending on the final application, may be a memory or microprocessor type. It is accessed either via physical contacts (e.g. a banking ATM or public phone) or through radio frequency modulations emitted by a remote reader. The former is the more widely used form and is referred to as a contact card, which accounts for 92% of all Smart Cards.

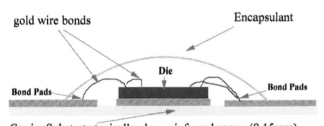

Module assembly.

The die, mounted and wire-bonded to a thin epoxy glass tape, is protected by an encapsulant. This assembly (module) is then embedded into a milled cavity in a plastic card (using cyanoacrylate adhesive or Hot Melt technology) to produce a Smart Card. Adhesives to attach the die, known as die attach adhesives, and encapsulants, are required for module fabrication. Loctite has developed products for both applications.

Die attach

Die attach adhesives may be conductive or insulating depending on the die function. Conductive adhesives are generally pure silver-flake filled, and require rapid polymerisation. Important features of die attach, which must be designed into the die attach adhesive, are viscosity, thixotropic index and working life. Thermal or electrical conductivity, electrical insulation and levels of hydrolysable ions (Na^+, K^+, Cl^-) are all important features of a die attach adhesive. Special features may include a degree of flexibility or low temperature cure. Special measuring techniques have been developed in Loctite Ireland to support the development of these products.

Encapsulation and protection

Liquid encapsulants, which are formulations of hardening adhesives/sealants, may be polymerised by heat or by exposure to ultraviolet (UV) light. The predominant chemistry is epoxy resin based. They offer low shrinkage, low moisture uptake and a tough barrier against the environment. The fine gold or aluminium wires in the module require a resilient coating to protect against damage in use. Liquid encapsulants require good flow characteristics to fully shroud the die and fine wires, ensuring a sealed, void-free assembly.

Heat hardened epoxy systems have traditionally been used for reliability. They are typically stored frozen in syringes and have a short pot life (8 hours) once thawed. They often require up to 16 hours at high temperatures (150°C) to polymerise. For this reason, UV hardening products have been developed in Loctite as an alternative. UV encapsulants are stable in darkness at room temperature and can be stored for up to 12 months, making product handling and shipment easier to manage.

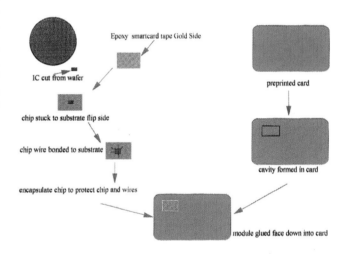

Smart Card assembly process.

Loctite's UV hardening products and light systems save time and cost compared to heat hardening technology. The die can be encapsulated and hardened within 30 seconds, thereby offering a fast throughput for high volume producers. Loctite's UV technology does not require any heat after light exposure to assist polymerisation, and this too saves time for the manufacturer. The more functional the die becomes, the more space it will require, and thus the greater the encapsulated area. However, given the restriction on encapsulant height (0.4mm), this poses a problem for high flow UV encapsulants.

Loctite is developing low flow UV products which are used to form a retaining "dam" around the perimeter of the wirebonded site. A low viscosity "fill" product is then used to fill the cavity created by the dam. The properties of the dam and fill must be compatible to ensure that wire bonds suffer no stress during hardening or when the card is in use. This "dam and fill" technology allows coverage of microprocessors (up to 5mm x 5mm). Today many European manufacturers are looking to meet all their encapsulation requirements with UV technology.

Contact: Loctite Research,
Development & Engineering Group,
Tallaght Business Park, Whitestown, Dublin 24;
Fax +353-1-4519073;
E-mail: helen.murray@henkel.de

MERCK SHARP & DOHME – A global company in a local community

A subsidiary of Merck & Co., Inc., one of the largest and most successful healthcare companies in the world, Merck Sharp & Dohme (Ireland) Ltd has been based at its manufacturing facility at Ballydine, Co. Tipperary, since 1976.

The Company's newly expanded facility at Ballydine is Merck's primary bulk manufacturing plant in Europe and represents the single largest capital investment by the Company outside the USA.

As the foremost facility of its kind within the European Union, the plant is heavily involved in the production of bulk active ingredients for a wide range of healthcare products. These ingredients are then exported to other Merck facilities where they are formulated into tablets and capsules and distributed to hospitals and pharmacies around the world.

An aerial view of the Ballydine plant.

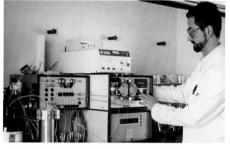

David Brazil, Senior Chemist, analysing a product sample in the Technical Operations Laboratory.

Production Manufacture
Ballydine is an Organic Synthesis Plant where production is carried out in two phases, "Wet" and "Dry". In the "Wet" section of the manufacturing building, liquid and solid raw materials are mixed and processed. The final product is then refined and packaged in the "Dry" or finishing section of the building.

A typical product manufacturing cycle includes chemical reaction, extraction, separation, distillation, filtration and isolation stages in the wet section, followed by drying, milling, blending and packaging in the dry area before dispatch to the finished product warehouse.

Control Room attendants Sean English and Tony O'Brien monitoring the Production Process in the Central Control Room.

Driven by Technology
As one would expect, advanced technology plays a critical role in the successful day-to-day running of Merck Sharp & Dohme's bulk manufacturing facility at Ballydine.

A state-of-the-art computer system, which relies on the very latest technology, is used to control all processes throughout the entire pharmaceutical facility. This ensures not only product quality, but also the plant's impeccable safety performance and environmental safeguards.

The highly skilled workforce at the Ballydine plant include process operators, technicians, chemists, engineers and IT specialists, serving the company's commitment to innovation.

Supporting the Local Community
Merck Sharp & Dohme may be part of a global company, but it has always believed that it has a corporate duty to the local community in which it operates.

As a result, the company is wholeheartedly committed to supporting an extensive range of local activities in areas as diverse as health, education and civic projects.

As part of a successful global company involved in the discovery and development of vitally important healthcare products, Merck Sharp & Dohme is acutely aware of its responsibilities towards the protection of the environment.

During the past ten years, Merck Sharp & Dohme has been officially recognised for the excellence of its environmental management systems and its commitment to clean production, receiving both the Good Environmental Management Award and the Clean Technology Award.

Facts and Figures	Merck Sharp & Dohme (Ireland) Ltd
Production Started	1976
Total Site Area	188 acres
Present Plant Area	50 acres
Capital Investment	£500 million
Annual Capital Expenditure	£20 million
Number of Employees	430
Annual Payroll	£14 million
Method of Production	Computer controlled batch processing
Water Consumption	Approximately 600,000 gallons are taken daily from the River Suir
Waste Treatment	Waste from the Plant is treated in an activated sludge waste treatment system
Raw Material Sources	World-wide but principally Ireland, Japan, USA, United Kingdom and Europe
Markets	The Plant exports to 30 countries. Main markets are Europe, USA, United Kingdom and Japan
Products	
MODURETIC®	– for treatment of High Blood Pressure
CARACE®	– for treatment of High Blood Pressure
INNOVACE®	– for treatment of High Blood Pressure
ZOCOR®	– for treatment of High Cholesterol
FOSAMAX®	– for treatment of Osteoporosis
SINGULAIR®	– for treatment of Asthma
AGGRASTAT®	– for treatment of Angina
MAXALT®	– for treatment of Migraine

Contact: John Condon, Merck Sharp & Dohme (Ireland) Limited, Ballydine, Kilsheelan, Clonmel, Co. Tipperary; Tel: 051-601000; Fax: 051-601241.

Laser cladding development for the gas turbine industry

SIFCO Ireland is involved in the remanufacture of turbine engine components, with three plants in Ireland located around Cork. In recent years SIFCO has devoted a large amount of resources to the research and development of new repair technologies for the gas turbine industry. One technology that is currently under development is Laser Cladding. As the name suggests, this process uses a laser to deposit a layer of material onto a substrate. The deposited layer can have a different composition, and subsequently properties, to the underlying material. This potentially has a range of applications in a number of areas, in particular the aerospace and automotive industries.

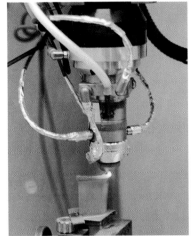

Figure 1. The nozzle plays a critical role in the deposition of the clad.

In the early stages of this project, SIFCO became involved in a technology transfer project funded under the E.U.'s "Innovation" programme. This has helped SIFCO to develop links with other institutions across Europe and allowed us to make use of a greater knowledge pool, including The University of Liverpool and the Instituto Superior Tecnico (IST) Portugal. These university research centres have established expertise in laser cladding at a laboratory scale. This project aims to transfer this knowledge to the industrial partners to develop production scale laser cladding processes.

The laser cladding equipment we have constructed in SIFCO during this project consists mainly of a 1kW CO_2 slab laser, a powder feed system, a custom built delivery nozzle and a CNC (computer numerical control) controller. Clearly one of the most important components of this machine is the nozzle (see Figures 1 & 2) which plays a critical role in the deposition of the clad. Within the nozzle, the gases, powder and laser converge to form a shrouded molten pool of material on the substrate. As the nozzle is co-axial, it allows the pool to traverse any direction, which was not the case with earlier nozzle designs.

Control and manipulation of this pool are the keys to producing a good deposit. The factors influencing the shape of the pool can be divided into four main categories, each of which has a distinct effect on the final clad:
1. beam energy,
2. powder characteristics
3. gas flows
4. part manipulation.

The beam energy at the substrate surface is dependant on the beam power density and the feed rate of the component. These settings are dependant on the materials being used. The powder flow rate, grade, composition and handling of the powder all have an effect on the microstructure of the clad formed. The gas flow rate through the nozzle has a particular impact on the aspect ratio of the clad. A low gas pressure gives a tall rounded clad whilst a high gas pressure produces a wide flat clad. Combining all of these parameters results in a deposit that can exhibit one of three general cross-sections (see Figure 3).

At the correct settings, the second structure is formed, which produces a uniform buildup of subsequent layers. If too little energy is available the result will be poor melting of the powder and/or poor adherence to the substrate. Too much energy results in alloying of the powder and the substrate and a poor layer buildup. As most of the components in a gas turbine are complex shapes, the contour followed by the nozzle must be precisely controlled. This is achieved by probing each component and adapting a stored contour for each profile type.

The powder and substrate materials that are used have a large bearing on the cladding. Aerospace components are typically manufactured from high temperature Nickel and Cobalt based superalloys which can be very difficult to work with. In particular, more modern aero components are now being made from directionally solified (DS – single direction grain growth) or single crystal (SC – no grain boundaries present) materials which require more difficult processing, such as inductive heating of the substrate during repair.

The high pressure turbine blade shown in Figure 4 is an example of a part whose tip was formed by the laser cladding process. Here the tip buildup material (Rene 142) and the blade substrate material (Rene

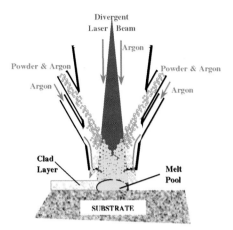

Figure 2. Nozzle arrangement.

125) are two different superalloys that are chosen for their high temperature mechanical properties. The blade material has excellent creep resistance and the tip material has very good wear resistance.

The advantages of this process over existing methods (e.g. TIG-welding) are improved turntimes, increased volume capability, excellent reproducibility and a cost saving. The mechanical and microstructural properties of the cladding are also of a higher quality than existing methods of repair which require a much higher heat input that can often result in distortion of the component.

In the near future it is hoped to move toward commercialisation of the laser cladding process. Work is currently ongoing to optimise the present process for blade tip repair, along with further mechanical and microstructural testing. Many other in-house applications for the machine have been identified and are also under development. New nozzle designs and control (vision) systems are also under investigation to increase the flexibility of the machine to process different components.

Contact : Neill Boyle, SIFCO Turbine Components, Carrigtwohill, Co. Cork;
Tel : 021-287300; Fax: 021-287301;
E-Mail: neillboyle@sifco.ie

Figure 4. High pressure turbine blade whose tip was formed by the laser cladding process.

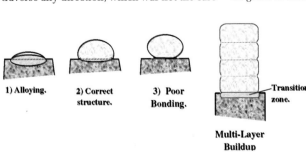

Figure 3. Deposit cross-sections.

Internetology

Nortel Networks is building a better Internet, one over which voice, video and data come together and are delivered with unprecedented speed, reliability and capacity. Much of the Research & Development that makes this possible takes place here in Ireland.

To find out more please contact Mike Gaffney in Galway, e-mail: **gaffneym@nortelnetworks.com** or Roger Johnson in Monkstown, e-mail: **rogerjo@nortelnetworks.com**

www.nortelnetworks.com

How the world shares ideas.

Nortel Networks, the Nortel Networks logo and "How the world shares ideas" are trademarks of Nortel Networks. © 1999 Nortel Networks.

Nortel Networks are at the forefront of everything that's exciting in the fast-moving world of telecommunications

Nortel Networks is one of the largest and most successful telecommunications companies in the world. It delivers value to customers around the world by being at the heart of the internet. Nortel Networks spans mission critical telephony to internet protocol optimised networks. Customers include: public and private enterprises and institutions; internet service providers; long distance cellular communications companies; cable TV carriers; and public utilities.

Our facilities in Ireland have responsibility for everything from R&D to product development to product marketing to manufacturing to logistics and distribution, to customer management. Moreover, this year, the corporation identified seven key sites worldwide to be "Global Systems Houses" – world-class facilities for building, integrating, and testing products and networks before shipping them to our global customers. Two of the global systems houses – Monkstown

Testing the latest generation equipment at Monkstown.

Nortel Vice-Presidents in Ireland: Liam Nagle, VP Europe, Middle East & Africa Operations, Enterprise Solutions Europe (heads Galway site); Pat Hobbert, VP Operations; & Peter Schuddeboom, VP International Optical Networks, Europe, & M.D. of NITEC.

Headquartered in Canada, and with over 77,000 employees worldwide, Nortel Networks' annual revenues total just over US$17bn dollars. Our employee base in Ireland now totals more than 2,000 people, and over US$2bn of our European business is supported from Ireland, north and south, making us one of the largest contributors to the island's economies. In Europe, Nortel has over 12,000 employees across some 35 countries, with revenues of US$5bn annually.

As a corporation, Nortel Networks is 104 years old: in Ireland, its Galway operation was established 26 years ago, in 1973, while in Monkstown (near Belfast), Northern Ireland, a Standard Telephone & Cables manufacturing facility (which later became part of Nortel Networks), opened in 1962. Monkstown's R&D lab, NITEC (the Northern Ireland Telecommunications Engineering Centre) came into operation in 1988.

We have an educated, young and hungry workforce in Ireland: Nortel Networks aims to enable those people to build their careers within a world-class telecommunications solutions enterprise.

With two world-class R&D facilities at Monkstown and Galway, Nortel Networks has the largest private sector telecoms R&D presence on the island. There are more than 600 engineers working on product development at our Monkstown and Galway sites, tightly integrated with our other research labs in Canada, the US, the UK and Europe.

and Galway – are located in Ireland. This decision recognizes the strength of Nortel Networks Ireland's expertise and commitment to world-class standards of research and enterprise. A key factor in the corporation's success has been the calibre of its people, their educational standards, and their commitment and determination. Over 50 per cent of Nortel Networks' employees in Ireland now have third level qualifications and our customer support organisation supports more than eight languages.

We have extensive and deepening links with universities, both in Northern Ireland and the Republic of Ireland. These links are strengthening Nortel Networks' involvement into the academic research environment and community. Just last year, Nortel Networks announced a £2.5m R&D investment link up – the Jigsaw project (http://www.optimise.net/jiqsaw/) – with the University of Ulster and Queen's University, Belfast, supported by a grant from the Industrial Research & Technology Unit.

As the world reorients itself around e-commerce and the internet, and as data and telephony merge, Nortel Networks is perfectly positioned at the heart of the emerging information society, with exciting, innovative solutions that add powerful value for its customers.

Contact: Mike Gaffney in Galway;
E-mail: gaffneym@nortelnetworks.com;
or Roger Johnson in Monkstown;
E-mail: rogerjo@nortelnetworks.com

View of Nortel Networks' Monkstown facility.

Forty years of experimental particle research at DIAS

This year sees the completion of nearly forty years of research in experimental charged particle physics and astrophysics by the authors and colleagues at the School of Cosmic Physics of the Dublin Institute for Advanced Studies (DIAS). Earlier distinguished cosmic ray work by Janossy, McCusker and others was continued by O Ceallaigh, who took over in 1953 and abandoned cloud chambers in favour of the latest nuclear emulsion detection techniques. When we arrived at DIAS in the early sixties, O Ceallaigh was already well known internationally and, through his influence, facilities at CERN (the European Organisation for Nuclear Research in Geneva) were made available to us as part of the European K-meson Collaboration, which included UCD as well as laboratories in Brussels, Warsaw, Berlin, Prague, Bristol and London. This accelerator-based work resulted in significant contributions to the physics of K-mesons, hyperons, nuclear fragmentation and nuclear structure. In particular, the work of this period was a major contribution to the physics of hypernucleus production and decay. (A hypernucleus is an atomic nucleus in which one of the neutrons has been replaced by a hyperon, which is a short-lived particle with mass considerably greater than that of a neutron). This research resulted in the discovery of new hypernuclear species (a "periodic table" of hypernuclei) including the world's first double hypernucleus which enabled the lambda-lambda hyperon interaction to be measured for the first time.

In the mid-sixties, we became interested in damage trails in solids caused by individual charged particles. This led to a joint research programme between DIAS and the General Electric Research Laboratories in Schenectady, during which a new technique of charged particle identification in solid state nuclear track detectors was discovered and then developed. The widespread application of such detectors by many laboratories soon followed. One application was Ireland's first direct space science involvement when Berkeley-DIAS detectors were exposed on the Lunar surface during the Apollo 16 and 17 Missions in the early seventies, resulting in the first measurements of heavy cosmic ray nuclei outside the Earth's magnetosphere.

DIAS-Bristol balloon launch in South Dakota.

The Berkeley-DIAS detectors on the lunar surface.

Another application during the seventies was a very successful series of ten massive high altitude DIAS-Bristol balloon flights to study the charge and energy spectra of heavy and ultra heavy cosmic ray nuclei and to search for examples of the very rare actinides (extending the "periodic table" of cosmic nuclei). The eighties saw Ireland's first Earth orbiting space science experiment (a further application of the above detectors) when DIAS, in collaboration with ESTEC (the Space Science and Technology Centre of the European Space Agency - ESA), designed and built the single largest experiment for the NASA Long Duration Exposure Facility (LDEF) with the authors as joint project leaders. In 1984 the hardware with its huge particle collecting area (twenty square metres) was deployed in Earth orbit via space shuttle,

Deployment in Earth orbit of the DIAS-ESTEC experiment on LDEF.

retrieved from orbit in 1990 after a six year space exposure, and returned to DIAS for analysis. This experiment yielded the world's largest (by a factor of ten) sample of relativistic cosmic nuclei with charge greater than 70 and the world's first statistically significant

Continued on next page

Armagh Observatory — Simon Jeffery

Cannibal white dwarf creates trembling giant!

For thirty years an extraordinary star has baffled astronomers. Known as V652 Her, it has a surface made almost entirely of helium. Stars create helium in their centres, but it is rarely exposed on the surface and then only with other products of nuclear burning such as carbon. But V652's surface doesn't have any other nuclear waste. Sometimes helium stars form by losing material to a companion star - but V652 has no companion. V652 Her shows another remarkable property. Every two and a half hours it expands and contracts by over 15,000 km (about 1% of its size). These pulsations enable us to measure V652's mass, which is essential if we want to know how it was formed. Moreover, the pulsations are speeding up, which tells us that V652 Her is shrinking.

Now, Armagh astronomer Simon Jeffery has measured the composition of V652 Her in detail and proved that the surface helium is the product of hydrogen-burning alone. He has also determined the mass more precisely then ever before. These observations helped Jeffery and Japanese colleague Hideyuki Saio explain how V652 Her might have formed. V652's life-line is shown in the *Figure*. Long ago, before the formation of the solar system, two Sun-like stars formed and evolved first to become red giants and then helium white dwarfs. Billions of years later, they spiralled into one another until one was sucked onto the surface of its companion. With a new energy source, the merged star expanded to become a bright giant. Unusually, the new star started to burn from the outside in. Computer models predict just the right properties for V652 Her.

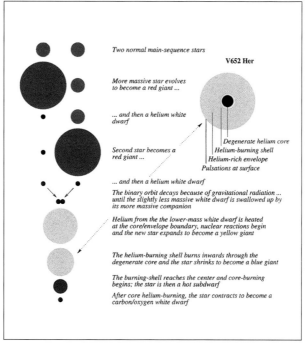

Stars like V652 Her are rare, but not unique, and recently another trembling helium giant was discovered. The Armagh group predicts that more of these engorged cannibals will be found, and will make extensive searches for them in the coming year.

*Contact: Dr C. Simon Jeffery,
Armagh Observatory, College Hill,
Armagh BT61 9DG, Northern
Ireland; Tel: +44-1861-522928;
Fax: +44-1861-527174;
Email: csj@star.arm.ac.uk;
WWW: http://star.arm.ac.uk/~csj/*

Dublin Institute for Advanced Studies — Denis O'Sullivan & Alex Thompson

Continued

sample of cosmic actinides. Analysis of the vast amount of experimental material involved has extended to the late nineties. Interest in this field will continue into the Space Station era via continuing involvement with Berkeley and NASA.

During the eighties and early nineties solar system studies employing semiconductor detectors for the real time measurement of charged particles (electrons, protons and light ions) in space were also undertaken. With collaborating scientists at NUI Maynooth, the Max Planck Institute for Aeronomie and other European Institutions, a major success was achieved with Ireland's first experiment on a European Space Agency scientific mission, the Giotto mission to Comet Halley. Dramatic measurements of particle energy spectra and flux densities were made during the very close flyby of Halley's nucleus in 1986. Analogous investigations were made during an Earth swingby (the first ever) in 1990, when the mission was extended to encounter Comet Grigg-Skjellerup, where the Irish-German equipment, still operational, made unique observations in 1992. In another major success using a similar detector system, Ireland's first experiment on a Russian space mission, the twin spacecraft Phobos mission, obtained pioneering charged particle measurements during 1989 while in orbit around Mars.

More recently one of the authors is co-ordinating a large European study of galactic and solar cosmic rays at aviation altitudes in the Earth's atmosphere. It involves nine laboratories including CERN and employs techniques developed for both ground based and space investigations, with detectors operating on many air routes and mountain tops to study cosmic radiation effects on aircrew.

Simulation of cosmic ray field at CERN in 1999.

*Contacts: Professor Denis O'Sullivan,
Professor Alex Thompson,
Dublin Institute for Advanced Studies,
5 Merrion Square, Dublin 2;
E-mail: dos@cp.dias.ie, at@cp.dias.ie*

GREENMOUNT COLLEGE OF AGRICULTURE AND HORTICULTURE, ANTRIM — JONATHAN MCFERRAN, NIGEL MURPHY & CIARAN HAMILL

The Internet – learning for life

Greenmount College of Agriculture and Horticulture is embracing new technology as a means of communicating information and providing access to learning material for farmers and growers in Northern Ireland (NI). The College Internet Website was launched in December 1996 and has developed steadily since, both in the content and in the use made of it by the farmers and growers in NI. It is one of the premier agriculture websites in the British Isles with more than 225,000 hits per month on the www.greenmount.ac.uk site.

There are a number of objectives for developing the Website.

1.
The first is to supply farmers and growers in NI with timely, effective technical information to help them manage their businesses. Information is provided covering all livestock, arable and horticultural enterprises and also on business management and information technology (IT).

2.
The second objective is to supply farmers with interactive decision support programs and "benchmarking" data. Performance of College farms enterprises and student-led projects such as CREAM (30 cow herd of high genetic merit Holsteins), LIMO (50 Pure, threequarter and half bred Limousin cows producing quality beef), RAMS (60 ewe pedigree texel flock) where students are involved in "learning by doing", is made available to farmers via the Internet. Information is also collated from farms participating in lifelong learning and systems development projects. The interactive programs provide farmers and growers with the opportunity to model the impact of various management decisions on performance and, using benchmarking data, allows them to compare their performance with others. This helps highlight strengths and weaknesses in their technical and management competence. Weaknesses can be addressed, with the help of material available via the Internet.

3.
The third is to promote all of the activities of the College and provide information to potential customers about the services offered by Greenmount College and associated staff throughout NI. Information is available to potential students on the courses offered at the College, including being able to apply on-line for any course offered. Services provided by Business Development Advisers in all areas of Agriculture and Horticulture and the work of the Technology and Business Divisions is outlined.

Embracing new technology.

The Internet allows learning material to be made available to farmers and growers to back-up other forms of delivery, and allows them to progress through modules at a time and at a pace that suits them. They can follow structured, modular programs and undergo assessment leading to nationally recognised qualifications.

Our target is that by 2002 30% of education & training, business development and technology transfer services will be delivered electronically.

Contact: Ciaran Hamill;
Tel: +44-(0)1849-426741;
Fax +44-(0)1849-426777;
E-mail: ciaran.hamill@dani.gov.uk

ROYAL COLLEGE OF SURGEONS IN IRELAND — HANNAH MCGEE & CIARAN O'BOYLE

Psychosocial contributions to illness and healthcare

Health Psychology has rapidly developed in the last decade as a new specialisation which focuses on psychosocial aspects of physical health, use of health services, and behaviour of health professionals and the public. The Department of Psychology, RCSI, is the first Department to specialise in this work in an Irish medical school.

One focus of study is investigation of psychological factors as causal mechanisms in disease. A multidisciplinary project with the Departments of Cardiology, Beaumont Hospital, and Clinical Pharmacology, RCSI, studies the role of psychological stress in triggering cardiac events. In experimental studies, increased levels of acute stress have been associated, for the first time, with reductions in levels of PAI-1/tPA complex. PAI-1 (plasminogen activator inhibitor-1) is an inhibitor of tPA (tissue plasminogen activator) activity and is hypothesised as a pre-thrombogenic agent. This continuing work may provide explanations for a link between stress and fibrinolytic and coagulation processes triggering cardiac events.

A second focus of our research is development and evaluation of psychometrically adequate tools for the measurement of quality of life (QoL). QoL is an important outcome for both patients and healthcare providers. Many instruments are available, but their design has until recently reflected the concerns and values of health professionals, with little patient consultation. The Department has developed an individualised yet quantifiable measure of quality of life: the Schedule for the Evaluation of Individual Quality of Life (SEIQoL). This allows individuals to outline QoL in terms important to them, and allows sensitive evaluation in diverse settings, e.g. orthopaedic surgery, palliative care, HIV, motor neurone disease, hormone replacement therapy, and chronic pain. As one of a small number of QoL instruments listed by the World Health Organisation, it is being widely applied in international settings. [See: Joyce C.R.B., O'Boyle C.A., McGee H.M. (eds), *Individual Quality of Life – Approaches to Conceptualisation and Assessment*, Harwood Academic Publishers, Reading, 1999.]

A challenge to healthcare providers is to ensure the provision of evidence-based interventions. To do this, measurement instruments must be capable of demonstrating clinically important outcomes, i.e. be responsive. Nine international instruments are being evaluated in a single large study to identify the most responsive tool for evaluation of QoL in cardiac rehabilitation programmes. Findings from this will inform Irish and UK selection of instrument(s) for validation of programmes and centres.

Contact: dhevey@hotmail.com

Contact: coboyle@rcsi.ie

Contact: hmcgee@rcsi.ie

Science and Technology come to Dun Laoghaire

This year has witnessed the arrival of science and technology education and scholarship to Dun Laoghaire Institute of Art, Design and Technology (DLIADT). The Institute, on Kill Avenue near Baker's Corner, incorporates a School of Art Design and Media, which has offered innovative and award-winning courses for over two decades. This history has contributed to the mission and ethos of the new Institute, which sees itself serving the knowledge, media and entertainment sectors of the Irish economy with a range of educational and training programmes, at various levels, as well as applied research and consultancy. We particularly want to broaden access to higher education in the way envisaged by the Green Paper on Adult Education.

The institutional mission gives a particular flavour to the new School of Science and Technology that can be seen in the unique portfolio of courses offered. The inaugural programme is a computer programming course with an emphasis on multimedia. The Irish multimedia industry is booming and media-savvy computer programmers are in great demand. Not surprisingly, the course, which had its first intake in October 1998, is also in great demand. We have recruited a high-calibre lecturing team with a broad set of skills and experience, both academic and industrial, to support this course.

The technological basis of the media and entertainment industries is undergoing a rapid transformation, with the shift from analogue to digital media, rapid drops in the costs of entry, and the proliferation of the channels of distribution. A factor holding back the industry is the shortage of staff who can combine current technical skills with the capacity to adapt quickly to the emerging technologies. Identifying this niche led us to develop a unique course in audio-visual media technology, which admits its first students in September.

The Executive of the new School of Science and Technology at DLIADT: Bryan Maguire (Head of Science), Ian Hughes (Head of School) and Mark Riordan (Head of Technology).

At the same time as the School of Science and Technology, the Institute established a School of Business and Humanities. From the beginning we sought areas of collaboration between the two schools, and the first fruits of this work is Ireland's first undergraduate course in Electronic Commerce beginning in September 1999. The experience of developing this unique combination of business and technology education to prepare students for careers in the network economy is also influencing the way we develop our other courses.

Science and technology are transforming Irish society, but there is a widespread concern about the breadth of understanding of science and technology in Ireland. The School is tackling this issue in a number of ways. A key means of bringing science to a wider audience is through the school system. We have introduced a programme of courses in science education for primary and post-primary teachers. These courses run in evenings and during school holidays. The development of these courses coincides with the updating of the curricula in primary and post-primary schools, and there has already been a lively demand for places.

For over a year we have been actively supporting the Southside Partnership in their bid to establish a Children's Museum, which will constitute an informal educational centre in which children and their parents can encounter, explore, understand and critique the new technologies which are shaping our society. The direct contact with a cross-section of the public, which this kind of public access facility offers, is central to our vision of opening up higher education institutions. People who would otherwise never venture onto a campus will accompany their children to the Museum. In this way the facility also provides a human laboratory for research into the impact of new technology.

We plan to offer new courses at undergraduate and postgraduate level which will directly take up the human responses to new technology. The core discipline for these studies will be psychology, where we intend to launch a first-degree course in 2000, but many other disciplines help interpret the impact of the information society on life, learning and work.

In all our courses we are striving to model participative and active learning. By learning how to learn while they study with us, as well as the disciplinary content of their specific programmes, students will develop the habits and attitudes which will enable them to sustain their own learning throughout their careers.

All of these activities are housed in a new building that was opened in January 1999. As well as lecture rooms, laboratories and staff offices, this fine, airy building will house the institute library and a range of student services. The past year has seen science and technology make an impact on DLIADT – in the years to come we hope to see DLIADT make an impact on the world of science and technology in Ireland.

The team which put together the first year of academic activity in Science and Technology at DLIADT also put together an apparatus for filling a champagne fountain without touching the bottle at an end-of-year staff development programme! (From left; Bryan Maguire, Kevin McDaid, Paul Comiskey, John Dempsey, Marion Palmer, Ian Hughes, Cliona Flood, Pamela Gaynor).

Contact: Dr Bryan T. Maguire,
Head of Department of Science,
Dun Laoghaire Institute of Art, Design and Technology;
E-mail: b.maguire@iname.com

Ecotoxicology research at Athlone Institute of Technology

Ecotoxicology involves the scientific study of the fates of pollutants in the environment. Ecotoxicity tests can be used to monitor the pollution potential of waste effluent discharged to the environment and to predict whether biota in the receiving environment will be adversely affected.

Algal enrichment (eutrophication) is one of the major consequences of freshwater pollution in Ireland. The eutrophic potential of effluents can be modelled using algal bioassays. Macro-invertebrates such as daphnids are a major source of food for commercially important fish such as salmonids. Any negative effect on these organisms can have significant implications for the receiving environment. Such effects can be screened for using the *Daphnia* bioassay. Another bioassay, the fish avoidance test, uses rainbow trout as the test species. This bioassay is based on the ability of salmonids to migrate away from sub-lethal concentrations of chemicals.

Research projects undertaken by the Ecotoxicology Unit at Athlone Institute of Technology include (i) the ecotoxicological assessment of several pesticides used in the Irish mushroom industry (ii) the microbiological degradation and detoxification of harmful chemicals and (iii) the ecotoxicological assessments of industrial effluents.

Agri-chemicals are widely used in the mushroom growing industry in Ireland. The industry is worth £77 million pounds annually and employs over 9,000 people. As this very successful industry continues to expand, the safe disposal of the concomitant spent mushroom compost (SMC) is becoming increasingly problematic. The Irish mushroom industry generates 272,000 tonnes of SMC annually which has to be disposed. A significant proportion of this spent mushroom compost is disposed of by landfilling at a considerable cost. While this organically rich SMC has many desirable properties for use in soil amendment, these benefits may be negated by the potential presence of pesticide residues. Researchers in the Ecotoxicology Unit have developed bacteria which are capable of detoxifying some of the pesticides found in the spent mushroom compost, which should enhance its beneficial properties, while eliminating many of the environmental concerns.

Contact: afogarty@ait.ie

Disposal of spent mushroom compost.

Cleaning validation, how clean is clean?

Cleaning validation is a process to ensure that equipment cleaning procedures are removing residues to predetermined levels of acceptability. Although "equipment cleaning" is part of current Good Manufacturing Practice requirements the term "cleaning validation" was not popular until late 1980s. The need for a systematic approach to proving the effectiveness of all the cleaning procedures was achieved in 1993 with a revised Food and Drug Administration Inspection Guide on Cleaning Validation.

Pharmaceutical products and active pharmaceutical ingredients (APIs) can be contaminated by other pharmaceutical products, by cleaning agents, micro-organisms or by other material (e.g. air-borne particles, dust, lubricants). Further sources of contamination might be raw materials, intermediates, auxiliaries, etc.

In many cases, the same equipment may be used for processing different products. To avoid contamination of the following product, adequate cleaning procedures are essential.

High risk products, such as penicillins, are a major concern in the cleaning validation field. Sensitive sampling methods require development and must be applicable to each specific piece of equipment used. Due to the possibility of inter-product contamination, highly sensitive analytical methods such as Liquid Chromatography are required for trace level analysis.

These detection methods should be specific for the target analyte, sensitive for trace and ultra-trace analysis, and be sufficiently able to separate and quantify the target analyte from potential interference. Currently there are two sampling methods in use.

Pharmaceutical process equipment in operation.

The direct method, incorporating a swabbing material, is favoured over the indirect rinsate analysis approach. However caution must be taken when choosing the correct swab, as factors such as recovery, background contribution and particle generation can hinder residue determinations. Therefore each should be evaluated independently and an overall correction factor applied to the swab. Swab recoveries may be determined using spiking studies incorporating coupons of equipment surfaces.

Rinsate analysis is a useful sampling tool for equipment such as blenders and reaction vessels. The theory is that by analysing an aliquot of rinse water, the total quantity of analyte residue can be estimated. This method however assumes that the residue is uniformly removed from the equipment and also presumes that if the rinsate is clean then the equipment is clean.

Probably the most important aspect of cleaning validation programmes is establishing predetermined levels of acceptability. In order to establish Acceptable Residue Limits (ARL), various product and equipment attributes are evaluated. This leads to wide variations in ARL values between different product trains and different manufacturing facilities. The most important aspect is therefore proving that the ARL values determined in the cleaning validation programme have been established using a sound scientific rationale.

Contact: School of Science,
Athlone Institute of Technology;
Tel: 0902-24453; Fax: 0902-24492;
E-mail: adrianr@ait.ie and jroche@ait.ie

Developing the Irish Business Excellence Model for an educational environment

The Irish Business Excellence Model (IBEM) was introduced by Excellence Ireland to enable organisations to achieve Business Excellence. It embraces key concepts such as Customer Focus, People Management, Continuous Improvement, Leadership, Benchmarking and Best Practice.

The IBEM may be used as a diagnostic tool for assessing the current economic health of the organisation, assisting it to balance its priorities, allocate resources and generate realistic business plans. It may also be used as a framework to aid the development of the organisation's future mission and goals. Through Business Excellence Awards, the model allows national and international recognition of successful organisations and promotes them as role models of Excellence.

Business Excellence enables a comprehensive, systematic and regular self-assessment of an organisation, to assess progress along the path to excellence, and culminates in planned improvement actions under nine key headings:

MANAGEMENT PRACTICES
Leadership – How the actions of the management team inspire, support and drive the pursuit of excellence by providing appropriate resources and assistance to all staff.

Strategy and Planning – How the organisation's visions, policies, strategies and mission are formulated, communicated, realised, reviewed, and updated.

People Management – How the organisation plans for, and realises, the full potential of its personnel resources; how targets are agreed, communicated and performance reviewed; and how personnel are involved, empowered, developed, and appreciated.

Resources – How effectively and efficiently the organisation manages and deploys its financial and information resources; supplier relationships and materials; and physical, technological and intellectual assets, in support of strategies.

Quality System & Processes – How the organisation identifies, manages, evaluates and improves key processes.

RESULTS
Customer Satisfaction – What is achieved in relation to external customer satisfaction.

People Satisfaction – What is achieved in relation to employee self-esteem.

Impact on Society – What is achieved in relation to the local, national and international community (as appropriate).

Business Results – What is achieved, in financial and non-financial terms, in relation to planned business objectives and satisfying the needs and expectations of all stakeholders.

The Irish Business Excellence Model.

The universal requirement for best practice in education is making the task of developing quality systems and processes increasingly important for educational institutions. Academic Quality Assurance is gaining increasing importance, particularly within the Institute of Technology sector, where it is being driven by the quest for self-validation and self-awarding powers.

Athlone Institute of Technology (AIT) is adopting the IBEM with an ultimate goal of improving academic quality within the Institute and of indicating areas where ongoing and future improvements can be made.

Contact: Mary Doyle,
Quality Development Officer, AIT;
Tel: 0902-24551; E-mail: marydoyle@ait.ie

Clean up of oil spillages using biosurfactant

Hydrocarbons, which include petroleum products, oil products and halogenated compounds, form an important class of pollutants on a global scale. The presence of these hydrocarbons in the environment is of considerable public health and ecological concern, because of their persistence, toxicity and ability to bioaccumulate. Different technologies are employed by companies to clean up contaminated sites, including various chemical and physical methods such as excavation, thermal evaporation, soil flushing, pump and treatment of the groundwater, and soil vapour extraction. However, more recently, one technology which is receiving increasing attention is bioremediation. This is a natural process by which microbes modify and breakdown contaminants into other less toxic compounds, and ultimately to CO_2 and water.

One of the problems associated with the biodegradation of hydrophobic compounds, which include petroleum hydrocarbons, is that they bind to soil particles and have limited solubility in water, resulting in limited availability to soil microorganisms, which in turn can retard and/or stop the degradation process. It is generally assumed that, if you can increase the availability of the hydrophobic compounds to the microorganisms, you can increase the rate of bioremediation.

One method which has been investigated to address the problem is the use of surfactants and emulsifiers. These chemicals increase the solubility and dispersion of hydrophobic compounds. To date the majority of studies carried out have used synthetic surfactants and emulsifiers. A class of surfactants which has received limited attention

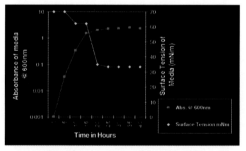

Growth rate of an isolated bacteria versus surface tension of the growth media.

are those produced naturally by microorganisms: they are called biosurfactants. Biosurfactants, as they are produced naturally, may be more ecologically acceptable and, unlike many synthetic surfactants, they are biodegradable.

Biosurfactant has been produced in our laboratory at Institute of Technology, Sligo, using a bacterium found in a local soil sample. Biosurfactant production occurred in a special medium – Proteose peptone glucose ammon-ium salts.

Preliminary leachate studies have been carried out comparing the ability of these biosurfactants to synthetic surfactants at removal of diesel from a sand column. The biosurfactant removed approximately 1.5 times more than the most effective synthetic surfactant and almost four times more than water alone.

At present, bioremediation enhanced processes, using biosurfactants, have relied on the direct introduction of biosurfactants into the contaminated site. The advantage of *in situ* production (growth in the site of contamination) is that it would be more cost effective, visibly and ecologically more acceptable with less transport/labour.

Further work in our lab will continue to investigate the potential for stimulating *in situ* production of the biosurfactant by, for example, bioaugmentation and soil amendment.

For further information contact:
Dr Michael Broaders; E-mail: broaders.michael@itsligo.ie;
Gary Canny; E-mail: cannyg@students.itsligo.ie

Biotechnological and Environmental Sciences at IT Carlow

Left: Researchers at IT Carlow - back row: Orla Sherlock & Hanane El Baz; front row: Amanda O'Brien, Guiomar Garcia-Cabellos & Dominic Garvin.

The Institute of Technology at Carlow offers flexible modular course programmes in the Biotechnological and Environmental Sciences, progressing from Certificate (two-year academic programmes), to Diploma (one-year post-certificate programmes), through to primary degree (one-year post-Diploma programmes) and onwards to postgraduate degree qualifications (M.Sc. and Ph.D.).

The B.Sc. in Industrial Biology at IT Carlow has a particularly impressive record in terms of graduate employment. 96% of Industrial Biology graduates are currently employed as Food Scientists, Biotechnologists, Pharmaceutical Analysts, Quality Assurance Inspectors and Managers, Microbiologists and Consultants in the Food, Healthcare, Pharmaceutical and Biotechnology industries. The remaining 4% are engaged in higher education studies at IT Carlow and other third-level Institutions. The success of this degree programme in terms of graduate employment has been largely due to the applied nature of the course, the direct relevance of the course content to modern industry, and the six-month work placement element at the end of the four years of study.

Building further on the Institute's strengths in the Biological and Chemical Sciences, IT Carlow recently introduced a new B.Sc. in Industrial Environmental Science. This was in response to the developing skills shortage in the Environmental area. Graduates of the course are prepared to work in areas such as Environmental Management, Pollution Control, Environmental Impact Assessment, Environmental Modelling, and Integrated Pollution Control Licensing. This B.Sc. degree, which also contains a six-month work placement module, is open to National Diploma in Science graduates from Chemistry, Biology, Analytical Science and related backgrounds.

Graduates from both B.Sc. degree programmes have joined the Institute's postgraduate research programme to further their studies to M.Sc. and Ph.D. level. The Biotechnology and Environmental Research Team at IT Carlow currently consists of over 30 research scientists and postgraduate students. While many of the current postgraduate researchers have transferred from degree programmes at IT Carlow and other Irish third level institutions, several researchers have come from further afield, having transferred from Scotland, Spain, France and Morocco. IT Carlow is collaborating with both national and international academic and industrial research partners in a number of multi-disciplinary project areas including:

- Bioremediation and Environmental Biotechnology,
- Biofuel Production,
- Biocatalysts for Clinical & Food Diagnostics and Production,
- Electrochemistry for Clean Technologies,
- Development of Probiotics,
- Microbial Production of Flavour Compounds,
- Assessment of the Environmental Impact of Current Agricultural Practices.

This research is funded by competitive research-grant awards from National and EU sources and relevant industries, and is carried out in collaboration with research teams in Ireland, Denmark, Spain, Germany and the USA.

For more information contact: Dr Patricia Mulcahy, Head of Department of Applied Biology and Chemistry, Institute of Technology, Carlow; E-mail: mulcahyp@itcarlow.ie

Biochemistry and biotechnological applications of amino acid NAD(P)$^+$-dependent dehydrogenases

Amino acid dehydrogenases catalyse the reversible NAD(P)$^+$-dependent oxidative deamination of amino acids to their corresponding keto acids. [NAD(P) is nicotinimide adenine dinucleotide phosphate]. Because of their physiological and biotechnological importance, NAD(P)$^+$-dependent amino acid dehydrogenases continue to serve as a fertile field for enzymological research. For example, amino acid dehydrogenases can be used in biosensors or diagnostic kits to screen blood serum for elevated levels of free amino acids associated with certain types of diseases. Furthermore, a range of current or developing industrial processes rely on NAD(P)$^+$-dependent amino acid dehydrogenases for final product formation (such as synthesis of L-amino acids for inclusion in foods and feeds, or as building blocks for the synthesis of therapeutics, herbicides and insecticides). A third application of NAD(P)$^+$-dependent amino acid dehydrogenases is in the development of coenzyme regeneration systems for industrial bioprocesses. Developments in this latter area is ensuring that NAD(P)H-dependent enzyme catalysed synthesis is gradually entering production scale and industrial practice.

Detailed structural studies and biotechnological application of the NAD(P)$^+$-dependent amino acid dehydrogenases is often hampered by the requirement for large quantities of highly purified enzymes with appropriate kinetic and stability properties. One approach to overcoming restrictions inherent in the use of conventional protein purification procedures is the development of highly specific

Researchers Julie Tynan and Laura Oakey at work in the Protein Technology Laboratory at IT Carlow.

bioaffinity chromatographic systems capable of purifying the target enzyme in a single chromatographic step with yields approaching 100%. The bioaffinity systems should also have a very high capacity for the target enzyme so that "large-scale" purifications can be achieved on the laboratory bench. Such bioaffinity systems are being developed at IT Carlow, and current research efforts are concerned with further development of the technologies involved through synthesis of immobilised ligands with improved chemical properties for protein purification applications, and through resolving difficulties traditionally associated with the scale-up of bioaffinity-based separations.

The amino acid dehydrogenases under study at Carlow have been selected based on their varying cofactor specificity (NAD$^+$, NADP$^+$ or dual-cofactor specificity), other distinguishing kinetic characteristics (e.g., allosteric regulation), thermo-tolerance, high specific activities in crude cellular extracts, potential biotechnological applications, and/or potential value for detailed structural analysis. The NAD(P)$^+$-dependent amino acid dehydrogenase activities currently under investigation at IT Carlow include alanine, leucine, phenylalanine, glutamate, glycine, lysine, serine, tryptophan, and valine dehydrogenase.

For more information contact: Dr Patricia Mulcahy, Department of Applied Biology & Chemistry, Institute of Technology Carlow; E-mail: mulcahyp@itcarlow.ie

INSTITUTE OF TECHNOLOGY CARLOW — DINA BRAZIL

Antibiotic resistant microorganisms in the River Barrow

The river Barrow has been of importance to the inhabitants of Ireland for over five thousand years and has recently been targeted as a priority resource for tourism. One essential factor in the successful tourist development of the river is high water quality.

At the Institute of Technology, Carlow, two microbial aspects of water quality are being investigated: the presence of microorganisms that indicate recent faecal contamination, and the prevalence of microorganisms that are resistant to antibiotics used in animal and human medicine. The significance of the antibiotic resistant microorganisms is that they may constitute a pool of drug resistances that could potentially transfer into human and animal pathogens, undermining successful antibiotic therapy.

Our investigations of selected sites from the Barrow have indicated recent ongoing faecal pollution resulting in introduced populations of microorganisms originating from the mammalian gut. This investigation is ongoing and a more complete picture of the faecal contamination and the resulting river microflora will be available.

When levels of river microorganisms resistant to antibiotics were analysed, between 20% and 59% were resistant to three

Researchers Shona Stewart and Dominic Garvan in the Molecular Biology Laboratory at Carlow IT.

or more antibiotics used in human and veterinary medicine. In addition, it was found that a sample of the multi-resistant bacteria identified could be stably maintained in the absence of selective pressure, suggesting a stable pool of antibiotic resistance genes in the bacterial population of the river.

The possibility of these antibiotic resistances being plasmid borne and therefore transmissible to other microorganisms was investigated, and we have detected transfer of resistances both in *in vitro* experiments and in non sterile microcosms into the enteric bacteria *Escherichia coli*. The significance of these results is the demonstration of the presence of a stable pool of transferable antibiotic resistance genes both in the indigenous and in the introduced microbial population in the river Barrow that could potentially enter the mammalian food chain.

The role of prescribed veterinary antibiotics entering the river in the establishment and maintenance of such populations is not known. In the US it has been suggested that antibiotics used in farming and aquaculture have become widely disseminated in waterways and sewage systems, resulting in multidrug-resistant organisms.

Our investigations are continuing with the aims of determining the levels of antibiotic resistant introduced and indigenous microorganisms in the Barrow, and of identifying possible hotspots for plasmid transfer in the water system.

Contact: Dr Dina Brazil; Department of Applied Biology and Chemistry, Institute of Technology, Carlow, Kilkenny Road, Carlow; E-mail: Brazild@ITCarlow.ie

---

INSTITUTE OF TECHNOLOGY CARLOW — DAVID DOWLING

Directed evolution of bacteria for the detoxification of pollutants

The goal of the research group in Carlow is to understand the scientific basis and develop the potential applications of using bacteria that grow and colonise plant roots (rhizobacteria) to breakdown harmful pollutants such as polychlorinated biphenyls (PCBs). One aspect of the research is focused on development of improved PCB degrading bacteria as a possible solution for the removal of PCBs from the environment.

The *bph* operon contains 11 structural genes that encode enzymes for growth on biphenyl and co-metabolism and breakdown of polychlorinated biphenyls. A critical point in the degradation pathway is the opening of the chemically stable aromatic ring. This ring cleavage step is catalysed by the *bphC* gene product 2,3-dihydroxybiphenyl 1,2-dioxygenase (DHBD).

Our objective is the directed evolution of *bphC* encoded enzymes with improved biological properties, such as higher activity in the presence of the inhibitor 3-chlorobenzoate (3CB). This compound is an end product of PCB transformation and a potent inhibitor of DHBD activity.

Physical/chemical mutagenesis of a rhizobacterial strain F113pcb, and subsequent selection in the presence of 3CB, was used to generate potential mutants with altered DHBD activity. PCR primers specific for *bphC* were used to amplify these putative mutant alleles, and the amplicons were cloned in plasmid vectors. These clones are currently being characterised with respect to enzyme activity and their DNA sequence. This will further our knowledge of the structure and function of this key enzyme, and generate novel/mutant genes with improved properties for PCB degradation and bioremediation.

Another approach under investigation is the transcriptional re-regulation of the *bph* operon using promoters that are activated in the rhizosphere environment. This will allow optimised gene expression of the PCB breakdown genes when the bacteria have colonised the plant root and are metabolically active.

Ultimately this work aims to use plants, such as *Medicago* sp. or *Salix* sp. and microorganisms, as a partnership to detoxify contaminated sites and soils as a novel cost-effective approach to bio-remediation.

Schematic ribbon drawing of subunit of the DHBD enzyme showing location of amino acid residue changes in one of the higher activity mutants.

Contact: Dr David Dowling, Department of Applied Biology & Chemistry, Institute of Technology Carlow; E-mail Dowlingd@pat.itcarlow.ie; http://www.itcarlow.ie/staff/DDHome.html

Sustainable development in Ireland – all talk and no action?

At a recent conference on environmentally superior products, held in Dublin, Dr Hans van Weenan of the United Nations Environment Programme – Working Group on Sustainable Product Development, stated that he had the impression that:

> . . . *(in Ireland)* . . . *concerning sustainable development there are policies but there is no action, there are plans but there is no implementation. Ireland is not doing much on sustainable development.*

For the past eight years, the Clean Technology Centre at CIT has been carrying out research in all aspects related to sustainable development. Furthermore, we have been working with industry to help industry to improve its environmental and economic performance. The results of this work have enabled CTC to evaluate the current world and Irish situation – so that we can attempt to throw some light on Dr van Weenan's statement.

Of course, Ireland has a number of policies and strategies. We are fully committed to the tenets of Agenda 21 of the World Summit at Rio de Janeiro, and we have perhaps the best environmental regulatory system in the world with Integrated Pollution Control licensing. The country also has a number of schemes, aimed at encouraging industry towards proactive environmental improvement. These include grants for environmental auditing, environmental management systems, environmentally superior products, etc. There are also moneys available for recycling, raising public awareness, and other similar measures. But is this enough?

Currently, the world's richest 20% consume 80% of the resources. To redress this imbalance it will be necessary for the developed world to reduce its consumption and waste generation to 25% of its current level. This is termed "Factor 4". At the same time, the current world population of approximately five billion is expected to increase to 10-11 billion within the next few decades. This leads to the concept of "Factor 10" if the already overburdened ecosphere is not to be further compromised.

These concepts are already written into national policies and plans in the more environmentally enlightened countries. Their incorporation can be explicit (as in Austria, for example) or implicit (such as in the Dutch voluntary targeted reductions of 75% and 90%). In Ireland, as yet there has been no public debate on the policy implications concerning either Factor 4 or Factor 10.

Nevertheless, many good things have happened in this country over the past 10 years. The Clean Technology Centre has researched the factors that promote or inhibit the uptake of innovative environmental improvements, particularly by adopting a preventive approach to environmental issues. As part of EU funded projects, we examined the dominant factors in the processing industrial sectors, e.g. fine chemicals, brewing, dairy products, etc. In separate projects, we looked at the effects of the local socio-economic situation and the importance of different agents in the adoption of this approach. From this, we showed that, while the picture is complex, and unique to every locality and firm, some common threads emerged. Promoting a preventive approach can not be achieved by a narrow or limited focus. A "programme of activities" rather than a number of "projects" is needed. Time is needed – Rome wasn't built in a day. The activities must inter-link and re-enforce each other. Firms need a different approach, depending on their size, sector and economic standing. Key individuals, whether in firms, local authorities, or the wider community, have an important role to play.

The CTC has actively supported companies. We have conducted waste minimisation assessments, technology transfer, environmental management system installation, training in ways to reduce waste and emissions, as well as operating an environmental management system or integrated pollution licence. This has brought us into contact with operating, administrative, managerial and technical staff. With our help, companies have identified improvement opportunities and implemented them, saving both money and the environment.

In addition to the private reports prepared for companies and contracting research agencies, our findings have been published by Cork County Council as "How to prevent waste and emissions from your company: a self-help guide", by the EPA as "The use of cleaner production technologies in the metal finishing and electronics industries", in a book to be published by Earthscan, and by ourselves as "Co-operation for environmental improvement: case studies and guidelines for industry/regulator co-operation towards sustainable production programmes", and a CD-ROM training tool for environmental management systems.

Ireland has policies and the Clean Technology Centre has provided many of the actions to achieve sustainable development. Our approach has been a model of partnership. In conclusion, we agree that prosperity for Ireland demands this attention, because again quoting Hans van Weenan:

> *The world is at the beginning of a reconstruction of its production base. Unsustainable enterprises will be ended, partly sustainable enterprises will be changed, and wholly sustainable enterprises will take the lead and their establishment will be stimulated.*

Contact: Dermot Cunningham,
Clean Technology Centre, Cork Institute of Technology, Unit 1, Melbourne Business Park, Model Farm Road, Cork;
Tel: 021-344864; Fax 021-344865;
E-mail: ctc@cit.ie

Probing the hearts of quasars

Quasars are the most powerful objects in the universe, emitting the energy equivalent of 1000 billion suns, or more, every second. Efforts to understand them have gone on since their discovery in 1963. The most widely accepted model invokes material falling into a supermassive black hole at the centre of a galaxy. As the material spirals inwards, it becomes superheated and incredibly luminous. For reasons which are not at all clear, approximately 10% of quasars, the so-called radiolouds, eject some of this material in powerful jets. In the optical, radiolouds are revealed by large-amplitude and often rapid variations in their light output.

A significant step forward in our understanding of quasars would result from finding objects which harbour very weak jets ("radioweaks") – these may represent the poorly understood early stages of jet formation. One of the research projects at present being pursued by the Astrophysics Group, CIT, involves examining a recently discovered sample of radioweaks to determine if their optical behaviour is similar to that seen in the powerful radiolouds, and whether the jet's internal structure evolves with jet power.

The approach involves determining the variations in the optical brightness of the radioweaks. Observations are made primarily during week-long observing runs with telescopes on the Canary island of La Palma, and at Calar Alto, Spain. The technique involves taking

The Nordic Optical Telescope at La Palma (photo by Niall Smith).

repeated images of a quasar and the stars around it with a state-of-the-art charge-coupled device (CCD) camera. By comparing the brightness of the quasar relative to the stars (which are assumed not to vary), we can produce a very accurate graph of the brightness variations of the quasar itself, even if the sky conditions are not ideal. Since we expect weak jets to produce only low-amplitude variability, we have developed a novel data reduction technique which optimises the photometric accuracy of the lightcurve. Further, since the number of CCD frames can be large (over 2000 in each observing run) we have developed sophisticated automatic routines to reduce the raw data.

Our group has observed 20 objects to date. A few show low-amplitude brightness variations, suggesting that jet activity in these sources is either absent, or the outflow in the weak jet is not highly turbulent. The most significant variations happen on nightly timescales, which is fast enough to place strong constraints on theoretical models. However, new sources are being observed all the time and the picture of what is going on in radioweaks is slowly becoming clearer.

Contact: Dr Niall Smith, Physics Department, CIT;
E-mail: nsmith@cit.ie

"Smooth & Shiny"

A maintenance-free finish. Guaranteed noticeable results in minutes. No, this is not an advertisement for the latest revolution in hair care, rather an introduction to a new research group in The Chemistry Department, Cork Institute of Technology (CIT), under the direction of Dr Rosamund Hourihane. The research interest is in the area of Electropolishing with Surface Analysis and involves collaboration with The Centre For Surface and Interface Analysis (C.S.I.A.), CIT, directors Dr Liam McDonnell and Dr Eamon Cashell.

What is Electropolishing, I hear you ask? Electropolishing or electrolytic polishing is the improvement in the surface finish of a metal by making it anodic in a suitable solution. The continued solution of the metal occurs in such a way that irregularities on the surface are removed and the surface becomes smooth and bright. Careful selection of conditions such as electrolyte, current density, temperature and viscosity are required.

Many advantages exist for electropolishing: for example it produces a surface which is more passive and decidedly more corrosion resistant than a mechanically polished surface. It is a simple technique which gives reproducible results whereas, with other polishing techniques, the degree of finish obtained depends to a large extent on the skill of the operator. It is versatile, where small areas can be polished as well as large articles with complex shapes.

An ideal polishing process should "smooth" by elimination of large scale irregularities above micron size and "brighten" by removal of smaller irregularities (nm size). A distinguishing feature of an electropolishing process is that it combines both functions.

How do you assess how smooth and bright a surface is? Smoothness or Roughness is assessed by determining an Ra value for the surface using, on this occasion, Atomic Force Microscopy (AFM). Brightness is determined using Specular Reflectance.

AFM images have highlighted one very interesting fact: the relevance of height based parameters such as Ra. This much-used parameter is insensitive to surface wavelengths and in consequence surfaces of different spectral content can have the same roughness value – that is to say two very different surfaces can have the same roughness value.

Research student Mary Oldham using Atomic Force Microscopy.

Funding for this research was secured under the Graduate Training Programme. O.M.C. Engineering Limited, Limerick, have endorsed the commercial logic of this project – they have agreed to partner the Chemistry Department CIT in an Enterprise Ireland Applied Research Project to develop this idea.

This project facilitates industry by not only investigating an alternative polishing technique, which may be a process option for them, but also provides an expert in the area of electropolishing who may advise and provide technical support.

Contact : Dr Rosamund Hourihane,
Chemistry Department,
Cork Institute of Technology;
E-mail: rhourihane@cit.ie

CIT develops virtual reality milling machine for interactive training

The increasing importance of training and updating skills has lead to the use of new interactive technologies in the development of more effective training tools. "VRmill" is a prototype training tool which has been developed at the Mechanical and Manufacturing Engineering Department at Cork Institute of Technology (CIT), in collaboration with Solas Data, a small software house operating out of Ballincollig, Cork.

The prototype is now the subject of a £40,000 research project funded primarily by Enterprise Ireland through the Applied Research Program (ARP), with a 25% contribution from Solas Data.

"VRmill" consists of two computerised windows applications which together constitute a fully interactive model of a Computer Numerical Control (CNC) milling machine. The first application provides a 3-D virtual world, containing a fully realised CNC milling machine around which the user can navigate in real time. The second application is an on-screen emulation of a Farco control panel, which gives the user full interactive access to programming and manipulating the virtual machine. At present, the control panel emulation is sold as a controller for miniaturised desktop milling machines, used primarily for training in technical schools and colleges.

The aim of future work in the project is to develop the prototype virtual milling machine into a commercially viable alternative to the miniature desktop machines, fully integrated with the Farco control panel emulation. This would provide a low cost application involving minimum space and hardware requirements.

"VRmill" is written using a virtual software application called Superscape, which allows the program developer to define how objects interact with each other, for example, how they move relative to each other. This enables the virtual world to provide continuous updating of the viewers' perspective of the objects they are viewing.

To be effective, "VRmill" must be able to "read" and interpret a CNC program in order to carry out the cutting instructions contained in the program. This is done using a piece of object oriented software (called the code converter) developed for the project, and a technique called Dynamic Data Exchange (DDE) whereby two separate windows applications can communicate with each other. The prototype "VRmill" is at present capable of executing the linear cutting actions of a milling tool, visually simulating the progressive machining of a workpiece as a program is executed *(see illustration)*. The new development program will enhance the capability of the virtual machine to include circular cutting actions and as many of the standard functions of a CNC milling machine as possible.

Contact: Mr Daithi Fallon;
Tel: 021-326213; E-mail: dfallon@cit.ie

Electronic analysis and design for industry

In the Department of Electrical and Electronics Engineering at the Cork Institute, various projects are underway designing new electronic components for both the computer, telecommunications and process industries. The efforts are in both the analysis and understanding of circuits as much as in their design and fabrication. The fields are varied and include chip design, predictive control, noise cancellation systems, motor characteristics identification and power supplies. A few examples are given.

Work is at present being undertaken for ArteSyn Technologies Limited, Youghal, Co. Cork, in the design of low wattage power electronic converters. Both the telecommunication and computer industries require highly accurate power supplies at various output voltage levels, irrespective of the load and input voltage levels. They also require the power supply to sense an over current or an over/under input voltage condition and for the circuit to self protect itself in such an eventuality. In essence, a high performance and robust control strategy is required for the power converter.

Another major thrust is to reduce the size of power supply units to allow for smaller units and, consequentially, allowing the use of automated manufacturing techniques. This entails a very focused study of the losses in the power converter and how the control strategy can help in increasing the overall efficiency and reducing down the losses. A Ph.D. student is looking at the temperature stress of the individual components of a typical power converter to see which items can be engineered in such a manner that the overall converter can operate at a higher temperature. The result of this will be to have higher losses per volume or, put another way, a greater through-put of watts per volume to produce the losses. It is hoped that the final converters will be in a dual-in-line package, allowing for ease of final product assembly. A cellular arrangement will be necessary with these converters to allow for paralleling of them for greater current output. This will require current sharing topologies between the converters.

In another related area, chip design has been undertaken for Silicon Systems Design Limited, an innovative silicon intellectual property company. The project focused on full custom design of phase lock loop systems for clock recovery and synthesis. Among the deliverables of this project was a full custom delay lock loop. This was fabricated under the Europratice program.

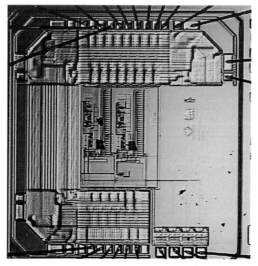

A photo-micrograph of the delayed lock loop fabricated for clock recovery and synthesis, which was designed at the Cork Institute of Technology.

Contact: nbarry@cit.ie or jhoran@cit.ie

Research Colloquium at Letterkenny Institute of Technology

A very successful Science and Computing Colloquium was held in the Institute of Technology, Letterkenny, from Wednesday 26 to Friday 28 May, 1999. This major annual conference is held in a different Institute each year.

The Colloquium was attended by approximately 140 academics and research students from all the Institutes of Technology in the country. The delegates were welcomed to Letterkenny by Mr Paul Hannigan, Director of the Institute. The keynote address was given by Mr Paddy McDonagh, Assistant General Secretary at the Department of Education & Science, who spoke of the importance of the Institutes of Technology in contributing to a regionally dispersed Science and Technology infrastructure, and their major role in the deepening of national Research and Development.

Mr McDonagh outlined the plans of the Department of Education & Science, in co-operation with private industry, to make a huge investment in research during the next five years, emphasising the importance of this investment in enabling the country to maintain its economic growth and provide the high-tech, highly-paid jobs which the young educated work force requires.

He stressed the critical importance of research in providing a catalyst for the type of innovation which an economy such as ours must have in order to sustain and expand economic growth.

The main business of the Colloquium started on Thursday, with the presentation of papers by researchers from all of the Institutes of Technology. Papers were presented on many varied topics, and included descriptions of a survey of Zebra Mussels in Lough Key, the use of waste microbial biomass as a source of oxido-reductases, the use of streamed multimedia to enhance the virtual classroom, developments of novel methods for estimation of histamine in foods, and for creatinine in blood, etc.

The conference dinner was held in the Mount Errigal Hotel and was addressed by Dr Eileen Stewart, a former student of the Letterkenny Institute and now Researcher and Lecturer at Queen's University, Belfast. Dr McDaid, Minister for Tourism, Sport & Recreation, attended the dinner. In his speech he acknowledged the importance of the Institute of Technology in the Letterkenny area, and confirmed that an additional 2.6 million pounds was being made available to complete Phase II of the College development.

The Conference closed at 1.00 p.m. on Friday with the distributions of prizes for the best research papers. The main sponsors of the event were AGB Scientific Limited.

A group of staff and students from Tralee Institute of Technology at the Science Research Colloquium.

Contact: Dr John Hines, Head, School of Science, Letterkenny Institute of Technology, Port Road, Letterkenny, Co. Donegal; Tel: 074-64129; Fax: 074-64107; E-mail: john.hines@lyit.ie

The use of Middleware in E-Commerce applications

With the increase in popularity of the Internet, and indeed E-Commerce, Middleware has become indispensable in the development of web-based applications. Many of these require continuous complex interactions with a number of databases. Middleware is employed to enable the seamless integration of the applications with the various types of databases. Increasingly however middleware is used in the development of applications which communicate across platforms as they traverse the Internet. Typically these applications are being developed using the Java Programming Environment. In choosing the type of middleware used for these applications, consideration must be given to a number of factors including: the type of application being developed, the need for security, quality of service and broadcast services, etc.

It is for this reason that research is currently being undertaken in Letterkenny Institute of Technology into the capabilities of the various types of middleware technology. The categories of middleware technology being analysed include:
1. Database Middleware (e.g. JDBC).
2. Remote Procedure Calls (RPCs) based middleware.
3. Transaction Processing (TP) Monitors.
4. Message Oriented Middleware (MOM).
5. Object Request Brokers (ORBs) including OLE/COM/DCOM.

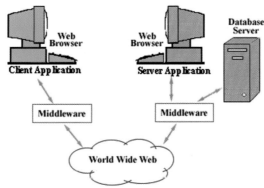

Middleware & E-Commerce application interaction.

Consideration must be given to the services provided by the various middleware technologies during their comparison. As such it may be found that no single middleware solution will be the perfect middleware technology to replace all others. Indeed, a number of middleware technologies may be required in conjunction with each other to provide the required solution for any individual application. As previously stated, each application must be assessed for its specific requirements before proceeding with the selection of the middleware technology.

It is envisaged that the development of an advanced prototype E-Commerce application will provide a platform for testing the qualities of the various types of middleware technology. The application created may be used on the server side and will link to a backend database representing a production Mangement Information System. The link to the database server software will be provided through the use of middleware *(see illustration)*. The application developed will be tested for factors such as: speed, flexibility and error rates in accessing various data types using a number of the middleware technologies.

Contact: Ruth Lennon, Computing Department, Letterkenny Institute of Technology, Port Road, Letterkenny, Co. Donegal; Tel: 074-64313; E-mail: ruth.lennon@lyit.ie

Research strategy and expertise at Dundalk Institute of Technology

Dundalk Institute of Technology – a centre of innovative research and technology enterprise support

Serving the North-East region since 1971, DKIT is one of the longest established Institutes within its sector, drawing the majority of its students from counties Louth, Meath, Cavan, Monaghan and North County Dublin. It has a student population of 2,500 full-time and 400 part-time, pursuing third level programmes at National Certificate, National Diploma, Degree and Post Graduate levels in Engineering, Science, Computing,

Dundalk Institute of Technology - strategically located to service the research and development requirements of the rapidly expanding industry base along the Dublin-Belfast economic corridor.

Business and the Applied Humanities. DKIT's Academic and Campus Masterplan envisages a growth in student numbers to over 3,000 by 2003, and the Institute is currently engaged in a £16 million Campus development and building investment programme.

DKIT, through the Research and Development activities of its Academic Staff, has developed a strong R&D reputation and is recognised as a centre of excellence in:
- Food Technology
- Software Development
- Enterprise Development and Innovation
- Product Design Engineering
- Electronics
- Building Surveying

The evolution of this range of research expertise has a direct bearing on new course and staff development, and on the relevance of graduate provision to the skills needs of the Regional and National economy. An Annual Research Forum provides Research Staff an opportunity to disseminate information on their activities, and these events have been highly successful in stimulating joint research projects with industry.

The Institute was one of the first to establish a Campus based Innovation Centre (the Regional Development Centre) with incubator facilities and an Industrial Services Office providing a range of services to Industry and business in the region. The Centre draws strongly on the knowledge base and expertise of the Institute and has become one of the most developed incubator units in the Institute of Technology sector.

DKIT is an active participant in the Enterprise Ireland Applied Research Programme, working in partnership with small, medium and large companies in the region, and sectoral industry groups, to develop innovative solutions to design, process, and manufacturing challenges.

Participation in a number of EU funded R&D projects under programmes such as ADAPT has brought a wider dimension to the Institute's activities, particularly in the areas of Innovation and Enterprise Development.

The Institute has now developed a Research Policy and Strategic plan which encourages research in key strategic areas critical to the development of the Institute, its region, and also relates to the demands of the regional economy for graduates and post graduates.

Contact: Colm Piercy, Industrial Services Officer, Dundalk Institute of Technology; Dundalk, Co. Louth; Tel: 042-9331161; Fax: 042-9351412; E-mail: iso@dkit.ie

---

Low fat poultry products

In recent years, there has been an increased consumer interest and demand for low fat food products. This interest stems from health concerns over conditions such as obesity, coronary heart disease and colorectal tumours. The development of these conditions has been attributed to high fat intake in the diet and, in particular, saturated animal fats. The outbreaks of Bovine Spongiform Encephalopathy (BSE) or mad cow disease in cattle has enticed consumers to look for safer or less risky types of meat for consumption. Hence the observed increased sales of poultry, pork and other meat products, and a corresponding decrease for beef.

Dr Sam Alwan with M.Sc. student Mark Wolfe at the Instron Texture Analyser workstation.

Comminuted meat products normally contain high levels of fat in their composition: for example, beef burgers normally contain 20 to 30% fat, pork sausages contain 25 to 50% fat, and frankfurters 30 to 35% fat. Fat, in these products, contributes to flavour, texture, juiciness and some resistance to abusive cooking. Fat reduction in meat products may be achieved by fat removal and replacement with added water and other fat replacing ingredients. Therefore, the aim of this research programme was to investigate the economic development of low fat (< 5% fat), poultry products of similar characteristics to the full-fat traditional products. This research work was required to overcome the problems of flavour and texture associated with many low fat meat products, such as a "cardboard" taste, other objectionable flavours in grain meat products, lack of juiciness, and increased toughness.

The food ingredients utilised as fat replacers included: Hydrocolloids such as seaweed extracts (carrageenan, alginate); Gums such as xanthan (XG), guar and locust bean gum; Non-Meat Proteins such as soy protein isolate (SPI), sodium caseinate (NaCas) and whey protein isolate (WPI); others such as microcrystalline cellulose (MCC) and pectin (Slendid). The poultry meat types investigated were chicken breast and thigh meat, singly and in combination. Chicken skin pasteurised or in an emulsion form was included.

Two new graduates with B.Sc. degrees in food science & technology (UCC) were selected to carry out the investigations. Ms Helen Smith has already completed her research programme and was awarded the M.Sc. degree in Food Science in 1998. Mr Mark Wolfe is expected to be awarded his M.Sc. degree in 1999.

Contact: Dr Subhi (Sam) Alwan, Department of Experimental Science, School of Science, Dundalk Institute of Technology, Dundalk, Co Louth; Tel: 042-9370277; Fax: 042-9327151; E-mail: subhi.alwan@dkit.ie

DUNDALK INSTITUTE OF TECHNOLOGY & THE QUEEN'S UNIVERSITY OF BELFAST   K.A. MURRAY (DKIT), D.J. CLELAND (QUB) & S.G. GIBERT (QUB)

The behaviours of edge panels in reinforced concrete flat slab structures

A typical flat slab structure is a reinforced concrete slab supported directly by concrete columns without the use of intermediate beams spanning between the columns such as are used in a traditional beam and slab structure. The bulk of research to date has tended to concentrate on the behaviour of flat slabs in the vicinity of interior columns as opposed to exterior columns. From the point of view of design, the edge column is perhaps more important than the interior column. This project aims to provide a better understanding of the behaviour of flat slabs in the vicinity of edge columns.

A series of one-third scale, two-column models has been tested in the laboratory at The Queen's University of Belfast (QUB). This type of model was selected as the best simulator of the prototype. A significant aspect of this project is the novel use of closely spaced internally strain gauged reinforcing bars, which was developed by Scott and Gill (1987) at the University of Durham. As a result of this research, a picture of the distribution of curvature and bending moment across each of the slabs was obtained for the entire loading history of the test programme.

The project has developed a computer model to simulate the experimental behaviour of the slabs. In order to simulate the behaviour of the slab over its entire loading history, the model has been split into four different phases. This stepped analysis is an attempt to model how cracking in both the slab and the column affects its behaviour.

As a result of this work, it should be possible to devise a more accurate set of design guidelines for the prediction of moment, shear and deflection in flat slab edge panels. This should result in a less conservative analysis and design process, leading to economic savings in the construction of flat slabs.

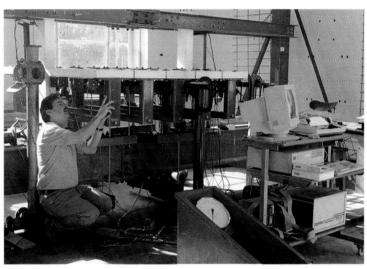

Flat slab test model.

Contact: K.A. Murray, Dept of Infrastructural and Environmental Studies, Dundalk Institute of Technology; Tel: 042-9370200
D.J. Cleland & S.G. Gilbert,
School of Civil Engineering,
The Queen's University of Belfast.

DUNDALK INSTITUTE OF TECHNOLOGY   TONY LENNON & DANIEL O'BRIEN

A chilled fibre optic humidity sensor

A Chilled Fibre Optic Humidity Sensor was developed, intended for use in mushroom growing tunnels and in the curing of meats, where medium to high values of relative humidity are required.

One method employed to measure relative humidity is to cool a surface and to determine at what temperature dew begins to form on it – known as the Dew Point temperature. The traditional way of achieving this is to cool a mirror to determine the temperature at which light is scattered from its surface as the dew begins to form.

Any material on the outside surface of an unclad optical fibre can cause a reduction in the light transmitted through the fibre. If an unclad optical fibre is chilled below the Dew Point, condensation will form on its surface allowing light to "leak out", and the intensity of transmitted light will be reduced. The change in light intensity can be used to detect when dew starts to form, so the Dew Point temperature can be measured, compared to room temperature, and so the relative humidity determined.

The effect can be maximized by introducing light into the optical fibre at an angle of approximately 27° off-axis. This causes a cone of light to be emitted from the opposite end of the fibre. As the fibre is chilled, the cone of light becomes diffused and less intense.

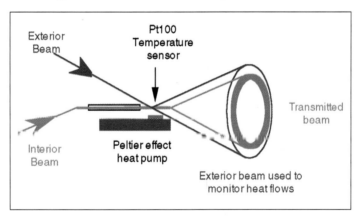

Introducing light into an optical fibre at an angle of approximately 27° off-axis causes a cone of light to be emitted from the opposite end.

In order to examine the time response of the sensor, a mathematical model of the thermal flow in the cross section of the fibre was developed using finite difference techniques. The model predicted the flow of an upward thermal wave around the circumference of the fibre as it was chilled. This mathematical model was verified through experimental results.

A Peltier effect heat pump was used to chill the base of the fibre. Platinum resistors were used as temperature sensors, while measurement and control was performed using a data acquisition board and personal computer.

It was found that the intensity of the light changes with the temperature of the unclad fibre even in the absence of moisture.

The final version of the sensor will compare the signals of a clad and an unclad fibre, both exposed to the same temperature excursion, in order to compensate for changes in light intensity due to the temperature.

For further information contact Tony Lennon, Department of Science, Dundalk Institute of Technology, Dundalk, Co. Louth; Tel: 042-9370200; Fax: 042-27151; E-mail: lennona@sci.dkit.ie

Supporting knowledge-based enterprise in the North-East Region

Established in 1989 by Dundalk Institute of Technology, the Regional Development Centre acts as the Institute's Innovation Support and Technology Transfer organisation. The Centre manages a range of Entrepreneurial Development programmes, and houses a variety of start-up high technology and knowledge based companies in its 1800m² incubation facilities. Services including Research and Development, Technology Transfer and Customised Training Programmes are also co-ordinated by the Centre. So far, the Centre has assisted 85 entrepreneurs on a variety of specially designed enterprise development programmes and has accommodated 24 new Innovative start-up businesses, 15 of which have now relocated throughout the North-East region.

The Regional Development Centre at Dundalk Institute of Technology is celebrating its 10th year of operation.

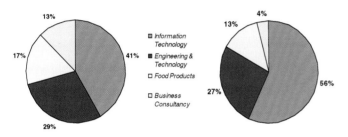

*Left: Summary and Sectoral Category of Projects.
Right: Sector v. New Employment Generated.*

Campus companies that have located at the Centre since its establishment have been involved in a wide range of sectors, with IT and Engineering/Technology businesses being the most dominant. A recent analysis of these businesses demonstrates the distribution of businesses across these sectors:

Assisted companies now employ 101 people in a variety of disciplines – the majority being at graduate level. These businesses have helped in establishing the North-East Region as a positive environment for knowledge based enterprises, and have been cited as a factor by State Agencies in attracting inward investment projects. The North-East can now truly be identified as a region with a growing capacity to support R&D intensive entrepreneurs, and the Institute and Centre continue to be pro-active in supporting the growth of this Science, Technology and Innovation infrastructure, and as contributors to continued Regional Economic Development.

High technology Campus Companies at DKIT

Incubation facilities at Dundalk Institute of Technology have provided an ideal support environment for the development of new high technology enterprises over the past 10 years. The following is an overview of just some of these:

RFT Vision Systems

Develops customised vision recognition, measurement and control systems, with applications across a wide spectrum of industrial sectors. The Company has been commissioned by a number of multinational clients to develop bespoke vision solutions in plant-wide process control, measurement areas, and pharmaceutical quality assurance.

Mr Mike English and his team at RTF Vision Systems are developing innovative vision-based production and quality control systems for the semiconductor and pharmaceutical industries.

Programmable Systems

Formed by DKIT Electronics Lecturers John Hanratty, Liam O'Gogain and Kevin Starrs, the company provides design, troubleshooting and support services to industry in the area of control and automation.

To address training needs in programmable devices, they have developed a self-tuition system, PATH, which provides the necessary hardware and software for Electricians, Maintenance Staff and Managers who need to update or maintain their knowledge and understanding of programmable control systems.

Foodlink R&D Europe

This is a new company offering innovative training and research and development consultancy services to the food manufacturing and service sector. The company operates mainly in the Republic and Northern Ireland, and has recently won its first contract in mainland Europe.

Foodlink's clients include large multinationals as well as SMEs, Enterprise Ireland, FÁS, and various County Enterprise Boards. As the Irish food industry enters a new paradigm of fewer and more powerful customers, growth of central distribution and the arrival of e-commerce, Foodlink works with clients to improve their competitiveness in this changing environment.

Danu Industries

Has developed an encrypted IP driver that secures all content of TCP/IP communications over a network. The product is based on a closed user group using symmetric private keys to encrypt and decrypt data. It can be used in conjunction with any third party application to provide secure multimedia communications.

Danu Industries is developing a range of high level encryption technologies which can be used to secure data communications over public communications networks.

Currently the Company is integrating its technology into a software-based solution for Voice over IP Telephony using PC to PC communication. Another project will integrate the technology into a hardware-based solution for encryption of voice and data communications over standard telephone lines on the PSTN.

Other activities which have been accommodated in the Centre include:
- Innovative food products development
- Multimedia games development
- Industrial control systems
- Project and business management consultants
- Precision engineering and design
- Advance facilities for inward investment Multinationals
- Enterprise Development Agencies.

---

*For further information contact: Gerry Carroll, Head of Development;
Colm Piercy, Industrial Services Officer;
Kate Wiseman, Business Programmes Co-ordinator;
Dundalk Institute of Technology; Dundalk, Co. Louth;
Tel: +353-42-9331161; Fax: +353-42-9351412;
E-mail: iso@dkit.ie; WWW: www.dkit.ie/rdc*

Supporting entrepreneurship at Dundalk Institute of Technology

Possibly the most fundamental question asked in the study of entrepreneurship is whether entrepreneurs are born or made?

Does the entrepreneur have a totally different psychological profile from the rest of the population or can anyone become an entrepreneur, given the right training, encouragement and resources? The Regional Development Centre at Dundalk Institute of Technology (DKIT) has been active in providing training, advice and assistance to aspiring entrepreneurs through its specially designed enterprise development programmes.

The Coca Cola National Enterprise Awards

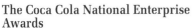

The Coca-Cola National Enterprise Award targets young Irish graduates with a business idea and offers them the opportunity to develop their ideas into meaningful commercial propositions. The Programme offers graduates the opportunity to compete for an award from a prize fund totalling £18,000.

The programme aims to encourage and develop entrepreneurial talent through training. In order to qualify for entry, an applicant must be a graduate of a third level educational institution, hold a diploma, degree or higher qualification, and have a business idea at concept or feasibility stage.

The project combines structured training with financial reward, and a recognised qualification, all within a competitive framework. Managed by DKIT's Regional Development Centre, the scheme comprises a series of six intensive business training modules, individual assessment sessions, and each participant is linked with an experienced business mentor. A comprehensive training manual was developed to support the business training modules delivered by DKIT Business School. Training is delivered in Dublin, and supported by regional mentoring and assessment sessions.

Through the initiative, a three-way partnership comprising of Industry, Higher Education and a State Agency was formed, which provides comprehensive support to aspiring entrepreneurs. The programme was jointly designed with Coca-Cola Atlantic, Drogheda, and funded by the U.S. based Coca-Cola Foundation, while Enterprise Ireland offers its network of experienced business mentors to support the participants in the development of their business proposals.

To date over 60 graduate entrepreneurs have been assisted on the programme. Recent research indicates that 18 participants' businesses had reached set-up stage, with the majority in full-time employment and/or pursuing further development paths. A wide range of projects has been supported – including designer knitwear, a shell-fish hatchery, a CD-ROM image bank, environmental services, and a residential artists print workshop.

Technology Enterprise Programme

The Technology Enterprise Programme (TEP) was first developed in 1992 as a cross-border enterprise support programme designed to assist those with technology based product or service ideas. The Programme targets both individual entrepreneurs and small existing companies seeking to develop new products or services, in an effort to encourage the development and growth of indigenous technology industry in the North-East region of Ireland.

TEP spans the border between Northern Ireland and the Irish Republic and is jointly managed by the Regional Development Centre at DKIT and Formanagh Business Initiative in Enniskillen. Entrepreneurs selected to participate in the Programme are offered a wide range of valuable support services to help them through the early stages of business development – including:

- comprehensive training in the area of business planning
- business counselling
- access to the Institute's library, laboratories, computers and equipment
- networking opportunities
- access to marketing and technical consultancy/mentoring
- fully serviced office facilities and secretarial support.

The training provided by the programme includes market research, business planning, legal issues, financial planning and sales skills. This training is designed to position technology entrepreneurs to build their businesses, their industries and, in due course, to contribute in real ways to economic growth.

Since 1992, TEP has supported 44 technology-based projects, resulting in the creation of 26 innovative businesses, with a total of 56 new jobs generated. Projects developed through TEP are both employment generating and attractive to potential investors. Examples include multi-media games software, quality management and financial software, consumer food products, and an electronic golf training product.

The Programme partners have recently launched the fifth Technology Enterprise Programme.

Promoting best practice in enterprise development

UNISPIN is a special collaborative project aimed at universities, colleges, development agencies and enterprise centres, interested in setting up their own regional spin-off programme. The programme offers a series of workshops, which offer the distilled practical experience of some of Europe's most successful spin-off programmes. Best practice models which are presented at the workshops include the Technology Enterprise Programme, along with University of Twente TOP programme, Holland, and the University of Linkoping ENP programme, Sweden. UNISPIN-CR aims to assist enterprise agencies and universities in the Czech Republic to design and establish successful programmes for supporting technology entrepreneurs.

The role of enterprise training in encouraging innovative enterprise

It has been recognised that a combination of factors determines the success of new business ventures. An innovative, market driven business proposal, and the entrepreneurial traits and abilities of the promoter, will be critical to the successful development of any new venture. Adequate grant aid and financial investment at the early stages of business set-up is also significant; however the benefits of non-financial support, such as business training, mentoring and advice provided by enterprise development programmes, is often underestimated.

For further information on enterprise support please contact:
Ms Kate Wiseman, Business Projects Co-ordinator, Regional Development Centre, Dundalk Institute of Technology, Dundalk, Co. Louth;
Tel: 042-9370413; Fax 042-9351412;
E-mail: kate.wiseman@dkit.ie

Mr Clifford Kelly, Chairman, Board of Governors DKIT; Margaret O'Shea, Second Prizewinner; Mr Jim Higgins, Director & Vice-President Concentrate Manufacturing, The Coca-Cola Company; Bernard Conlon, Overall Winner; Ms Mary Harney TD, Tanaiste and Minister for Enterprise & Employment; Mr Denis Murphy, Acting Director DKIT; and Barry Finnegan, Third Prizewinner.

Pervaporation for waste management

Strategies for dealing with industrial liquid wastes involve reducing both the quantity produced and their environmental impact. Recent Irish and EU statutory provisions require consideration of waste minimisation and environmental performance. Innovative solutions are required for reduction, recovery and recycling of hazardous liquid wastes in particular.

We have been using pervaporation for the separation of liquid solutions. This process has advantages – including improved selectivity, reduced energy consumption, and separation of azeotropic mixtures and mixtures of components with close boiling points. (Azeotropic mixtures are those whose composition does not change on boiling and thus the components cannot be separated by simple distillation.)

Pervaporation is a membrane separation process, which uses a dense polymeric membrane for selective permeation of one or more components from a liquid mixture. In the process, a concentration/vapour pressure gradient is established across the membrane, and one component preferentially permeates the membrane. A vacuum applied to the permeate side is coupled with immediate condensation of the permeated vapours.

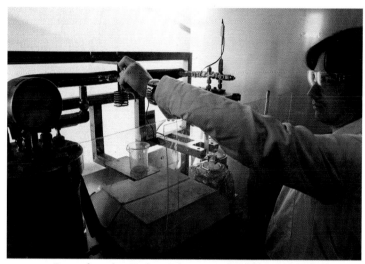

Research student Rodney Lakes operates a laboratory scale pervaporation unit at DIT.

The choice of membrane is most important for selective separation of specific components. We have used hydrophilic membranes for the separation of azeotropic solvent mixtures containing water. Hydrophobic membranes can be used for the preferential permeation of organic solvents from organic/water solutions or organic/organic mixtures.

Experiments are performed on a laboratory-scale unit incorporating a membrane cell. Samples of permeate and retentate are analysed using gas chromatography and Karl-Fisher titration. The variations of permeate concentration and flux with feed concentration and temperature are studied in order to generate data to be used for the design of pervaporation and hybrid pervaporation/distillation systems for the separation, recovery and recycling of organic solvents.

Contact: Paul Ashall,
School of Chemistry, DIT;
E-mail: paul.ashall@dit.ie

Dublin Institute of Technology Spectroscopic Facility

The Spectroscopic Facility at DIT is a collaboration between the Schools of Chemistry and Physics to address common needs for spectroscopic services and promote interdisciplinary research at undergraduate and postgraduate levels. Any one modality of spectroscopy can only probe a limited range of physical properties, and therefore a range of techniques is essential for a comprehensive research programme. Currently, the facility houses the following suite of instruments.

Shimadzu UV-2101PC Absorption Spectrometer – a double beam, direct ratio photometric measuring system providing a range 190–900 nm.

Perkin Elmer LS50B Luminescence Spectrometer – a ratioing luminescence spectrometer with the capability of measuring fluorescence, phosphorescence and luminescence. The source is monochromated in the scan range 200–800 nm. The luminescence can be scanned over the range 200–900 nm.

Mattson Infinity FTIR Spectrometer – a single-beam, Michelson interferometer-based, Fourier transform infrared spectrometer, with an operating range in the mid and far infrared, 200–5000 cm^{-1}, with a resolution of 0.5 cm^{-1}.

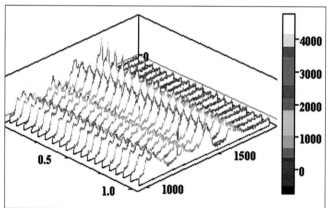

Scan using the Labram Raman imaging microscope system.

Instruments S.A. Labram 1B – a confocal Raman imaging microscope system. It has a range from 150–4000 cm^{-1} in a single image, or greater resolution (1 cm^{-1} per pixel) in a combination of images.

The facility provides support for research projects and collaborations with other national and international institutions, as well as a specialist service for industry.

Current research activities include fullerene thin films, single wall carbon nanotubes, organic polymers for light applications, organic photopolymers for holography, ferroelectric crystals, liquid crystal displays, defects in silicon wafers, whole cell studies in biological systems, photoactive donor-acceptor complexes, detection systems for pollutants, chiral selection of amino acids using cyclodextrins, and self-assembled monolayers for detection of biomolecules.

Contact:
Dr Hugh Byrne or Dr Mary McNamara,
Schools of Physics and Chemistry, DIT;
E-mail: hugh.byrne@dit.ie, mmcnamara@dit.ie

Centre for Industrial and Engineering Optics

The Centre for Industrial and Engineering Optics in the School of Physics, DIT, specialises in optical solutions to engineering problems, and provides technical services to industry. Activities include applied research, practical problem solving, and system design and consultation services.

The Centre develops systems for a range of optical non-destructive, non-contact measurements in a variety of industrial applications. These include surface profiling and surface roughness measurement by optical interference methods, measurement of surface displacements for stress/strain analysis, and defect detection using Electronic Speckle Pattern Interferometry (ESPI).

ESPI is a PC controlled laser interferometric technique for measuring small displacements of rough surfaces. It is not limited by the shape of surface under test and can be used to examine complex shapes in moulded plastics, as well as regularly shaped objects such as blocks of masonry, metal tanks and pressure vessels. A recent project to analyse building materials, in collaboration with DIT's Department of Engineering Technology, compared results using the Centre's ESPI system and the strain gauge technique. The results show that, whereas strain gauges can measure strain at specific points on a surface, ESPI has microstrain sensitivity over the whole surface. It is a non-contact method and can be used to examine delicate surfaces.

The image shows the ESPI patterns on a sandstone block subjected to a compressive load. The bright and dark fringes are contours of equal displacement of the surface. A continuous pattern of this type would indicate a smooth distribution of strain. The cracks can be seen as discontinuities in the fringe pattern. The cracks themselves were not yet visible to the naked eye.

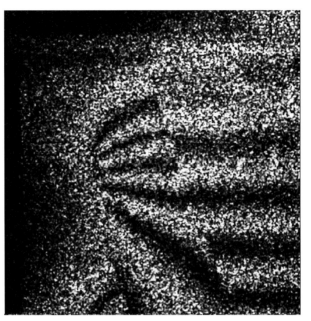

Growth of a crack in a sandstone block.

Contact; Drs Suzanne Martin & Clodagh Feely, Industrial and Engineering Optics, DIT; E-mail: suzanne.martin@dit.ie, clodagh.feely@dit.ie

Optical sensor research

Dr James Walsh and his group in the School of Physics at Dublin Institute of Technology are doing optical sensor research and development for a range of applications. An example is a novel fibre-optic sensor system for in-vitro and in-vivo analysis of mitochondrial redox reactions. The work of optimising the sensor's spectrometer is underway.

Cellular respiration involves the oxidation of pyruvate to carbon dioxide and water to generate cellular energy in the form of ATP. The energy is generated by the passage of electrons via Fe^{+++} and Cu^{+++} ions, FeS clusters, and a series of organic electronic carriers. Measurement of the redox kinetics of these components provides information on the respiratory electron transport.

The state of oxidation of respiratory substrates may be studied by measuring the absorbance difference, over the range 250-650 nm, between fully oxidised and fully reduced mitochondria. But the samples are very small, highly scattering and turbid. They need arrangements to maximise the incident and transmitted light. The system will allow study of cellular changes in-vitro and in-vivo.

The project is in collaboration with Professor Matt Harmey, Botany (UCD), and Drs Michael Farrell and Orla Hardiman, Pathology (Beaumont Hospital).

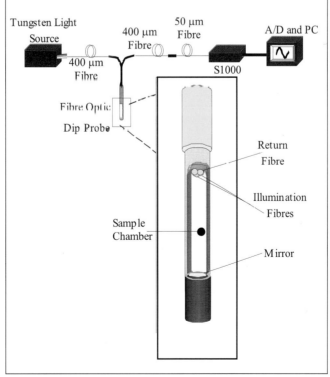

Redox microspectrophotometer system with detail of fibre-optic front-end probe.

Contact: Dr James Walsh, School of Physics, DIT; E-mail: james.walsh@dit.ie

Power estimation at high levels of integrated circuit design

Until recently, area and speed were the main factors of concern to integrated circuit (IC) designers. Portable devices – notebooks, cellular phones – were the only applications where the power consumption was of major interest. Higher integration densities and increased speed have led to real time applications. But the power consumption is directly proportional to the speed of computation. Thus, it is vital to monitor the power consumption at the earliest stages of the design cycle.

The Very Large Scale Integration (VLSI) research group at DIT has developed a tool, PowerCount[1,2], capable of estimating the power consumption at the circuit description level. This tool has several advantages over existing tools. It is fully incorporated into the normal IC design cycle. Only one additional file is needed for the computation of the power consumption, and it can be generated automatically within the design environment in seconds. PowerCount uses a simple user-

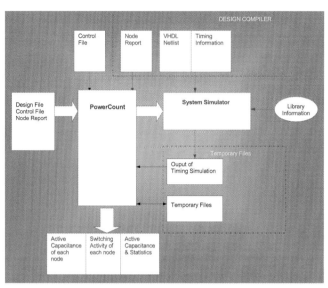

PowerCount in the IC design environment.

friendly interface. It allows the testing of the design at the initial stage when a redesign is easily done, and also a fast simulation time. Better productivity and a less costly design process result.

In the second phase of development of PowerCount, very large designs can be simulated within hours. For example, an image processing chip with more than 1100 nodes was tested in less than eight hours, 25% better than the previous version.

References:

1 A.Th. Schwarzbacher, P.A. Comiskey & J.B. Foley, "PowerCount: measuring the power at the VHDL netlist level", *Electronic Devices and Systems Conference*, Brno, Czech Republic, June 1998.

2 A.Th. Schwarzbacher, P.A. Comiskey & J.B. Foley, "High level power estimation with PowerCount", *Irish Systems & Signals Conference*, Dublin, June 1998.

Contact: A.Th. Schwarzbacher,
School of Electronic & Communications Engineering, DIT;
E-mail: aschwartzbacher@electronics.dit.ie

Identification and control of delayed systems

A delay in a system is the time interval between the start of an event at one point in a system and its resulting action at another point in the system. Delays arise in physical, chemical, biological and economic systems: for example, a delay of approximately 0.2 seconds arises in the light reflex action of the pupil of the human eye.

The research carried out in Dublin Institute of Technology, Kevin Street, looks at both the identification and automatic control of processes with delays. In identification, research has been carried out into open loop identification in both the time and frequency domains, and into closed loop identification in the time domain. In automatic control, the use of dedicated delay compensators has been researched in detail. Research group members are Ms Morena Stolfa, Ms Ruiyao Gao, Dr Susan Carr and Dr Aidan O'Dwyer. Current and future research activities include:

- Identification and Proportional Integral Derivative (PID) autotuning control of delayed processes, in continuous time.
- Identification and time delay compensation of delayed processes, in discrete time.
- Robust controller design for delay systems using Linear Quadratic Gaussian and control techniques.
- Analytical determination of performance and robustness criteria for PI and PID control of delayed processes.
- Performance and robustness issues in the use of time delay compensator strategies.

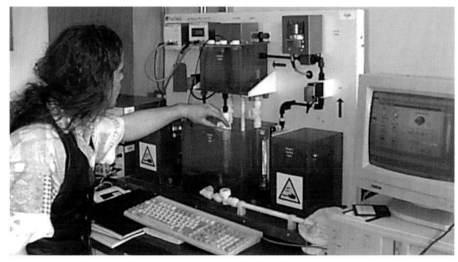

Controlling the pH system - Jim Condron adjusts a valve.

Acknowledgement: The author wishes to express his appreciation of the support provided for the above projects by the seed, strategic research and scholarship funding schemes of the Dublin Institute of Technology.

For further information, please contact:
Dr Aidan O'Dwyer,
School of Control Systems and Electrical Engineering,
Dublin Institute of Technology, Kevin Street, Dublin 8;
E-mail: aodwyer@dit.ie

Making sense of water pollution

Research in the area of infrared optical sensors began in LIT in 1997. Preliminary work funded by the Forbairt Strategic Research Programme, 1997, and Graduate Training Programmes, 1997 and 1998, has led to a number of advances in the environmental sensing area.

Chlorinated hydrocarbons are major pollutants of ground, surface and drinking water. Their origin can be traced to their use in industrial plants, household cleansers and research laboratories. Research into optical sensing devices at LIT includes the development of a novel sparging (air purge) sampling system for determination of chlorinated hydrocarbons. The sparging method in use at LIT involves purging a sample with air to remove the analyte, and transfer of the analyte onto a polymer film. The polymer film is attached to an attenuated total reflection (ATR) element (a zinc selenide crystal) which is mounted in a Fourier transform infrared (FTIR) spectrometer. The polymer films are designed to enable selective and rapid enrichment of the analyte(s) of interest. Simultaneous analysis of six chlorinated compounds has been possible using a poly (isobutylene)-coated ATR element. These sensing devices are robust and reusable. This work is a collaborative venture with Waterford Institute of Technology.

Sparging set-up used by research student Ambrose Hayden, with ATR-FTIR spectroscopy.

Successful outcomes of the polymer-ATR work at LIT with on-going collaborative research with Dublin City University has led to development of fibre-optic sensing devices. These devices have shown the potential for determination of numerous chlorinated hydrocarbons simultaneously. The fibre-optic sensors will be tested in industrial waste streams for detection of alarm levels of chlorinated solvents.

Leakages in oil pipelines are hazards associated with storing, processing and transporting petroleum hydrocarbon products. Fast detection and location of leakages in installations are important to minimise the emission of hazardous chemicals into the aqueous environment.

Chemists at LIT are collaborating with physicist Dr J. Walsh, at Dublin Institute of Technology (DIT), to develop a modular infrared spectrometric instrument for determination of hydrocarbons. An interdisciplinary program of research will investigate the design and development of a sensing instrument, which can be used for analysis of petroleum hydrocarbons in the environment. The proposed sensing system will consist of an optical fibre configuration which will be suitable for *in-situ* analysis. Special emphasis will be placed on development of sensors for groundwater protection, due to increased vulnerability of groundwater in the vicinity of underground petrol tank facilities. The interdisciplinary nature of the research will mean that the sensing system will be characterised at LIT, while development of the modular instrument will take place in DIT.

For further information, contact:
Dr Fiona Regan, Department of Applied Science, LIT; Tel: 061-208333;
E-mail: fiona.regan@lit.ie

A framework for migrating to distributed object computing

In computing terms, heterogeneous environments have quickly become the norm rather than the exception. This gives IT managers major headaches and poses serious dangers to software interoperability. The dawn of the World Wide Web as a tool for electronic commerce has left many organisations and their legacy code as mere onlookers to this phenomenon. Many organisations can neither afford to redevelop legacy systems or afford to have them isolated from everyday use.

Organisations who find themselves in this quandary may well find comfort in a M.Sc. project that is currently being undertaken at Limerick Institute of Technology. The cornerstone of this project will be the development of a framework, which will map the transition from legacy applications to distributed objects. The project examines software interoperability and, in particular, how an organisation can migrate from a legacy system to a distributed object environment with minimal reengineering, which can be achieved by use of CORBA (Common Object Request Broker Architecture).

The project is broken into two strands. Strand one is the development of a framework, which will outline how an organisation can integrate its existing base of installed software (legacy systems) with newer distributed objects. Up to a number of years ago, this would have required a significant amount of reengineering. But nowadays, with the help of technologies such as CORBA, disparate systems can be integrated seamlessly, and in a cost and time efficient manner. The framework will embrace topics such as technology transfer, reengineering, systems integration, object wrapping, data migration, etc. The framework when completed will, like CORBA, not be tied to any particular operating system or programming language.

The project is actively supported by two companies who have vast experience in the fields of systems integration, reengineering, and systems development using middleware.

The second strand is the development of software. This demonstration software will support the findings of strand one of the project. The software will take form of a Java applet located on a web server communicating with a COBOL program located on the College's AS/400. Both programs will swap parameters via a CORBA architecture. The ORB used in the development will be OrbixWeb by Iona Technologies.

All told, this project should be of benefit to most IT organisations who wish to integrate existing software with newer distributed object systems. The approaches towards software evolution are changing at a phenomenal pace, and two approaches towards its evolution seem to be in decline. Firstly, it is not feasible to completely replace legacy systems and start from "scratch" and, secondly, it is becoming increasingly undesirable to continually reengineer legacy systems in the hope that they will turn into a maintainable asset. However, a middle ground does exist, and it is the aim of this project to examine and outline this "middle ground".

Contact: E-mail: brendan.watson@lit.ie

Marine microalgae as a source of ω3 fatty acids

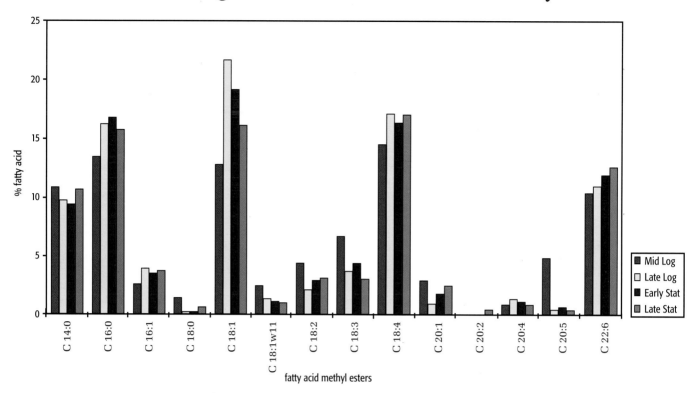

Fatty acid profile of Isochrysis galbana *clone T. ISO at log and stationary phases.*

Marine microalgae, or phytoplankton, provide the food base which supports the entire animal population of the open sea. Most algal classes are represented in ocean populations: the best known are the diatoms (Bacillariophyta), the dinoflagellates (Dinophyta), the green algae (Chlorophyta) and the blue-green algae (Cyanophyta).

Microalgae contain a wide range of fatty acids in their lipids. Of particular importance is the presence of significant quantities of the essential polyunsaturated fatty acids, ω6-linoleic acid (C18:2) and ω3-linolenic acid (C18:3), and the highly polyunsaturated ω3 fatty acids, octadecatetraenoic acid (C18:4), eicosapentaenoic acid (EPA, C20:5) and docosahexaenoic acid (DHA, C22:6).

Three species of marine microalgae – *Skeletonema costatum*, a diatom, and two flagellates, *Isochrysis galbana*, and *Tetraselmis suecica* are widely used as food for growing larvae in shellfish hatcheries along the Irish coast. The combination of the three organisms ensures a balanced supply of the ω3 and ω6 fatty acids for healthy growth and survival of the larvae, and several studies have shown that the critical factor in this diet is the content of EPA and DHA. Our current work in the Department of Life Sciences (GMIT) is concerned with the qualitative and quantitative analysis of the fatty acids of these microalgae, as no detailed study of the fatty acid composition of the indigenous organisms has been carried out to date.

STRUCTURAL FORMULAE OF ω3 FATTY ACIDS
(Note: C18 : 3 signifies 18 carbon atoms and 3 double bonds)

Formula	Code	Name
$CH_3\text{-}CH_2\text{-}(CH=CH\text{-}CH_2)_3\text{-}(CH_2)_6\text{-}COOH$	C18 : 3	Linolenic acid
$CH_3\text{-}CH_2\text{-}(CH=CH\text{-}CH_2)_4\text{-}(CH_2)_3\text{-}COOH$	C18 : 4	Octadecatetraenoic
$CH_3\text{-}CH_2\text{-}(CH=CH\text{-}CH_2)_5\text{-}(CH_2)_2\text{-}COOH$	C20 : 5	Eicosapentaenoic acid
$CH_3\text{-}CH_2\text{-}(CH=CH\text{-}CH_2)_6\text{-}CH_2\text{-}COOH$	C22 : 6	Docosahexaenoic acid

The microalgae are the primary source of EPA and DHA. Humans obtain their supplies indirectly by the consumption of oily fish in their diets, and there are also numerous products, including the traditional cod liver oil, available in healthfood stores. But the intake of fish or fish products is regarded as being grossly inadequate in the current Irish diet: the low incidence of heart disease among Greenland Eskimos, Japanese and Icelanders is attributed to the high proportion of marine fish in their diet. There is growing evidence that EPA and DHA play an extremely important role in the prevention and treatment of atherosclerosis, thrombosis and high blood pressure, of inflammatory conditions such as asthma, arthritis, migraine headache and psoriasis, and of cancer such as breast, colon and prostate cancers.

The next phase in our studies will involve screening and selection of microalgae for their EPA and DHA content, and the manipulation of their growth conditions to maximise the production of these important ω3 fatty acids. Declining fish stocks and the anticipated increase in human demand dictate that an alternative source of EPA and DHA will have to be found. The microalgae would seem to be the ideal source to fill this gap.

Contact: ide.nifhaolain@gmit.ie

The periwinkle, a perfect model for biodiversity studies

Conserving biodiversity is a high priority but we do not know that much about it. No one can yet explain, for example, why some environments support a much wider range of living organisms than others. To make conservation efforts more effective, we need to know what diversity exists not only between species also within a species, and how this diversity is generated, maintained or lost and, therefore, at what level this diversity should be conserved.

Most studies of marine biodiversity concentrate on one organism, using it as a case study. To many people the periwinkle might not seem the most exciting of marine organisms, but we have chosen it because we feel that it may be the perfect model to study the dynamics of biodiversity on rocky shores. The rocky intertidal presents a linear habitat subdivided into a number of patches. The pronounced and compressed environmental gradient from fully marine at low water to fully terrestrial at high water provides a unique continuum of physically changing habitat. Such an environment creates the selection pressures necessary to generate a high level of biodiversity, and this is seen in the periwinkle at two levels. Firstly, within a species, individuals show plenty of variation in characteristics such as shell shape, shell colour and, believe it or not, penis shape. Secondly, the group that we study, comprising some nine species of littorinids, is evolving rapidly and will eventually generate new, separate species.

Our main objectives are to determine the mechanisms involved in the transition from polymorphic interbreeding populations to those of species status, and to elucidate how such changes can be distinguished from phenotype plasticity. Results from our research will be used to produce operational concepts of biodiversity that can be applied to marine ecosystems in general. The research is EU-funded under the Marine Science and Technology Programme (MAST III), and involves the Galway-Mayo Institute of Technology, the University of Leeds, the Royal Belgian Institute of Natural Sciences, and the University of the Azores. The project, AMBIOS, uses a wide range of techniques, which include analysis of shell morphology using both traditional measurements and computer-aided image analysis, supplemented by transplantation experiments in the field. Starch and polyacrylamide gel electrophoresis (SGE and PAGE) are applied to screen enzyme variation, while PAGE and isoelectric focusing are used to investigate non-enzymatic protein variation. Genetic variation is further assessed by Random Amplified Polymorphic DNA (RAPD), microsatellite markers, single strand conformational polymorphisms (SSCPs), and sequence analysis of calmodulin introns, nuclear rDNA (18S, 28S, ITS) and mitochondrial DNA genes. To provide a phenotypic estimate of biodiversity at the physiological level, thermal tolerance of selected populations is measured. An important aspect of the project is that these different techniques are being applied to the same samples collected from selected key sites. This exploits the expertise of our research groups to the full, and allows us to assess the efficacy of particular techniques in the measurement of biodiversity.

Figure 1. Littorina saxatilis and L. tenebrosa; scale bars 4mm and 10 mm.

Classical taxonomic "species" boundaries only describe a limited part of biodiversity, so conservation efforts that are restricted to this level may lead to a loss of evolutionary relevant variation present at the intraspecific level. This problem can be circumvented by applying the "Evolutionary Significant Unit" (ESU) concept to, for example, some of the microgeographic genetically differentiated forms of Littorina saxatilis. This species shows enormous variation in shell morphology to the extent that several well-differentiated forms are considered as distinct species. An example of this is the case of Littorina tenebrosa, a small (< 6 mm shell height) snail that lives in non tidal lagoons on permanently submerged macro algae *(Figure 1)*. Its habitat is unique in that all other littorinids being investigated by AMBIOS live on rocks and boulders subject to the daily flux of the tide. The habitat of L. tenebrosa effectively isolates it from shore populations of littorinids. It is not surprising that this snail has been accorded species status. However, our enzyme and DNA results show that, while there is significant genetic differentiation between L. tenebrosa and L. saxatilis, this is not large enough for L. tenebrosa to be considered as a separate species. However, by considering L. tenebrosa as an ESU – an entity that contributes substantially to the overall genetic diversity of the species L. saxatilis – will ensure that it has conservation value.

As an extension of the ESU concept, AMBIOS is also endeavouring to identify "Evolutionary Significant Areas" (ESAs) by focusing on particular habitats, or on isolated or peripheral populations. In Ireland, a good example of an ESA is the lagoon habitat of L. tenebrosa (Figure 2). The application of the ESA concept provides direct guidelines for conservation actions, and represents therefore a potentially very useful and practical tool for protecting biodiversity.

Figure 2. The natural rock lagoon at Golam Head, County Galway, is about one hectare in area, and is only in contact with the sea on spring tides and during storms.

Contact:
Dr Elizabeth Gosling,
School of Science,
Galway-Mayo Institute
of Technology;
Tel: 091-753161;
E-mail: egosling@aran.gmit.ie;
http://www.leeds.ac.uk/biology/europe/ambios.html

Sewage detection in seawater – the detector needs a boost

Coastal waters have been a repository for domestic sewage for decades. Such water poses a hazard to human health as a consequence of bathing and by consuming contaminated shellfish. Gut-derived pathogens are generally present in low numbers in seawater, making detection by standard microbiological assay unwieldy. Instead, microbiologists have developed methods to detect other microorganisms associated with the intestine and which exist in larger numbers. The presence of these "indicator" microorganisms acts as an alert system that the seawater may potentially harbour pathogenic microorganisms.

Escherichia coli is a bacterium which inhabits the intestinal tract of mammals, including humans. With cell densities reaching 10^9/gram of faeces, *E.coli* is considered to be readily detectable in faecally contaminated environments. It is the faecal indicator organism most commonly assayed for in putative sewage contaminated waters. Microbiological media have been developed to enhance the selective growth of *E.coli* from water samples while inhibiting the growth of indigenous organisms which would otherwise overwhelm the relatively few *E.coli* cells present and prevent their detection. In addition to selective media, the water samples are subjected to standardised processing protocols.

The standardised processing protocols can detect *E.coli* at levels of one cell per 100 ml test volume. Protocol validations, however, use metabolically active cells. Do these really mimic the physiological state of *E.coli* in seawater? Within any environment external to the mammalian intestine, *E.coli* is considered to be allochthonous - i.e. a transient occupant of an alien ecosystem. In coastal waters, factors including temperature, pH, protozoal predation, salinity level, lack of nutrients, light, and oxidative stress will influence the capacity of *E.coli* to grow on standard selective media. Comparisons of selective media counts with counts determined by intricate viable staining procedures indicate that as little as 1% of the viable *E.coli* numbers in seawater may be capable of growing on selective media!!!

Investigations are currently ongoing in the Department of Life Sciences, GMIT, to devise sample processing regimes and resuscitation media which could boost the percentage recovery of *E.coli* from seawater without compromising selectivity and differentiation. Factors being investigated include sample handling and storage, diluent composition, medium composition, nutrient levels,

Post-graduate research student Niamh Gray investigating novel E.coli *recovery media.*

nutrient ratios, antioxidant additives, osmoprotectants, radical scavengers, sonicated bacterial extract inclusions, and sample incubation parameters. In addition, biotypically distinguishable *E.coli* isolates are being studied to determine whether individual strains respond differently to stress exposure and resuscitation components. This would aid in devising protocols and media which could be universally applicable towards boosting the recovery of *E.coli* strains in general.

This project received funding from the Institutes of Technology Graduate Training Programme.

Contact: Dr Vincent Jennings or Dr Declan Maher, GMIT, Galway; Tel: 091-770555; Fax: 091-751107; E-mail: vjennings@aran.gmit.ie or dmaher@aran.gmit.ie

Meat quality assurance in GMIT

Over the past ten years a major research project has been underway in GMIT where a systematic analysis has been carried out on processed beef. This project has been jointly funded by Forbairt, with a substantial financial contribution from Rangeland Meats, providers of beefburger to Supermac's and McDonald's both at home and abroad.

The spoilage of burgers is very complex and the team, under the direction of Dr Mary Ui Mhuircheartaigh, set about investigating the factors which influence meat stability at freezer temperatures. Chemical, biochemical and microbiological processes can be involved in spoilage: the exact pathway to off flavour development depends on the contribution of each. Initial high levels of spoilage organisms (Microbacterium thermosphactum, pseudomonads and Lactobacilli) dramatically shorten the shelf life. They do not multiply at –18°C but contribute large quantities of lipases, peroxidases and hydroperoxidases, and proteases. The lipases and peroxidases continue to catalyse fat degrading reactions that contribute to taint, even at –18°C.

Fat in beef can be divided into two types – the triglycerides/neutral fat and the polar fat/phospholipids, a component of cell membranes. Typical burgers have a total fat content of 20%, of which 3-4% is polar lipid.

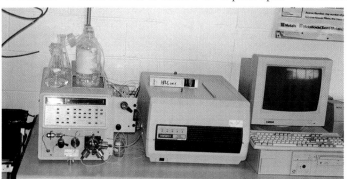

HPLC apparatus for testing the rancidity of meat.

The latter is present in all meat and is composed of relatively unsaturated fatty acids that readily undergo oxidation, producing aldehydes and ketones with "rancid" off flavours. Ironically, low fat meats develop taint much more rapidly than high fat products.

The expertise in GMIT across different scientific disciplines has been an essential factor in the success of this project. Fat extraction methods, spectrophotometric methods, distillation, gas liquid chromatography, high pressure liquid chromatography (HPLC), and Kjeldahl equipment have been of use in this study. The use of GC-Mass Spectrometry to identify the oxidation products is currently being investigated.

Prevention of lipid oxidation is best managed by dietary manipulation of the cattle feed before slaughter. If this is not possible, maximum attention to hygiene and fast processing is recommended. The addition of antioxidants in legally permissible levels can prolong the shelf life of cheaper burgers if added before the production of hydroperoxides.

This project, in addition to giving full-time and summer employment to our graduates, has helped develop staff skills, with downstream benefits to our undergraduates, at no cost to the Institute.

Contact: Dr Mary Ui Mhuircheartaigh, Galway-Mayo Institute of Technology, Dublin Road, Galway; Tel: 091-753161; Fax: 091-751107; E-mail: muimhuir@gmit.ie.

Mathematical modelling in science and engineering

It is a well known fact that many students studying science and engineering courses are finding some degree of difficulty with the mathematics module they are required to take. In some cases the students are unable to relate the mathematics they are studying to their chosen discipline and therefore loose confidence. In order to address this problem, we have examined the mathematical content of a number of science and engineering courses at GMIT. We then developed mathematical models which use computer graphics to help the students get a better understanding of the underlying mathematical theory. We feel the results of our findings will be of great benefit to all THIRD LEVEL institutions and therefore we summarise them below.

Production/Mechanical Engineering

(a) In order to make the learning of FOURIER series more enjoyable, we developed an analog to digital converter. This routine was written in Visual Basic and uses Microsoft Excel as the backdrop for visual graphics. Each harmonic could be drawn on-screen and the student could readily interface with output by simply clicking a command button.

(b) The application of LAPLACE transforms to the solution of differential equations lends itself well to a graphics treatment. We developed a number of models of vibrations which allows the student to easily change boundary conditions and see a new plot of the solution.

(c) 3D parametric mapping was easy to model and it allows the student to easily input new data to the matrices and re-plot the new image.

Biology and Physics

(a) The natural model to develop here was population growth and decay. We therefore focussed on the single module model for fish and were able to easily demonstrate to the student such factors as saturation coefficient, equilibrium stage L, and harvesting factors. The model was based on a first order differential equation with variables separable, and again Visual Basic was employed to allow for user friendliness.

(b) In physics we chose models of particles on springs, impulse and momentum which allowed the student to see the mathematics in action.

(c) Transformation of matrices (rotation, translation and scaling) allowed the student to see why the order of matric multiplication was important.

Conclusion

The models can all be used with Microsoft Excel 97 with Visual Basic as the driver. Most students use Microsoft Office and therefore could easily use these models. We are developing more applications in other areas of mathematics which we hope will benefit other course structures.

Contact: Dr Robert Loughnane; E-mail: robert.loughnane@gmit.ie

---

Minimax & Applications

The minimax (or maximin) criterion is a widely used criterion for parameter optimization and more generally as a solution criterion for problems in many areas of pure and applied mathematics.

The criterion is best introduced through an example: We consider a wager on a race with n runners at winning odds $r_i:1$, $1 \leq i \leq n$. We are allowed to stake a total of £1 in any combination of win bets. Thus any possible bet is represented by a vector $u=\{u_1,u_2...u_n\}$, $0 \leq u_i \leq 1$ for $i=1$ to n where $\Sigma u_i = 1$. If $f(i,u)$ is the net loss when horse i wins with bet u, the minimax criterion chooses the bet u which minimizes the maximum losses – i.e. we consider for each bet u the worst possible result $\max_i f(i,u) = \hat{f}(u)$ and then seek the bet u which minimizes this. Thus we seek

$$\min_u \hat{f}(u) = \min_u \max_i f(i,u)$$

In general an "optimal parameter" value u_0 is sought which minimizes

$$\hat{f}(u) = \max_{x \in X} f(x,u)$$

where u ranges over a "parameter" space U, a convex subset of a Banach space, and X is a compact "index" set. Thus we seek u_0 for which a

$$\min_{u \in U} \max_{x \in X} f(x,u) = \hat{f}(u_0)$$

is achieved. The theoretical results concerning the minimax solution, such as the Von Neumann's saddle-point theorem, rely on stringent global conditions of convexity and/or concavity being imposed on f and assume a saddle point solution as in *Figure (i)* ($X=[-1,1]$, $U=[-1,1]$) which, as *Figure (ii)*, showing two anti-inclined ridges, demonstrates, is generally not necessary. In both cases the function $\hat{f}(u)$ is obtained as the upper envelope of the f surface in three dimensions when viewed from the positive x axis direction, and thus is minimized in both cases at $u=0$ with $\hat{f}(0) = 0.5$.

Under quite lax regularity assumptions the author has produced general necessary and sufficient criteria for the existence of a minimax solution.

These criteria have been in turn successfully applied to a range of problems to provide alternative proofs of some well known results such as Von-Neumann's Minimax Theorem of Game Theory, Helly's Theorem and an extension Kuhn Tucker's nonlinear optimization results. They are also used to extend and unify the First Frobenius Theorem for positive matrices and Jentzsch's Theorem for positive integral operators, while also providing a new short proof of the foundation existence result for Jentzsch's Theorem. They further provide a new generic theorem of Chebyshev approximation theory which unifies many existing results in that area. Finally, they can be used to provide a sharp lower bound of the second derivative of a function of one variable on a finite interval, and to prove the existence of a weight matrix in robust regression.

Contact: e-mail: mfitzger@tinet.ie

Analytical probes for the future

Antibodies were first discovered at the turn of the century. They are proteins produced by the body's immune system in response to foreign material entering the body. They elicit their protection through selective binding to the offending material, facilitating its neutralisation and destruction.

These abilities led Paul Ehrlich to coin the term "magic bullet" with regard to the potential ability of antibodies to selectively target disease causing molecules/cells. This specificity in recognition can also be exploited in the use of antibodies as powerful analytical and/or diagnostic reagents - where they have made enormous contributions to modern biology and medicine.

Developments, particularly in the 1970s, led to an explosion of interest in antibodies and potential applications when techniques to enable large-scale production of reproducible antibodies were first developed. Today, attention is focussed on genetic methods for the generation of recombinant antibodies.

As part of our research at IT Tallaght, we have been developing recombinant ScFv (single chain variable fragment) antibodies with specificity for astrocyte cell surface domains, and with possible applications in the analysis of controlling factors for cell division.

The growth and division of normal cells within the body is a highly regulated process. Within tissues, cells will normally only divide to occupy the available space. They then undergo a process known as contact inhibition that results in the cells entering a condition of stasis. The biochemical processes involved in this are the subjects of intensive investigation, not least because some of these processes become defective in cancerous tissues and lead to the unregulated growth and proliferation of these tissues.

The same processes also apply to astrocytes, a major population of cells within the central nervous system. Astrocytes do not normally divide *in vivo* due to the presence of an auto-inhibitor. However, following injury to the central nervous system, astrocytes increase in size, number and fibrous appearance. Rapid call division is promoted by chemical triggers, which convert astrocytes from a resting to a reactivated state. The rapid growth of the astrocytes leads to a condition known as glial scarring, the effect of which is to block the regeneration of the pre-existing neural pathways in the central nervous system.

The inhibition of astrocyte proliferation is made by a variety of molecules, including a complex carbohydrate termed glycosidic mitogen inhibitor (GMI). Little is known of its structure, origin and mechanism of action. Knowledge in these areas may lead to a better understanding of growth control of astrocytes, provide a means to reduce/prevent nerve scarring, and facilitate the regeneration of neural pathways. *In vitro* studies also indicate that the inhibitor affects other rapidly dividing cells, including cancerous cell lines of the nervous system. Thus, an understanding of the control of glial cell growth may provide clues to the control of growth of certain brain tumours. Antibodies produced here have applications in the affinity isolation and purification of GMI(s), and the analysis of GMI function and mechanism of action.

Recombinant antibodies also have potential applications in the development of novel diagnostic reagents for clinical and veterinary practice.

Research conducted by Kevin Chestnutt under the supervision of Dr Ken Carroll and Dr Jeremy Bird (Institute of Technology, Sligo). Contact: E-mail: ken.carroll@it-tallaght.ie

Ultrasound propagation in wood – developing quality assessment systems

Wood is a material which is light, strong and easily worked. It is a natural material with a variability which adds greatly to its charm and beauty. However, from an engineering point of view, the variability of wood presents a major challenge. Great effort has been expended on developing testing and evaluation techniques for wood in order to ensure that particular specimens will be fit for their task.

In this work, the elastic structures of certain species of wood are examined using material wave or ultrasound propagation techniques. The aim is to produce a non-destructive evaluation of the quality of the wood being examined. Ultrasonic techniques have been found to work well when determining the quality of metals, crystals and man-made composites. However wood presents a much more complex challenge since it is a naturally grown material, not a manufactured one. It is also anisotropic and heterogeneous, yielding ultrasonic signals which are complex and often difficult to interpret.

Some of the interesting work of the group to date has shown how the annual ring structure of wood can have filtering effects on material waves passing through wood.

Researcher Michael Pedini carries out measurements in the forest.

This modelling and measurement of the influence of heterogeneity on ultrasound in wood represents an advance on previous work which considered wood as a homogeneous material. In other work, neural network classification techniques have been used to identify which species of wood a particular received signal comes from. The current project aims to extend this work to develop an assessment technique for assessing the quality of wood in standing trees. The potential benefits of such a technique include targeting of high quality stands of trees for intensive and expensive maintenance, as well as ensuring buyers and sellers of stands have a better picture of the quality of timber they are buying.

This work has been carried out with the financial support of Enterprise Ireland and Coillte. It has been carried out with the input of Dr D. Thompson (Coillte), Dr J. Evertsen (Enterprise Ireland), Dr R.C. Chivers (University of Surrey), Mr R. Jordan (IT Carlow) and Dr J. Keating (NUI Maynooth).

*Contact: Barry Feeney,
Science Department, IT Tallaght, Dublin 24;
E-mail: barry.feeney@it-tallaght.ie*

WATERFORD INSTITUTE OF TECHNOLOGY & LIMERICK INSTITUTE OF TECHNOLOGY   PETER MCLOUGHLIN (WIT), PADRAIG KIRWAN (WIT), RHONA HOWLEY (WIT), FIONA REGAN (LKIT) & AMBROSE HEYDEN (LKIT)

The diffusion of gases into polymer films

Regulatory requirements placed on industry require the rapid and sensitive determination of residual solvents in many areas. Residuals are quantified in occupational exposure studies and routinely quantified as part of US Food and Drug Administration regulation of pharmaceutical preparations. Therefore there is an increasing drive in research into rapid analysis techniques such as spectroscopic analysis, in particular infra red techniques.

At our laboratory in Waterford Institute of Technology we are investigating the use of a polymer modified Fourier Transform Infra Red (FTIR) method for the quantitation of gas phase solvent residues. Industrial partners (Clonmel Healthcare) and research partners at Limerick Institute of Technology (Dr Fiona Regan) support this work. The system design used in this experimental work is illustrated in *Figure 1*.

The purpose of this research is to use the advantages of ATR-FTIR spectroscopy,
* speed,
* specificity,
* non destructive, and
* limited sample preparation requirements,

coupled with the use of polymer films, to aid sensitivity and exclude unwanted background effects, to quantify residual solvents.

The approach to this research is twofold. Firstly the system design is optimised with respect to polymer type, cell design and flow. Secondly a mathematical model is used to determine a rate constant for the diffusion of selected solvents into a number of polymer films. The diffusion of small molecules into polymer films has been shown to follow a Fickian type mathematical model. The use of the diffusion model allows us to quantify system changes and define optimum polymer types for the analysis of selected classes of solvent species.

The use of polymer modified ATR FTIR is an important methodology for investigating the possible applicability of selected polymers for use with optical fibre techniques. Both ATR and fibre optic FTIR techniques are based on the internal reflection of electromagnetic radiation and use the evanescent wave phenomenon to obtain the absorbance spectrum of analyte species. Therefore the use of polymer modified ATR FTIR can be used to evaluate polymers for use with optical fibre materials in order to develop new sensor systems. A number of research groups have investigated the applicability of polymer modified silver halide fibres as a means of residual solvent analysis[1-3].

Work at WIT has specialised in the use of polymer modified ATR spectroscopy for the analysis of gas phase organic residues. To date the system designed has allowed rapid, sensitive and reproducible analysis of a number of commonly used chlorinated solvents[4].

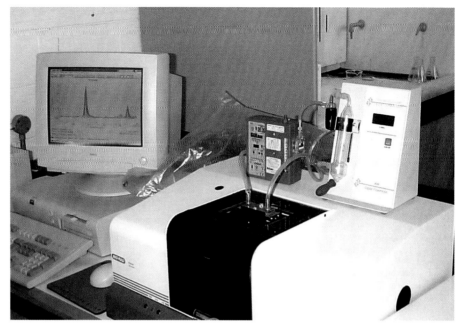

Figure 1: Polymer modified Attenuated Total Reflectance Fourier Transform Infra Red (ATR FTIR) system for residual gas phase analysis.

Diffusion into the polymer film is found to be rapid, reproducible and reversible. A typical diffusion profile is illustrated in *Figure 2*. A comparison of our experimental diffusion data to the model data is also illustrated in this diagram. It is possible to improve the sensitivity of this technique by the use of extended diffusion times.

The future of this research will involve further investigation into system design, as well as a study of novel materials for use as pre-concentrating ATR coatings.

For more information contact:
Peter McLoughlin,
Department of Chemical and Life Sciences,
Waterford Institute of Technology;
Tel: 051-302056; Fax: 051-378292;
E-mail: pmcloughlin@wit.ie

References
1. Walsh, J.E., MacCraith, B.D., Regan, F., Vos, J.G., Meaney, M., Lancia, A. & Artjushenko, S., Mid-infrared fibre sensor for the *in-situ* detection of chlorinated hydrocarbons, *Proceedings SPIE*, **233**, 2508, 1995.
2. Regan, F., MacCraith, B.D., Walsh, J.E., O'Dwyer, K.J.E., Vos, J.G. & Meaney, M., Novel Teflon-coated optical fibres for TCE determination using FTIR spectroscopy, *Vibrational Spectroscopy*, **14**, 239, 1997.
3. Krska, R., Rosenberg, E., Taga, K., & Kellner, R., *Applied Physics Letters*, **61** (15), 1778-1780, 1994.
4. Hayden, A., McLoughlin, P. & Regan F., Gas phase study of chlorinated organic compounds using polymer-coated ATR spectroscopy, *28th International Symposium on Environmental Analytical Chemistry*, 1998.

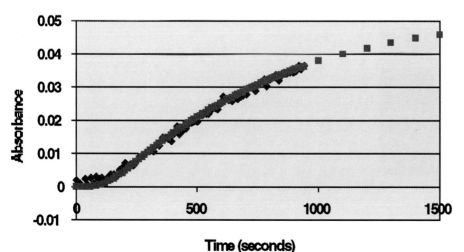

Figure 2: Theoretical and experimental results mapping the diffusion of tetrachloroethene into Polyisobutylene.

Institute of Technology, Tralee – a Centre for Applied Research and Product Development

The Chemical and Life Sciences Department of the Institute of Technology, Tralee (ITT), is a Centre for Research and Development on what can be loosely describe as the National Resources Area. Surrounded as we are in the southwest with what are substantial natural land and marine resources, it is perhaps not surprising that the Institute has found lots of opportunities to develop its expertise and research infrastructure in this area.

From research into environmental issues, waste management, food, agriculture and marine resources, the emphasis has been of an applied nature and often involves significant levels of product development and commercialisation. Product development and the industry focus of ITT activities are also reflected in its undergraduate courses, where for example a new BSc in Analytical Science with Product and Process Development has been launched.

Research undertaken by ITT is supported through a diverse range of grants, including Enterprise Ireland Applied Research Programme grants and the Marine Institute Marine Research Measures Awards.

Located close to the sea, in a region with significant marine resources by international standards, research into fisheries and aquaculture resources figures highly. Investigations of the importance and impact of fisheries, aquaculture and sea-angling resources have recently been undertaken. Development of aquaculture technology is included here, with two projects aimed at assisting salmon farming at sea. One project involves the design and development of a submerged fish cage system, while another research initiative will see the production of a devise to allow fish cage nets to be cleaned in situ, with major benefits to the environment and profitability.

Farming on the waves may become farming below the waves.

Marine algae resources represent a particular area of research where significant successes are being achieved. Kerry has a large coastline and extensive seaweed resources. Research into the extraction of alginates has lead to projects on the use of

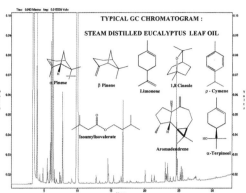

Waste from the foliage industry can be used as a source of valuable essential oils.

extracted gels in the development of hydro-seeding technology. This can be used to facilitate the recolonisation of disturbed and poor soil terrain with grasses to prevent erosion, and has a wide range of applications from restoring overgrazed mountain pastures to reinstatement of land works associated with mining and major road developments.

In Tralee, algal gels are being used in the development of food ingredients and also in the medical devices area, where they have uses in the treatment of bedsore, in prostheses and in cosmetic surgery. All these areas are leading, or already have led, to commercialisation.

The Institute has expertise in animal breeding programmes, particularly where it comes to aquatic species, and has been successful in getting funding for salmon breeding research. Another aquaculture species, the mussel, is also the subject of a study. The commercial value of mussel stocks in Tralee Bay is suffering because of the presence of a pest, known as pea-crab. The project will identify methods of relieving the negative impact of pea-crabs on the mussel stock.

As well as having significant marine potential, the southwest has also benefited from its unique climate, which has favoured the growth of a new industry, namely the production of foliage for the bouquet business. ITT has a number of applied research projects that are addressing the new industry's research priorities in the areas of propagation, breeding, cultivation, pest controls, quality and harvesting. Not only are cultivated varieties being studied but also new product development initiatives are being pursued using wild grown varieties such as Lodgepole Pine, Holly and Rhododendron.

The environment continues to be the focus of much attention in the area of applied research. Oil pollution control technology developed over the years is now being commercialised after earlier R&D identifying the oil absorption properties of waste bird feathers and a method of using feathers for oil spill control. Continuing the theme of environmental protection, the Institute is introducing the "Constructed Wetland Systems" approach to wastewater management. This follows years of R&D into biofiltration systems for waste-water from industry and domestic sources.

Wetland plants can control wastewater in an attractive arrangement of ponds.

The theme of pollution control is also the central theme to the R&D project on the identification of new locally sourced media for use in air bio-filtration used in the control of off-gases from waste-water treatment.

As well as developing food ingredient applications for seaweed extracts, applied research relevant to the food industry includes three other studies currently underway. The development of tests for freshness based on Histamine determination should yield a new rapid text kit that can be commercialised. In-process real time determination of glucose is being researched in a project where a new electrode sensor is being developed for the food industry. A project to find added value uses for yeast left over from brewing is underway also.

The Institute's successes, as well as being supported and facilitated by national funding sources, have grown substantially out of the co-operation with private sector enterprises which confidently provided the matching funds necessary to allow grants to be won.

For further information, contact:
Dr N. Mulligan; Tel: +353-(0)66-714-5600;
E-Mail: noel.mulligan@ittralee.ie

INSTITUTE OF TECHNOLOGY, TRALEE — MARTIN MCINTYRE & MICHAEL HALL

The SHaPE Centre - tomorrow's sports stars today

The SHaPE centre is a Sport, Health and Performance Evaluation centre based at the Institute of Technology, Tralee. It specialises in exercise physiology, fitness testing and human performance, and is one of only three human performance laboratories in the country. The Centre was established three years ago and was primarily developed for student practicals and tutorials. More recently it was opened to the public, and has been in huge demand ever since.

The Centre is run by professional staff qualified in the areas of fitness testing and sports science. It posses a wide range of equipment, which allows the quantification of a number of variables in relation to an athlete's performance, and is in great demand from high performance athletes, clubs, teams and the general public. The centre also offers sports specific testing, such as research into the physiological profiles of elite Gaelic footballers.

The ITT's Senior Football Team completed a Sigerson Cup three-in-a-row in 1999. Football in the country, indeed, countrywide is very healthy. From a scientific perspective, however, there is a distinct lack of literature available on relevant physiology. Put simply, what are the physiological attributes of an elite Gaelic footballer?

Taking the Sigerson team as test subjects, several of whom are inter-county footballers, research, leading to an MSc, into Gaelic footballers is being carried out in the Institute. Excellence in sport these days requires planning, dedicated players, coaches and access to services that provide support and feedback on training, technique, diet as well as physiological and performance characteristics for athletes. At the SHaPE centre, the Gaelic footballers are being profiled by modern analytical equipment, which carries out:

- Respiratory-gas analysis during various levels of exercise on the bicycle ergometer, rowing ergometer, or treadmill
- Blood analysis (haematocrit, haemoglobin and lactate levels)
- Anthropometric analysis (body fat, flexibility, body weight, posture, force from isometric muscle contraction, reaction times, etc)
- Lung function analysis
- Pulse analysis during training sessions.

In obtaining a physiological profile of Gaelic football players, it may be possible to deduce which attributes are desirable for good players, how specific training exercises affect group or individual profiles, and how these attributes differ in other sports.

Other related topics of research at the SHaPE Centre include:
- Kinesiological (muscle mechanics) investigation of sit-up variations using surface EMG
- Postural deviations in college students
- Exercise-induced asthma
- Physical activity in primary school children
- Isokinetic strength analysis
- The use of isokinetic dynamometry in the assessment of hamstring/quadriceps strength ratios
- The use of isokinetic dynamometry in the establishment of prevalence and degree of injury
- Validation study on blood lactate analysers.

Jimmy McGuinness, 1999 ITT Captain and member of all three Sigerson winning teams.

For further information, contact the SHaPE Centre, Tel: +353-(0)66-714-5600 or E-mail: martin.mcintyre@ittralee.ie

---

DUBLIN CITY UNIVERSITY — KIERAN NOLAN & DERMOT DIAMOND

Molecular recognition and chemical sensors

Chemical sensors have become an ideal tool for performing real-time monitoring of both chemical and biochemical species. A chemical sensor incorporates the usage of various technologies from synthetic chemistry to surface analysis technologies. In order for a sensor to be of use it must have a particular selectivity to a specific species, without which it would not be able to differentiate between different chemical moieties. Such specificity can be achieved using the principles of molecular recognition, based upon the concept of a host molecule/guest molecule interaction. In designing host molecules, a series of criteria must be considered in order for effective binding to take place. The key criteria are:

1. The host must contain functionalities which can electrostatically attract a charged species.
2. The host must also have the appropriate preorganization which can selectively hold the guest molecule.
3. The binding event must be detectable by either spectral and/or electrochemical techniques.

Our group has concentrated much effort on developing new ion selective electrodes which use macrocycles known as calixarenes, which normally contain four phenolic groups. The calixarene macrocycle is shaped like a conical basket where it possesses a lower rim and an upper rim. The lower rim of the basket contains polar hydroxy groups while the upper rim normally contains non-polar hydrophobic tertiary butyl groups. It is the lower rim of the macrocycle which is of interest in host/guest interactions. The lower rim hydroxy groups are oriented in such a way that they can encapsulate various charged species via polar interactions.

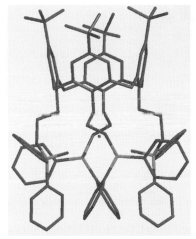

A 3-dimensional model of tetra(diphenylphosphineoxide)calix[4]arene.

Selectivity for specific host species can be achieved by selectively functionalizing the lower rim hydroxy groups. A series of various classes of functionalities (e.g. esters, amides, phosphines and phosphine oxides, etc.) have been introduced into the lower rim with each class of functionality demonstrating specificity for target cation hosts.

Of particular interest to our group are the phosphine oxide modified calixarenes (see Figure). We have successfully demonstrated their specificity for the actinides and lanthanide metals, and these compounds have now found application in the extraction of plutonium from nuclear wastes. Recently we have discovered that these compounds have excellent selectivity toward both lead and mercury cations, making possible the development of new sensors for the online detection of these environmentally hazardous pollutants.

Our attention is now focussing on the synthetic development of new macrocyclic receptors for the specific detection of phosphate and nitrate anions which are the pollutants responsible for algae blooms that lead to the eutrophication of natural water systems in Ireland.

Contact: Dr Kieran Nolan; Tel: 01-704-5610; E-mail: nolan_kieran@hotmail.com

Rapid antibody-based analytical methods: tools for detection of food, drug and environmental contaminants

The utility of antibodies for diagnostic and therapeutic applications depends primarily on their affinity, kinetic and stability properties. We employ Enzyme Linked Immunosorbent Assays (ELISA) and state-of-the-art biosensing technology to select these molecules, define their characteristics and develop novel diagnostic assays. Our current areas of interest are food and environmental contaminants and development of novel assays for illegal drugs. This research has wide applications, at present, considering the media coverage given to outbreaks of *E. coli* 0157 and *Listeria*, overuse of pesticides and antibiotic residues, and the wide availability of illegal drugs in Irish society.

The ELISA technique is based on the premise that antibodies specifically bind to their target antigens. This interaction allows the development of quantitative and qualitative assays for target compounds. In comparison to analytical techniques such as High Performance Liquid Chromatography (HPLC), ELISA offers rapid analysis without the need for expensive equipment and highly-skilled personnel. In addition, portable formats can be developed such as the commonly available pregnancy tests.

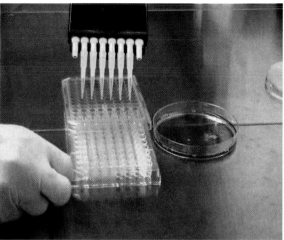
Production of a monoclonal cell line.

molecular weight affinity recognition molecules.

At present we are collaborating on a number of European funded research projects between industrial and academic partners. The FAIR project aims to develop rapid immunoassays for the detection of pesticides in wheat and antibiotics in milk. Since its initiation in 1996, a number of assays have been developed for commonly used pesticides such as Chloropyrifos, and antibiotics such as Cephalexin. In addition, broad-specificity antibodies have been developed which can identify a number of pesticides and antibiotics simultaneously. The resultant assays have broad commercial applications as they offer a rapid, sensitive detection method for these contaminants.

Another project funded by the INCO COPERNICUS programme has the overall objective of developing rapid, specific test methods for the detection of *Listeria monocytogenes* in foods. Specific antibodies have been produced by raising them against multi-antigenic peptides (MAPs) selected from a pathogenicity protein unique to *L. monocytogenes*. The binding characteristics of these antibodies to the MAPs and whole *Listeria* cells will be evaluated. Antibodies possessing suitable binding characteristics will be employed in the development of novel biosensor-based immunoassays.

Detection of illegal drugs has primarily focused on analysis of confiscated material. It is becoming increasingly important to be able to detect the presence of drugs in the workplace and in motorists, as performance may be seriously affected by recreational drug use. We are currently involved in research projects aimed at drug detection in these circumstances. In conjunction with a number of research groups in DCU, led by Prof. D. Diamond and Prof. B. McCraith, we are developing an assay for Heroin. In addition, we are collaborating in an EU funded project to detect Amphetamines, Cocaine and other drugs. The concentration of drugs in saliva closely correlates the concentration in blood and is not as laborious to collect and analyse as blood or urine. The final format will allow road-side testing of suspected drug abusers, as the project involves development of a novel saliva collector and rapid assays specific for each drug.

As part of an Enterprise Ireland Strategic Research Grant we are developing assays for illegal hormones in meat, in conjunction with Enfer Scientific. One of the main limitations of illegal hormone detection is the ease with which the parent compound can be altered to develop equally effective secondary compounds. Our project aims to develop antibodies to the basic structure of commonly used illegal hormones such that all secondary structures can be detected. At present it is necessary to develop novel assays for each new secondary structure, as it becomes available.

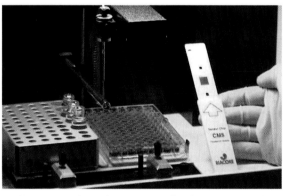
The autosampler and CM5 sensor chip: key features of the Biacore 1000 System.

"Real-time" biomolecular interaction analysis (BIA) using the BIACORE™ system enables us to rapidly optimise conventional immunoassay formats and to develop novel analytical strategies. Immunoassay development using BIA technology obeys many of the principles applicable to ELISA, but does not require labelling of the antibody or the antigen. Moreover, the biointerface can be regenerated using pH shock, thus allowing the ligand-coated sensing element to be used repeatedly. Our biosensor development program, funded by Enterprise Ireland, has successfully produced new "real-time" biosensing devices. In collaboration with Élan Pharmaceutical Technologies we aim to further develop these technologies to yield a highly sensitive biosensor for direct monitoring of interactions involving low

Other research is focused on the development of analytical methods for the measurement of coumarin-derived compounds (e.g. aflatoxins and warfarins) and for their metabolites. This involves the use of HPLC, Capillary Electrophoresis and electrochemical approaches. The latter is carried out in conjunction with Prof. Malcolm Smyth and Dr Tony Killard at DCU. In addition, we have established a novel method for developing recombinant antibodies. These offer further advantages to our analysis, as they result in assays of comparable sensitivity and specificity to traditional monoclonal and polyclonal antibodies but can be developed more rapidly and cost effectively.

The vast experience and expertise garnered from these research projects is unique in Ireland. We are currently setting up a company that will offer a Research and Development facility to Irish companies requiring assay development and optimisation and also offer analytical services.

For details contact:
Research Director:
Prof. Richard O'Kennedy;
E-mail: okennedr@ccmail.dcu.ie;
Assay Development:
Dr Bernadette Manning;
E-mail: manningb@ccmail.dcu.ie;
Analytical Services: Dr John Quinn;
E-mail: johngquinn1@excite.com

Solitons in field theory

Solitons play an important role in a wide variety of phenomena. They appear in benign form as solitary waves in canals, or threaten the Japanese west coast as tsunamis, giant waves which can propagate over great distances. As optical solitons, they increase the capacity of fibres for telecommunications and, as sine-Gordon solitons, they occur in condensed matter. In all its manifestations, a soliton's hall mark is its stability.

Whereas the solitons mentioned so far travel in one space dimension, the solitons studied extensively in field theory during the last 25 years live in two or three space dimensions. Magnetic flux tubes in superconductors are one example of solitons in two space dimensions; magnetic bubbles in ferromagnets and vortices in He^3 and He^4 are others. These extended objects are very important for the performance of the corresponding materials. Not of practical, but of fundamental importance for our understanding of the universe, are cosmic strings and monopoles.

The stability of these extended objects is due to their nontrivial topology far away from the centre. Although the bulk of their energy is concentrated near the centre, the shape of the fields at infinity is very important. Their nontrivial shape can be explained in terms of a strip twisted by 180° and then glued together end-to-end. Such a strip cannot be deformed into an untwisted strip without tearing it. In a similar fashion, the fields of our extended objects are twisted at infinity. To unravel them would require an infinite amount of energy.

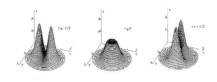

Energy density of Abrikosov vortices for $t = -1/2, 0, +1/2$.

Whereas topology explains the existence of solitons in field theory, nonlinear partial differential equations provide the proper mathematical framework for their description. The mathematical problem is to find solutions to the relevant equations which describe the properties and the behaviour of the extended objects. The Differential Equations Group of the School of Mathematical Sciences in DCU is engaged in this project as part of its wider research efforts.

In some models, by a suitable choice of coupling constants, the forces can be made to balance out. For magnetic flux tubes (Abrikosov vortices), this occurs at the borderline between type-I and type-II superconductivity. This makes it possible to look for static vortex solutions, which simplifies the mathematical problem considerably. Even so, no explicit solutions of the equations have been found. Recently we constructed these solutions in terms of a series. Using the terms of lower order, the energy density of two Abrikosov vortices can be calculated. It is plotted in the figure for three different values of the separation parameter.

Slowly moving vortices are at all times close to static configurations such as those depicted in the figure. This idea led to a mathematical technique, called the slow-motion approximation, for certain models with wave-like propagation. Within this approximation, the dynamics of solitons can be studied. The surprising result is 90° scattering, also depicted in the figure if we read it left-to-right as an evolution in time. In this scattering process, the extended objects merge in a head-on collision and then emerge on the sides, to the utter bewilderment of snooker players.

Contact: Jurgen.Burzlaff@dcu.ie

The chemistry of pure hydrogen

The hydrogen atom is the simplest atom, made up of just one proton and one electron. The hydrogen molecule H_2 is a more complicated object, but again very simple as far as molecules go. It would seem that in pure hydrogen there is really no proper chemistry to speak of. But nothing could be further from the truth. In fact when a plasma is formed in pure hydrogen, a fascinating world of chemical possibilities opens up.

A plasma contains electrically charged particles in addition to neutral atoms and molecules. It is actually energetic electrons that break up the molecules and ionize the gas. The electrons in a plasma are often very hot and the plasma is far from thermodynamic equilibrium. All sorts of exotic reactions occur and a hydrogen plasma is typically composed of three types of positive ion (H_3^+, H_2^+ and H^+), one type of negative ion (H^-), electrons, and atoms and molecules in a whole range of excited states. It is a formidable problem to keep track of all these species and to predict the plasma composition.

At the Plasma Research Laboratory in the School of Physical Sciences at DCU, a detailed model of hydrogen plasma chemistry has been developed. By averaging over time and space in a reaction volume, the model can determine the composition in a few seconds on a PC. This is a substantial advance on previous models and it allows us to incorporate many reactions. Seventy nine separate processes are included, and the only problem that stands in the way of making the model more complete is lack of fundamental data on collision cross sections and wall reaction rates.

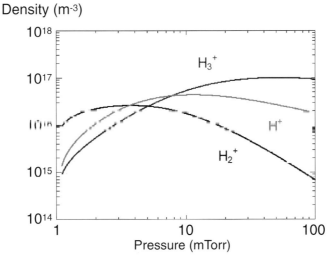

The concentration of positive ions in a hydrogen plasma.

The first applications for the new model are in predicting the production rates of negative hydrogen ions in the ion sources which are used for heating and re-fuelling plasmas in fusion experiments. Hydrogen plasmas are also used in many other applications. The highly reactive species produced in the plasma can gently remove corrosion from archeologically valuable objects, and countless other plasma cleaning applications exist. Adding a little methane (CH_4) can produce thin diamond-like films for wear resistant or biocompatible coatings. Adding a little silane (SiH_4) can produce films of silicon, such as the amorphous silicon films used in solar cells. There is no end to possible applications, but in each case it is important to understand the plasma chemistry. Our model is a valuable tool in this quest.

Contact: venderd@physics.dcu.ie

Creating a plasma

Plasma aided manufacturing encompasses a vast range of industrial applications. A plasma is essentially a gas containing electrons, ions and neutral particles. It is the presence of the charged particles that is the most important feature used in manufacturing.

An important issue in plasma processing is spatial uniformity. As processes increase in size, an even distribution of charged particles across the surface being processed is often difficult to achieve. It is therefore important to explore the physical processes that lead to the spatial structure of the discharge.

Charged particles are produced in several ways. The main production process in the argon plasmas we study is electron impact ionisation of a ground state atom. This is simply a collision between an electron and a neutral atom which results in a positive ion and two electrons. Plasmas used in important industrial applications are created in vacuum chambers by the application of electrical energy to a gas at low pressure. The plasma is in a "steady state" when the production rate of charged particles is the same as the loss rate. At low pressures, charged particles are lost mainly by diffusion through the gas to the chamber walls.

A variety of diagnostic tools have been developed and used by the Plasma Research Laboratory at DCU to investigate the fundamental physics of radio frequency plasmas. The most frequently used is a Langmuir probe, named after the scientist who coined the term "plasma". This probe allows spatially resolved measurement of the plasma parameters, such as electron and ion densities, and the electron energy distribution.

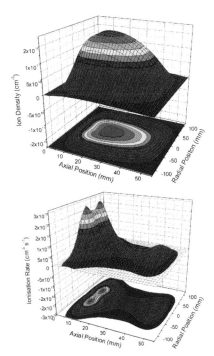

Measured density (top) and extracted ionisation profile (bottom) for an argon plasma at a pressure of 300 mTorr.

When the experimental data is compared with models and theoretical predictions of plasma behaviour, we can both understand the internal structure of the plasma and optimise the plasma conditions for use in industry. The Langmuir probe measures the density profiles in the plasma. This is important information in itself but, if we know the main production and loss processes, then we should be able to deduce precisely where the charged particles are generated.

A technique has been developed that extracts the two dimensional ionisation profile from experimentally determined density profiles. This technique can be extended to different gases or gas mixtures as well as to different chamber geometries. The ionisation profile for the plasma density profile shown in the *Figure* was determined by this procedure. Using this method, we are maximising the information obtained from our diagnostics in a way that enhances both our understanding and our ability to predict how plasmas will behave.

Contact: cmgd@physics.dcu.ie

Electronic Engineering research highlights at DCU

The School of Electronic Engineering at Dublin City University is heavily involved in cutting-edge research in a number of fields covering the entire spectrum of electronic engineering. With a faculty count of twenty four, together with a technical support staff of eight, and forty research postgraduate students, the School is one of the largest faculties in the Republic specifically devoted to electronic engineering.

Research activities fall into three broad categories: Telecommunications & Signal Processing (ten faculty), Microelectronics & Materials (four faculty) and Systems & Control (ten faculty). This work is supplemented by the presence of two government-sponsored Programme in Advanced Technology (PAT) centres: Teltec Ireland specializing in R&D for the telecommunications community, and PEI Technologies focussing on the development of instrumentation for the industrial sector.

A brief summary of some of our research highlights for the past year is given below:
- World's first use of synchrotron x-ray topography to evaluate the strain distribution in epitaxial lateral overgrowth of GaN on sapphire. GaN based materials are showing great potential for the development of blue and ultra-violet solid-state lasers for applications such as improved storage capabilities on DVDs.
- Silicon wafers, into which carefully controlled concentrations of impurities were introduced – known as defect engineering – have been examined for crystalline integrity using x-ray topography. This was part of a collaboration with researchers at Europe's premier particle science research centre (CERN) to evaluate detectors for the new Large Hadron Collider.
- Work on a new technique for completely characterising very short optical pulses (100 fs - 10 ps). The technique known as frequency resolved optical gating (FROG) combines the techniques of spectral and temporal analysis to obtain the complete electric field of an optical signal, and it is useful for designing high capacity optical communication systems.
- Development of a superior modelling strategy using artificial neural networks (ANNs) has resulted in the best results so far achieved for modelling the sun-spot cycle.
- The Control Systems Group was contracted by the ESB to develop a forecasting package for the National Control Centre to forecast 24-hour load profiles. This package will forecast up to three days ahead and will be important in the future deregulated Irish energy market.
- New low temperature production of thin films of beta-carbon nitride crystals (C_3N_4), a new synthetic material, which was postulated in the mid-eighties and predicted to have a hardness similar to diamond. Until now it was only possible to synthesise it at high temperatures.
- Establishment of the Performance Engineering Laboratory. This Laboratory is concerned with performance issues in telecommunications systems, including computer networks and software systems.

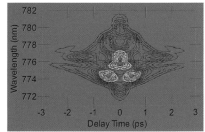

A FROG (frequency resolved optical grating) trace of an optical pulse.

Contact: info@eeng.dcu.ie

Dublin City University — Alan Smeaton, Sean Marlow & Noel Murphy

Indexing and retrieval of digital video

In four decades of video in the home, we have come from the ON/OFF button on the TV (we could choose *if* we watched) through multichannel TV (we could choose *what* we watched) to the VCR (we could choose *when* we watched). Now, with digital TV, low cost computing and gigabyte data storage, we are on the eve of the first major technological breakthrough in video access in two decades: we will be able to choose *how* we watch, and not just select, but *design* what we watch.

The digital video project at Dublin City University, sponsored by the National Software Directorate, is a research effort to develop technologies required for efficient management of video content. These include fully automatic video indexing processes that can break a TV programme, a movie, or an entire day's viewing into individual shots (a shot is a single piece of continuous video imaged by a single camera), and also include user interface features for efficient and easy video content navigation and browsing. Think of a movie or a news programme spread in front of you like a graphical web

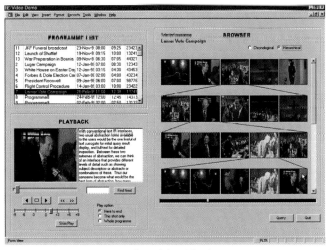
A screen shot of the video browser.

page. You can select which clips or which topics or which story you want to view.

Combining the expertise of the Multimedia Information Retrieval group in Computer Applications, and the Video Coding group in Electronic Engineering, the first objective of the project has been to produce a series of technology demonstrators of increasing sophistication which would act as nuclei for future product development. By combining recent DCU graduates working as R&D interns, with more traditional postgraduate and postdoctoral researchers, the project is succeeding in a second objective of the NSD sponsorship, which is to produce a pool of highly qualified R&D-ready talent with specialist knowledge and skills in this area.

A substantial difficulty for research in video indexing is the lack of a base-line against which to test algorithms. We have digitised eight hours of broadcast TV programming (18Gb of MPEG-compressed data) and individually logged *each* shot by when it starts and ends, how it breaks to a new shot (e.g. a cut, dissolve, wipe or fade) and by its content. We are building a demonstration system which selectively records, indexes and allows access to digital versions of broadcast TV programmes. This acts like a large communal VCR, but with vastly more powerful search and navigation than the traditional Fast Forward/Rewind buttons.

Contact: Prof. Alan Smeaton (01) 704-5262; http://lorca.compapp.dcu.ie/Video/

Dublin City University — Rory O'Connor & Tony Moynihan

Intelligent software project management

Due to the growing complexity of products and commercial systems, large software projects are facing more constraining production objectives in terms of time, cost, quality and risk. This evolution in the nature of projects being undertaken by software organisations results in increased difficulties associated with planning, managing and executing software development projects.

Software-intensive projects often fail because the project managers lack knowledge of good practices and effective processes which can reduce risk and increase the likelihood of success. Managers of projects need to know how to establish a set of processes which are tailored to a project's requirements in terms of functionality, time, cost, quality and their associated risks.

Although many project management software systems are currently available, the enormous scope and complexity of current software development means moving beyond the current state of practice, as such systems do little to support the "average" project manager. In addition, many of these systems fall short of supporting project managers in their decision-making processes and do not offer assistance in representing knowledge about plans and designs, or provide mechanisms for reasoning about plans and designs in a flexible manner. There are several areas in which an "intelligent project assistant" could be of benefit in assisting project managers in their decision making processes:

- The ability to reason about a project's plans, analyse alternatives and select the most suitable course of action.
- The ability to track the history of a project and capture knowledge gained, and reuse this knowledge as an aid to future project decision making.
- Assisting the project manager in adherence to standards, industry best practices and implementation of company policy.
- A facility to intelligently manage and analyse large amounts of project data.
- Suggestions are made which help the user balance cost, quality and time in making decisions about the use of project resources.

In the School of Computer Applications, DCU, the P3 project (Project and Process Prompter) is focused on addressing this

shortfall in existing tools, by providing an intelligent support tool which will increase the likelihood of success by helping project managers in their decision making, thus improving the current state-of-the-art in software project management tools.

The P3 project is funded by the Fourth Framework Programme of the European Commission as ESPRIT project 22241. It consists of five partner organisations - Dublin City University, Catalyst Software (Ireland), Intracom (Greece), Objectif Technologie (France) and Schneider Electric (France). The main deliverables of this project are a pre-commercial prototype decision support tool "Prompter" and a "Handbook and Training Guide".

Contact: Dr Rory O'Connor or Professor Tony Moynihan, School of Computer Applications, Dublin City University; E-mail: roconnor@compapp.dcu.ie or tonym@compapp.dcu.ie

Left: The P3 Project team. Front row: Philippe Boudot, Schneider Electric, Grenoble; Ioannis Nanakis, Intracom, Athens; Marty Sanders, Catalyst Software, Dublin; Annie Combelles, Objectif Technologie, Paris; Eamon Gaffney, DCU; Rory O'Connor, DCU. Back row: Tony Moynihan, DCU; Mark Johnston, Catalyst Software, Dublin; Robert Cochran, Catalyst Software, Dublin; Herve Le Corguille, Objectif Technologie, Paris.

Sensing knee injury

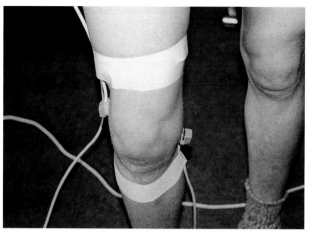

Knee ligament injury and associated instability is a common and serious clinical condition. A notable example is the cruciate deficient knee which affects many sporting enthusiasts.

Research currently being conducted in the School of Mechanical and Manufacturing Engineering at Dublin City University is focused on developing a reliable and clinically beneficial system for the successful diagnosis of the condition. At present, diagnosis is difficult without invasive surgery and is very often inaccurate. This project will attempt to map the micro-kinematics of the knee in normal patients and those with instability to identify specific patterns which can differentiate normal and abnormal knees. The system under development will incorporate a sensor system for measuring knee movement, and a computer interface for performing data capture and analysis.

This Knee Analysis System being developed seeks to overcome the limitations of existing systems by using the 3SPACE FASTRAK (from Polhemus Inc., a subsidiary of the McDonnell Douglas corporation) device and appropriate supporting software. The product line was originally developed for military applications (head up cockpit displays in jet fighters). The system is envisaged to provide a much greater degree of precision and accuracy than existing systems, while facilitating a wide range of exercise of the knee.

This product will be utilised to produce six degrees of freedom data from each of two sensors attached to one of the subject's knees. One sensor will be attached to the tibia and the other to the femur. The device operates by means of sensing a near magnetic field that is generated by a triad of electromagnetic coils within the fixed transmitter. The measurements will indicate the relative position and orientation of the femur with respect to the tibia. The measurements will be processed or analysed by the developed software, and various displays will be used to present characteristic information obtained from these measurements. This characteristic information presented by the application should then be of great use to a specialist in identifying abnormalities of the subject's knee. This system will be validated using trials conducted by a consultant orthopaedic surgeon.

Contact: Dr Brian McNamara,
School of Mechanical and Manufacturing Engineering, Dublin City University, Glasnevin, Dublin 9;
Tel: 01-704-5712;
E-Mail: brian.mcnamara@dcu.ie

The Centre for Health Promotion Studies, NUIG

The Centre for Health Promotion Studies at the Department of Health Promotion, National University of Ireland, Galway, was established around 1991 with the support of The Department of Health and Children under the academic direction of Professor Cecily Kelleher. Its primary aim is to be available as a national resource to conduct original health promotion research in an Irish context.

The Centre fulfils its mission through both research and consulting services, offering research skills and advice covering both qualitative and quantitative approaches across a wide range of activities, both at local and national level, in a variety of settings: lifestyle and nutrition surveillance, mental health promotion, workplace and occupational health, schools and young people, and social and primary care.

Three major projects being carried out exemplify the research currently undertaken by the Centre.

Funded by The Department of Health and Children are two comprehensive surveys of health related behaviours among adults, SLÁN, and school-going young people, HBSC. This comprehensive piece of research is the first ever undertaken in Ireland and was carried out in 1998.

MAIN FINDINGS OF THE SURVEYS:
- 33 per cent of 15-17 year olds are regular smokers
- 31 per cent of adults are regular or occasional smokers
- 27 per cent of men and 21 per cent of women take more than the recommended weekly units of sensible alcohol consumption
- 29 per cent of children report having had a drink in the last month
- 64 per cent of respondents reported consuming the recommended quantities of fruit and vegetables per day
- in all social classes, 35 per cent of 15-17 year old girls believe they need to lose weight
- one-third of those over 55 take no exercise in a regular week.

Professor Cecily Kelleher, Director of The Centre for Health Promotion Studies.

The Department of Justice, Equality and Law Reform has commissioned a similar survey on the general healthcare of the prisoner population. Its primary aim is to provide information in the planning and development of appropriate health policy, programmes and services in Irish prisons.

A study in the Agri-sector has been ongoing for the last couple of years, and its objectives are to develop and evaluate initiatives and to establish a model for health promotion practice in small enterprises and farming units in a rural setting.

The work of the Centre contributes and enriches the curricula of the Department with its substantial stratum of research, and it is guided by developing a research agenda that will integrate health promotion concepts into on-going empirical research.

Contact: Mary Cooke, Administrative Director, Centre for Health Promotion Studies, Block T, Distillery Road, Galway;
Tel: +353-91-750454(Direct); Fax: +353-91-750577;
E-mail: Mary.Cooke@nuigalway.ie;
Web Site: http://www.nuigalway.ie/hpr

Measuring water clarity from space

Has the water clarity of the Irish Sea deteriorated in the last few decades? This question has been posed by those involved in marine based activities.

A measure of the water clarity is the amount of suspended inorganic material (mineral suspended solids, MSS) in the water. Increased amounts of MSS alters light penetration in the surface waters affecting biological functioning, and causes the water to look "dirty" for people engaged in leisure activities in the coastal zone. The Martin Ryan Institute at NUI Galway, together with the University of Wales, Bangor, are investigating the problem through a project funded by the EU (Ireland/Wales) Interreg program, administered in Ireland by the Marine Institute.

To cover the whole Irish Sea region, the novel technique of measuring the water clarity from satellite sensors has been developed. The technique relies on the fact that the reflectance (R) of visible light increases with the amount of suspended sediment in the surface waters. The satellite sensor used is the Advanced Very High Resolution Radiometer (AVHRR), which has been in operation since the early eighties, allowing evidence for trends in Irish Sea sediment load over time to be assessed.

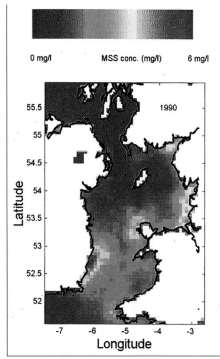

The highest values of suspended sediment in 1990 were found off Anglesey and St David's Head in Wales, and off the SE coast of Ireland in the region of the Arklow Banks.

Initially a relationship between R and MSS was found using field data collected on four research cruises – two using the Irish Research Vessel *Celtic Voyager*. To measure R at sea, an optical instrument was used which mimics the measurements made by the AVHRR.

Knowing the relationship between R and MSS concentration, the satellite images can be processed to make maps of sediment concentrations for the Irish Sea. The seasonal variation in MSS (about four times higher in winter than summer) is accounted for, and the final result is a map of annual mean MSS concentration based on a 6 km resolution grid (by averaging the individual 1 km resolution points). Between 40-60 images per year are used in the process and an example is shown in the *Figure* for 1990. The map shows that the highest values of suspended sediment are found off Anglesey and St David's Head in Wales, and off the SE coast on the Irish side of the Irish Sea in the region of the Arklow Banks. When all image processing is finished, it is hoped that trends in sediment load over time can be assessed, and the causes for any trend or changes in any particular year, established.

Contact: Martin White, MRI NUI, Galway; E-mail: martin.white@nuigalway.ie

Keeping an eye on the grass of the seas

Phytoplankton are microscopic organisms and the major form of plant life in the oceans. They are a key element in the global carbon cycle and are intimately connected with global warming and other climatic effects which may be caused by "greenhouse gasses". They also act as the "grass" of the seas and are thus the ultimate determinants of the size of fish stocks. Algal blooms can cause considerable environmental disruption and are a problem for the mariculture industry, destroying stocks or making them unfit for consumption. Studies in the south west of Ireland and elsewhere have shown that these blooms do not always grow *in situ* but may be moved into inshore areas from the open ocean by winds and currents.

Investigations in the Marine Microbiology section of the Martin Ryan Institute, NUI, Galway, supported under the EU's 4th Framework Programme, have focused on methods of monitoring the development and movement of phytoplankton populations. The seas and oceans represent a vast and constantly changing environment. Comprehensive monitoring of this environment by ships is difficult and costly.

The objective of the EU BIOCOLOR project

Lightfish is towed by vessels to provide details of surface optical properties over a wide area

is to provide a way of converting measurements of the optical properties of light in surface waters and that leaving the ocean surface to the types and quantity of phytoplankton which are present. In this way, optical instruments can provide measurements of phytoplankton populations, even in the absence of the obvious coloration of a severe algal bloom. Such instruments can be deployed in several ways. Satellites enable us to "look" at large areas rapidly and in considerable detail, optical moorings can be placed in critical areas to provide continuous measurements which are not curtailed by cloud cover, and sensors mounted on ferries or other "ships of opportunity" are a low-cost system with some of the advantages of both satellites and buoys.

In conjunction with the Marine Institute and several European partners, we are carrying out extensive field studies in the Baltic and the shelf seas to the south and west of Ireland. Comprehensive sampling of phytoplankton populations is coupled with determinations of the chemical, physical and optical properties of the water column. The latter is carried out by the use of satellite data and the deployment of optical instruments, both on buoys and from ships. The models presently being developed from this data will enable us to use optical methods to determine the distribution, quantity and type of phytoplankton in our seas with a higher level of precision than was previously possible. This capability provides an invaluable tool to assist in the sustainable development of marine resources.

For further information, contact: john.patching@nuigalway.ie or visit the Martin Ryan Institute website at http://mri.nuigalway.ie

MARTIN RYAN INSTITUTE, NATIONAL UNIVERSITY OF IRELAND, GALWAY STEFAN KRAAN & MIKE GUIRY

Growing a new sustainable sea-vegetable in Ireland

Intensive seaweed cultivation in China began in 1951 with the farming of an introduced kelp from Japan, Laminaria japonica (ribbon weed). In 1952, some 62 tonnes of kelp were produced, whereas in 1998 the harvest reached an incredible three million tonnes dry weight.

At present, large-scale seaweed mariculture is carried out only in Asia. Most cultivated seaweeds are grown as human food, although, notably in the Philippines, Taiwan and some African countries, seaweed is cultivated to be used for carrageenan and crude agar production. Cultivated species of global importance used in the food industry are Nori and different types of kelp (Laminaria japonica and Undaria pinnatifida).

A very popular sea-vegetable used in Asian cooking is the kelp Undaria pinnatifida or Wakame, with a harvest of over 450,000 and 500,000 tonnes in Japan and Korea, respectively. In Ireland we have a native Irish equivalent of Wakame, Alaria esculenta (literally, "edible wings"). Atlantic Wakame, as it is becoming known, has considerable potential as a foodstuff for human consumption for the home and export market, or as a sustainable, cheap and alternative source for protein, vitamins/minerals and biostimulants in animal feed for agriculture and aquaculture.

The Irish Seaweed Industry Organisation, in collaboration with the Martin Ryan Institute, and funded by the Marine Institute in Dublin via the Marine Research Measure, Operational Fund for Fisheries (1994-1999), a European Union research and development Programme, has started a cultivation project to grow and select strains of Alaria esculenta. Strains are crossed and seeded on a rope system using Chinese cultivation techniques. The microscopic gametophytes, part of the life cycle of A. esculenta, are sprayed on nylon twine, which, after appearance of small plantlets, are wrapped around a long-line and out planted in the wild. Using hybridization techniques, we have selected and established a premier-quality fast-growing, large and broad strain (see photograph) to be used in Irish seaweed aquaculture.

This is a further success story for the Irish Seaweed Industry Organisation in its quest for the enhancement of Irish seaweed aquaculture by the development of natural sustainable high-value-added sea vegetables.

Jim Morrissey of the Irish Seaweed Industry Organisation proudly shows the size of the field-grown Atlantic Wakame strain in comparison with the naturally occurring strain.

Contact: Stefan Kraan, MRI, NUI, Galway;
Tel: +353-91-524411 ext. 3199;
E-mail: stefan.kraan@seaweed.nuigalway.ie

MARTIN RYAN INSTITUTE, NATIONAL UNIVERSITY OF IRELAND, GALWAY BRENDAN BALL & ADRIAN LINNANE

Helping seabed life

Concern about the environmental impact of fishing has assumed a high profile in recent years. Commercial trawling along the seabed (demersal trawling) tends to affect the structure and composition of bottom-dwelling communities. In general, it appears that small, short-lived species are favoured, while longer-lived, larger species suffer the greatest damage and disturbance.

Scientists from the Department of Zoology Benthos Research Group and the Marine Fisheries Environment Unit of the Martin Ryan Institute (MRI), National University of Ireland, Galway, are heading up a 1.2 MECU multi-national EU-funded project to tackle this important problem. The project aims to reduce the adverse impact of demersal trawls on benthic marine organisms through changes in net design and alternative methods of stimulation. The Irish end of the work concentrates on otter trawls and is being conducted in association with personnel from the Marine Institute in Dublin, and in close co-operation with the fishing industry.

We believe that a unique rollerball modification system could be a solution to the environmental impact of otter trawls. While research has shown that they tend to have less impact than beam trawls, especially on soft grounds, reducing the degree of otter trawl contact with the seabed is still an important priority. The new rollerball

A unique rollerball modification system could be a solution to the environmental impact of otter trawls.

system has been devised by Mr James McDonnell of Gear Tech Ltd, Howth, Co. Dublin. It consists of swivelled rollerballs that run along the wings of the trawl. The rollerballs are size and number dependant and can be changed and modified on deck as required. A series of rollerballs are also to be incorporated on to the tickler chains. The aim of this design is to allow the trawl to move over, rather that through, the seabed. The net will also incorporate a square mesh panel of Ultracross knotless netting in the lower belly of the net, together with the possible inclusion of flotation aids to reduce net contact with the seabed.

As well as reducing mortality of benthic organisms, it is hoped that the modified design will also have a number of practical benefits for fishermen, including a reduction in hauling force, thereby decreasing the energy required to haul the net across the bottom, a reduction in sorting time, and a reduction of damage from debris to nets and to fish within the net. This new design, if successful, could become a new standard for trawl design throughout Europe.

For further information contact: Dr Brendan Ball;
E-mail: Brendan.Ball@nuigalway.ie; or Dr Adrian Linnane;
E-mail: Adrian.Linnane@nuigalway.ie

Laser Centre expands to meet new needs of Industry

The National Centre for Laser Applications (NCLA) is a centre of excellence in laser technology, that works closely with Irish industry on the development of new production tools and techniques based on laser and optical systems. NCLA laser systems can be used to drill, cut, mark, etch or weld a range of materials as diverse as glass, steel, ceramic or polymer with a high degree of precision.

The NCLA has worked with industry on projects at all levels of the R&D phase, from initial trials, through prototype fabrication to workstation design, assembly, commissioning and operator training. Recent industrial projects have seen laser processes based on polymer welding, miro-hole drilling and polymer etching installed into Irish manufacturing operations. For example, the NCLA recently concluded a project with CR Bard, a medical device manufacturer in Galway, on the development of a process for drilling tiny, accurately positioned holes in their angioplasty balloons with a precision unattainable with conventional techniques such as hot pin drilling. The project went from concept, through initial trials to installation and commissioning of a laser workstation in the space of 12 months. The system was subsequently transferred to the CR Bard manufacturing headquarters in New York where it is now in full production.

In response to the increasing product and process innovation now on-going in Irish manufacturing industry, and the increasingly diverse applications which clients are bringing to the NCLA, the Centre has expanded its laser facilities with the acquisition of two new state-of-the-art laser systems. The first is a compact, high-power diode laser with an integrated fibre-optic beam delivery system. This laser system will

Figure 1: Dr Richard Sherlock of the NCLA using a YAG laser system to cut steel.

be used in the development of laser welding and soldering applications. The second system is a UV excimer laser micromachining station with a range of features to allow rapid processing of micron-sized structures in a wide variety of materials. An example of the precision and cleanliness of this process can be seen in Figure 2, which shows a number of tiny holes drilled in a novel medical device for drug delivery.

These new laser systems will complement the existing laser facilities, which include high-power Nd:YAG and CO_2 laser systems with both galvo-driven scanning heads and static machining heads. These systems, integrated to CNC controllers and CAD/CAM software, can mark products directly with logos or serial numbers, can drill high aspect ratio holes, and can cut precise features in up to 5 mm thick stainless steel with a minimal heat affected zone. The NCLA now offers a range of high power lasers with operating wavelengths from the ultra-violet to the far infra-red, capable of processing the widest variety of materials. Laser based processes offer a number of clear advantages over conventional manufacturing processes; they are fast, efficient, they offer high precision, and are readily automated.

The NCLA also operates a number of laser and optics based characterisation laboratories featuring Raman and photoluminescent spectroscopy, and surface characterisation including atomic force microscopy (AFM) and white-light interferometry. Together these techniques give sub-nanometre resolution in lateral and vertical directions with a large range of measurable thin-film thicknesses and surface features.

Contact: Tony Flaherty,
National Centre for Laser Applications,
National University of Ireland, Galway;
Tel: +353-91-750364;
Fax: +353-91-525700;
E-mail: ncla@nuigalway.ie

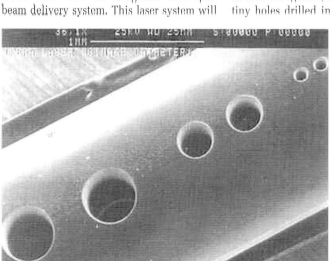

Figure 2: Small holes drilled in a medical device (catheter) using the excimer micromachining system.

Physics research at NUI Galway

The Department of Physics at NUI, Galway, has a wide variety of research activities – some specific projects are described below. Collaboration with other departments in NUI, Galway, and with other institutions in Ireland, Europe, and the USA, is very important. The main areas are laser spectroscopy and laser physics, astronomy and optical instrumentation, atmospheric and environmental physics, and, more recently, bio-physics and biological instrumentation.

Micro-Raman analysis is possible on solids, liquids, or gases, in microscopic quantities.

Laser Spectroscopy and the National Centre for Laser Applications (see previous page)

The National Centre for Laser Applications (NCLA), located in the Physics Department, specialises in the development of laser-based materials processing applications (welding, cutting, drilling, and marking), and laser-based materials analysis (spectroscopy, surface characterisation). Centre facilities include laser laboratories equipped with excimer, CO_2, Nd:YAG, dye, and high-power diode lasers, and a number of tuneable solid state sources. A primary aim of the Centre is to encourage and facilitate the use of optics and laser technology in manufacturing industry by providing laser-based manufacturing and research services, custom laser workstation development, and consultancy and training. Basic and applied research is on-going in the areas of laser-based materials surface modification, welding, drilling and marking. The Centre has a large client base in several sectors, including Medical Device & Healthcare, Microelectronics, Aerospace, Automotive, Pharmaceuticals, and Light Engineering. Using conventional high-power laser sources such as Nd:YAG and CO_2, Centre personnel have developed numerous applications for industry which are now in full production both in Ireland and abroad. The materials processing facilities have recently been expanded with the addition of a high-power fibre-delivered laser diode system and a dual-wavelength excimer laser micro-machining centre.

The Centre grew out of the activities of the laser spectroscopy group, and spectroscopic studies of doped insulators, low-dimensional structures in semiconductors, and thin films, are important elements of the Centre's current research programme. Much of this work is concentrated on GaInP semiconductor structures for Vertical Cavity Surface Emitting Lasers (VCSEL) and quantum dots (InP on GaAs substrates). Fluorescence, selective absorption, and lifetime measurements are carried out over a wide temperature range, using various laser sources for excitation. Other materials characterisation work includes research in Raman scattering, surface profilometry, and atomic force microscopy (AFM). Recent Raman scattering studies have centred on the detection of illicit narcotics: heroin, ecstasy, and cocaine have been detected and quantified in pure form and in mixtures, in concentrations as low as 10%. The technology is being developed to expand the range of drugs and to improve the sensitivity and specificity with which they can be detected. New miniaturised optical components are being used to develop a portable system for this application, which will also be used to measure doping concentrations and residual stresses in semiconductor wafers.

A separate **Biological Optics Laboratory** has also evolved from the laser spectroscopy group. This inter-disciplinary collaboration, between the Biochemistry and Physics Departments, studies biologically-active materials by optical techniques. Such non-invasive techniques, which are growing in importance in many areas of biology and medicine, permit rapid determination of diverse biological functions with minimal perturbation of the sample under study. Current interests of the group include the development and application of systems, based on laser fluorescent techniques, which can be used to monitor spatial variations of intracellular ion dynamics. Current projects include a study of the release of calcium in embryos following treatment with $Ins(1,4,5)P_3$.

Guide Star Adaptive Optics

By using adaptive optics (AO) on large telescopes, it has proved possible to remove much of the distorting effects of the atmosphere on astronomical observations – ground based observations can now be as good, or better, than Hubble images taken from space. However, in order to make corrections, a bright star needs to be in the field of view and this limits the technique to less than 0.1% of the sky.

The aim of laser guide star adaptive optics (LGS-AO) is to produce an artificial star anywhere in the sky, to be used as a point source for AO. We use a high powered laser tuned to 589.6 nm (sodium d2 line) which is scattered off naturally occurring neutral sodium atoms in the mesosphere at 90-120 km: this produces a fairly bright "artificial star" suitable for use as an AO guide star.

The 4W sodium LGS on the 3.5m telescope in Calar Alto (courtesy MPE, Garching).

NUI Galway is part of a Research Network under the Training & Mobility of Researchers programme in the IVth Framework, which studies the implementation of LGS-AO for large telescopes. We are specifically involved in monitoring mesospheric sodium in experiments at the Calar Alto Observatory (Spain) and at the Pierrelatte Nuclear Research Station (France).

Airborne Doppler Lidar Research

Atmospheric Doppler Lidar technology allows for remote measurement of wind speed by transmitting a short pulse of light into the atmosphere via a telescope, and measuring the Doppler shift in the light back-scattered by molecules and aerosol particles. Velocity accuracy better than 1 m/s can be achieved from a single pulse at ranges out to 5 km.

The European Commission funds research projects in the areas of aircraft safety and efficiency. The Applied Imaging Research Group (AIRG) is involved in one such project, which is developing an airborne, forward looking, pulsed Doppler Lidar system which will be capable of detecting turbulence hazards in the airspace in front of an aircraft. Among the hazards which this system aims to detect are wind shear and wake vortices.

Wind shear is a generic term describing any sudden change in wind speed or wind direction. Since 1964, wind shear has been a causal factor in at least 29 commercial aircraft accidents, causing 500 fatalities and over 200 injuries. Wake vortices, the horizontally-

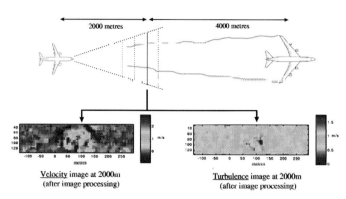

Simulation of airborne Doppler Lidar detecting wake vortices behind a Boeing 747 aircraft.

oriented, counter-rotating, "mini-tornadoes" shed from the wings of all aircraft, are a potential hazard to following aircraft. In order to avoid such encounters, airports schedule large separations between aircraft during take-off and landing. However, current landing separations are conservative, as wake vortices only rarely remain stable long enough to pose a hazard to a following aircraft. Given a reliable vortex detection system, a significant increase in peak time traffic capacity could be achieved at busy airports.

The AIRG has developed detailed simulations of the proposed system, and has used these simulations to develop the processing algorithms that will be capable of automatically detecting turbulence hazards. The *Figure* (above) illustrates one such simulation scenario. Field trials of a prototype system are scheduled for Toulouse airport in late 1999. If field trials are successful, it is envisaged that airborne Doppler Lidar systems will be available to aircraft manufacturers within five years.

Close up view of shore laboratories at Mace Head.

Atmospheric Research

Since 1958, the Department of Physics has operated an Atmospheric Research Station on Mace Head, near Carna, Co. Galway, on the west coast of Ireland. The Station has grown from modest beginnings in a small refurbished coastal look-out post to become one of the most important sites for atmospheric research in the northern hemisphere. Strategically situated with open exposure to the North Atlantic Ocean, it is in an ideal position to study changes in the global atmosphere. Its location facilitates the investigation of trace constituents, of both natural or man made origin, in marine and continental air masses. Long term measurement of such substances can detect trends in the background concentrations which may influence our weather and climate.

Atmospheric pollution is complex and transcends international boundaries, and, as a consequence, the research requires international collaboration and a sharing of resources and expertise. Mace Head is recognised world wide as a key location in the study of climate change, and its facilities are regularly used by scientists who visit the station for intensive long-term measurement campaigns, shorter-term projects, and international workshops. It is central to a number of international research networks and is a baseline station for the Global Atmosphere Watch of the World Meteorological Organisation since 1994. It is an important point of contact between European and American scientists and has achieved a high profile in atmospheric research.

Research activities at Mace Head have been concerned mainly with measurements of trace substances in air from the Atlantic Ocean which represent global background levels in the northern hemisphere. Since 1987, Mace Head has been one of five sites around the world studying the life times of chlorofluorocarbon (CFC) gases and their substitutes in the atmosphere as part of the Advanced Global Atmospheric Gases Experiment.

Air Quality Research

Occupational Hygiene is concerned with the recognition, evaluation, and control of environmental factors arising from work, which might adversely effect the health of people at work or in the community. For more than ten years, specialist courses in aspects of occupational hygiene have been provided within the Physics Department. Recently, technology transfer to industry and public bodies has been facilitated through the establishment of the Air Quality Technology Centre (AQTC). The Centre was formed in 1998 to address many air quality topics, including the growing concern about urban air quality in Ireland and its health implications. The role of the Centre is to provide expert services in the field of indoor and outdoor air quality for the purpose of complying with current health and safety and environmental legislation.

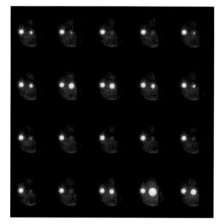

Time resolved images of the Crab pulsar. Sub-images represent about 1.3 msecs (or 4% of full-phase). The data was collected using TRIFFID mounted upon the Russian 6m telescope.

Astrophysics and Applied Imaging

The Physics Department, in close collaboration with the IT Centre at NUI Galway, has an active research programme in the fields of pulsar astrophysics and globular cluster photometry. These projects use data gathered using the unique camera, TRIFFID, developed at NUI Galway, which is used on some of the world's largest telescopes. TRIFFID produces images that can have a time resolution which varies from microseconds to hours. From this data set, very weak signals from optical pulsars can be analysed, and the effect of atmospheric turbulence can be reduced, thereby enhancing the ground-based images of globular clusters. The data sets from these observations require the use of large parallel computers to perform both the image analysis and the reduction of the pulsar time series. As part of this work, two new optically emitting pulsars (PSR0656+14 and Geminga) were discovered, as well as a new population of RR Lyrae type stars in the Globular Cluster M15.

One of the spin-offs from this work has been the application of Astronomical Image Processing techniques to Medical Imaging. In particular, digital medical x-ray images have been enhanced using algorithms developed for astronomical purposes. Another spin-off has been the development of a camera for the study of bio-luminescence, enabling biochemical reactions to be spatially resolved and tracked in time.

Further information can be obtained on the Physics Department web site http://www.physics.ucg.ie/

Biomedical Science and Biomedical Engineering at NUI, Galway

The broad multidisciplinary research areas of biomedical science and biomedical engineering contain some of the most challenging problems at the frontier of modern scientific research. They bring together researchers from many different disciplines in a shared, problem-centred approach to fundamental questions originating in biology and medicine, but seeking solutions using technologies, skills, and intellectual creativity from outside the traditional boundaries of biology and medicine. This activity has emerged as a major interdisciplinary research thrust at NUI Galway, and the University has recently established on campus a **Centre for Biomedical Engineering Science** to co-ordinate its diverse activities in this exciting research area. The Centre will act both as a local and a national resource, and will place Ireland in a positive strategic position internationally with respect to this emerging discipline.

The emergence of this research activity in Galway parallels the changing needs of society, reflected by the establishment of large centres of excellence dedicated to biomedical engineering science in many of the major universities abroad. As the European home of many multinational and national biomedical industries, and with a pool of dynamic and scientifically literate graduates, it is important that Galway and Ireland are positioned at the leading edge of this exciting field. The decision to focus on a targeted research programme is timely and appropriate, given the existing fundamental research strengths at NUI Galway on a broad range of topics, spanning disciplines from the life sciences, through the physical sciences, information technology and engineering, to clinical medicine. At NUI Galway we view biomedical engineering science as an approach rather than a distinct discipline, and regard it as a **problem-centred** multidisciplinary activity, located on the interfaces between medical, scientific, and engineering disciplines. Problems originating in medicine and biology are more likely to yield to the efforts of teams of talented researchers working together in a common approach. These teams are dynamic and will evolve in response to changing problems and to the skills needed to develop useful solutions.

Priority research areas have already been targeted for special focus as follows:

- **New measurement methodologies**
  The development of new and more accurate measurement methodologies, in particular those which allow non-destructive, non-invasive monitoring of living cells, tissues, and organisms.
- **Cellular and molecular bioengineering**
  Integration of advances in genomics and proteomics within a broader understanding of structure and function relationships in biological systems, cells and tissues.
- **Biofilms**
  Understanding, control, and exploitation of films of cells and other biological materials which develop on surfaces.
- **Biomaterials**
  Cell and tissue interaction with both biological and non-biological substrates, particularly on materials which can be manipulated to foster useful biological interactions.
- **Biomechanics**
  Physical forces and their measurement with particular reference to biology, including solid and fluid mechanics and computational models of biological problems.

The Centre brings together research personnel from a large number of departments in Science, Engineering, Medicine, and Information Technology. Support for their research activities has come from EU Framework Programmes, Enterprise Ireland, Health Research Board, Marine Institute, International Fund for Ireland, Wellcome Trust, and various industries. As a result, the University has excellent resources in terms of specific **technical expertise** (equipment and personnel) in disciplines which are essential to a basic research programme in biomedical engineering science. These include genomics and proteomics, cell and tissue biology, animal studies, microscopy, computational modelling, imaging, instrumentation, materials, lasers and laser processing, surfaces, solid/fluid mechanics, diagnostic/clinical. The research teams include members with dual appointments in NUI Galway and University College Hospital Galway. This ensures access to clinical material, specialised diagnostic facilities, and to a large and diverse patient population.

The Centre will be prominent in organising courses, workshops, and conferences on topics broadly within the scope of Biomedical Engineering Science. It will also co-ordinate and strengthen the industrial links to indigenous and multinational medical device and healthcare companies, and the national and international research links to other institutions in Ireland and abroad. In support of these activities, the University is committed to targeting new staff with a Bioscience/Bioengineering research bias, and to the development of interdisciplinary research groups on campus.

In addition to the research activities, related developments in this area, unique to NUI Galway, consist of new undergraduate programmes in Biotechnology (1991), Biomedical Engineering (1997), and Biomedical Science (1998), as well as an M.Sc. degree course in Biotechnology (1981). A new M.Sc. degree course in Biomedical Science will begin in 1999.

The proposal to develop biomedical engineering science combines a natural evolutionary process and a strategically-planned phased approach, deriving from the University's internal and external review process. It resulted in a successful application to the HEA Programme for Research 1998/2000 under the title **BioEngineering Science: An Integrated Approach to New Measurement Methodologies**. This proposal centred around an integrated suite of short-term multi-disciplinary research projects, which drew upon the collective strengths of participants from the faculties of Science, Engineering, and Medicine. Separately, these individuals reflected the University's acknowledged research expertise in the life sciences, physics and chemistry, medicine, and engineering while, together, they represent an emerging strength in the area of biomedical engineering science. The proposal also involved participants in partner institutions who brought expertise and resources not available within NUI Galway. The success of this proposal has added new personnel and equipment to assist in the expansion of the Centre.

Contact: Prof. Thomas Glynn,
Centre for Biomedical Engineering Science,
National University of Ireland, Galway;
E-mail: biomed@nuigalway.ie

Biomaterials science - a challenge for the future

Most people are aware of the fact that many of the largest medical device manufacturers in the world now have production and research facilities in Ireland.

They design a wide range of products for use in medicine, particularly for use in non-invasive surgery. Many of these devices are made from metals and alloys, whose stability in body fluid environments (blood, gastric juice) is of critical importance to the health of patients.

The need to ensure minimal corrosion/degradation has been the major determining factor in the selection of materials for use in these complex environments. Stainless steels have been used successfully as implants for many years. They owe their stability to the thin oxide layer on the surface which acts as a barrier between the material and the host solution. They have become particularly important in the construction of stents for use in the treatment of cardiovascular disease (blockage of arteries in particular). A stent is designed to serve as a temporary or permanent scaffold to maintain or increase the diameter of a vessel. Since blood contains around 0.9% sodium chloride, and because chloride ions can attack and destroy the oxide film on steels, attention must be focused on the design of corrosion test procedures for these devices. Any surface pretreatments which can improve the stability of the oxide can also be of considerable benefit. Research in these areas is currently being undertaken in the Chemistry Department of NUI, Galway.

For use in peripheral regions of the body (e.g. oesophageal stents), steels are not suitable because a blow to the neck may damage the stent causing a blockage. For use in these applications a very innovative "shape memory" material called Nitinol is used. Containing 50% Nickel and 50% Titanium, the alloy was discovered at the US Naval Ordnance Laboratory. After being subjected to suitable heat treatments, devices made from this material will return to their original shape after deformation. Flexible spectacle frames (as seen in TV ads) are made from this alloy. The heat treatments used to give these shape memory properties often result in the formation of less protective oxides on the stents, rendering them susceptible to corrosion.

Improving the corrosion resistance and hence the long-term stability of such stents forms the basis of another research effort underway in the Department. The next generation of implants will, it is hoped, have coatings which make them more biocompatible and acceptable in the body. These coatings will also be used to anchor drugs which will be released at the point of deployment. Work in this area is due to start shortly in the Department.

Michael Kelly and Eimear Costello compare test results.

Contact:
E-mail: william.carroll@nuigalway.ie

Air Quality Technology Centre

Clean air is a vital requirement for life and for many industrial processes. Inhalation is the principal route of entry into our bodies of pollutants in the form of gases, vapours and particles. Many modern industries, such as microelectronics, aerospace and health care products operate in stringent clean room conditions, while others, such as food preparation and pharmaceuticals, require increasingly demanding air quality standards. Concern for the environment requires studies of the effects of pollutants in the atmosphere on global climate change, damage to plants and property, public health and the quality of life in urban, rural, industrial, commercial and domestic settings. Hence the need to be able to establish the quality of the air for a given application and the concentrations of pollutants, gaseous and particulate, in it.

The Air Quality Technology Centre (AQTC) in the Department of Physics of the National University of Ireland, Galway, evolved from a long tradition of education and research on atmospheric pollution and occupational hygiene. It has benefited from funding from the European Union's STRIDE Environment sub-programme, Measure 1, for upgrading RT&D capability in the regions, and from Enterprise Ireland. It draws support from the technical and information resources of the University and the academic expertise available in the various departments to complement its own resources and experienced staff. The AQTC is now offering industry, commerce, public authorities and other groups a research and consultancy service on all aspects of air quality in outdoor and indoor settings.

Recent or current projects include a survey of PM_{10} particulate matter in Dublin city, mercury vapour levels in ambient air, thermal comfort surveys of offices, solvent levels in factories, pollutants produced by the laser processing of materials, and assisting compliance with occupational exposure standards. Specialised equipment available includes a portable mass spectrometer, optical particle counters, toxic gas detectors, a scanning electron microscope, a microbalance and a thermal stress meter.

The AQTC also contributes to education and technology transfer though short courses on health and safety at work, aspects of occupational hygiene and air pollution measurements, as well as a major input into the postgraduate courses for an M.Sc. and a Higher Diploma in Applied Science (Occupational Health and Hygiene).

Airborne particulate monitoring close to a construction site.

For further details contact
Prof. S.G. Jennings, Director AQTC,
Department of Physics, NUI, Galway;
Tel: 091-750364; Fax: 091-750584;
E-mail: gerard.jennings@nuigalway.ie

Secure File Transfer over the Internet (SFTI)

The concept of electronic commerce has been much hyped in recent years. The reason for the intense interest has been the upsurge in the use of the internet. Initially the commerce conducted over the internet was primarily business to consumer. One of the reasons for this is that business to business electronic commerce is long established in the form of Electronic Data Interchange (EDI). These systems enable the exchange of structured business documents such as invoices, purchase orders etc. over telecommunications lines. Traditionally the telecommunications method used has been Value Added Network Services (VANS). Such services provide reliable but expensive assistance in enabling the completion of the electronic commerce transaction. The reason for the expense is that the current price structure of VANS is volume based - you pay for each character you send.

When EDI was first developed, the internet was not a viable telecommunications infrastructure. However, with the advances in the internet, it is now technically feasible to send EDI messages over the internet rather than through VANS. Currently, EDI at Nortel is carried out over Value Added Networks. Using such telecommunications is expensive for both Nortel and their trading partners. The internet offers an alternative telecommunications method for the delivery of EDI messages. Such a communication method for EDI would be substantially cheaper than VANS, as the cost is not related to the volume sent.

An additional benefit of such as system is that it would enable Nortel to conduct EDI with additional trading partners, in particular with smaller trading partners. Currently, smaller trading partners are reluctant to adopt EDI due to the prohibitive costs of VANS. The ability to send EDI messages over the internet would lower the entry barrier for such firms.

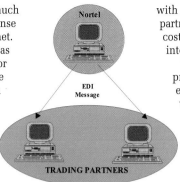

Link with Trading Partners.

THE AIM of the project is to build a software prototype that will demonstrate the viability of enabling the reliable exchange of EDI messages over the internet. Success in this, and the subsequent deployment of the prototype on a wide-scale level, will enable Nortel to reduce its use of VANS. Such a replacement would result in substantial cost savings for the company, thereby keeping aware of customer needs and facilitating product innovation.

THE OBJECTIVE of the project is to investigate and test the feasibility of transmitting EDI messages over the internet. As such, a software prototype will be developed which will be used to send test EDI messages between Nortel and one of its trading partners.

The scope of this project involves the transmission of the EDI message over the internet. As such, the EDI translation software required at either end of the transmission lies outside its scope.

The project kicked off on 1 April 1999 and is of twelve months duration. The partners are the Computer Integrated Manufacturing Research Unit (CIMRU), NUIG, and Nortel Networks. It is funded by the Enterprise Ireland Applied Research Grant Programme.

Further information regarding the SFTI project will be available shortly on the CIMRU homepage http://cimru.nuigalway.ie, or by contacting ingrid.hunt@nuigalway.ie

Rapid PDM

Product Data Management (PDM) is an enabling technology for distributed concurrent engineering and design for sustainability. Successful implementation of standards, such as STEP and ISO-9000, make the use of PDM unavoidable, because PDM is needed to manage all the product data and the processes in which the data will be exchanged. In the Extended Enterprise, PDM offers the enabling infrastructure for fast exchange of product data along the supply chain *(see Figure)*.

PDM vendors, CAD Vendors and independent systems integrators now offer the market an expanding range of electronic PDM systems. Implementation of a PDM system in an engineering organisation takes considerable time and effort for customisation and configuration of the package and for engineering process redesign. This effort often exceeds the purchase cost of the package by a factor of three or more. These high implementation costs create a significant barrier for enterprises, especially Small and Medium manufacturing Enterprises (SMEs), considering investing in PDM. The implementation process must be better understood, structured in manageable steps, and supported with proper tools to reduce the resources required for successful implementation and make PDM a viable option for SMEs.

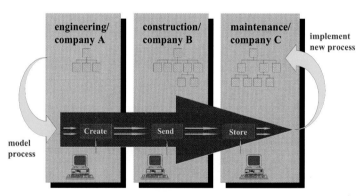

PDM in the Extended Enterprise.

The management of Product Data is a limiting factor in the competitive performance of manufacturing industry in Europe. This can be explained by the fact that products become increasingly customer specific, so that product and process specifications have to flow synchronously with the customer order. Only companies that have information systems which can manage and support this flow can maintain their responsiveness to customer needs. PDM systems offer such a solution, which however is typically complex and resource intensive in implementation, putting PDM outside of the scope of many SMEs.

This project, funded under the ESPRIT programme, aims to reduce the time, cost, risk and effort required to implement PDM successfully. The project will do so by developing innovative methods and tools that make implementation of PDM systems faster, easier and more effective. The primary objective of the Rapid PDM project is to develop a generic PDM implementation methodology that is supported with a set of IT-tools. This will address the needs of industry by supporting the implementation process for PDM.

The Rapid PDM consortium has been carefully constructed to combine excellent experience in PDM implementation (M.I.S., CIMdata), PDM software development (Cyco), PDM modelling research (BETA, Cranfield, CIMRU), product development process management (CEL) and ESPRIT project management (BETA, Cranfield, CIMRU).

*For further information contact
Dr Kate Goggin; Tel: +353-91-750-414,
Fax: +353-91-562-894;
E-mail: kate.goggin@nuigalway.ie*

Geoscience probes mantle secrets

Magmas formed beneath the East African continent ascend along NNE-SSW trending faults to create the Natron-Engaruka explosion crater area *(see illustration)*. Carbonated silicate magmas erupt to form melilitite composition tuff cones *(foreground)* and water in lakes reacts with molten rock causing explosions which result in maar craters *(middle distance)*. Carbonate magmas can build up impressive volcanoes such as Oldoinyo Lengai *(background)* which is erupting continuously, sometimes explosively. Its sodium-rich ashes turn water caustic and can burn the skin of the local Massai' livestock.

Although volcanic rocks of melilitite and carbonatite composition are rare, they have been found on every continent, from the Rhine Graben in Germany to the Amazon Group in South America[1]. The two rock types are intimately associated, both regionally and within volcanic complexes. This indicates that, ultimately, they have the same source region and form by similar processes. Their low silica concentrations negate an origin in the silica-rich crust, and they frequently erupt xenoliths (rock fragments) from the mantle. This is the region of the Earth that extends from between 10 and 40 to 2900km depth.

Magmas ascending from considerable depth will react with any mantle and crustal wall-rocks through which they pass. Volatile components become less soluble with decreasing depth and evolve as gases or escape to surrounding rocks in fluids. Reaction and exsolution alter the chemistry of the primary magma and the volcanic rock that is eventually generated can have a very different composition. Examination of such volcanic rocks alone can therefore give limited information regarding their origin.

As geologists cannot observe sub-surface processes directly, it is necessary to carry out laboratory simulations to understand mantle thermodynamics. Analyses of xenoliths carried to the surface in violent eruptions give the likely composition of the mantle. The results are used to mix synthetic compounds to form a powder approximating the mantle beneath old thick continental regions, where carbonatite volcanoes are generally found. This synthetic mantle is then loaded into high pressure and temperature apparatus in an attempt to "make" mantle rocks and magmas. If all the minerals of the mantle coexist with the magma at the end of the experiment, then the magma composition is possible. Further experiments show whether the primary magma composition can evolve into melilitite or carbonatite rocks.

Ongoing research at the Department of Geology, NUI Galway, in collaboration with European Union research facilities, specialises in:
- The temperature and pressure of formation of melilitite, carbonatite and intermediate composition magmas in the mantle.
- Variations within the carbonatite and melilitite rock groups as a result of mantle heterogeneity.

The Natron-Engaruka explosion crater area.

Contact: Dr Kate Moore;
E-mail: kathryn.moore@NUIGalway.ie

1. Woolley, A.R., 1989. The Spatial and Temporal Distribution of Carbonatites. In: *Carbonatites*, Bell, K. (Editor), pp 15-37, Unwin Hyman.

The shape on the map

Most people when they look at a map can easily recognise the objects being depicted. They do this through a combination of interpreting the form, size and shape of a feature plus the use of standard cartographic symbols and text annotation. To automatically process data for intelligent tasks such as in-car navigation, land use planning and market analysis, a computer system must itself interpret the map. Part of the automatic structuring (feature coding and object recognition) of topographic data, such as that derived from air survey or raster scanning large-scale paper maps, requires the classification of objects such as buildings, roads, rivers, fields and railways, partially based on their shape.

Recognition is based on the matching of descriptions of shapes. Numerous shape description techniques have been developed. The application of one of these techniques, Fourier descriptors, to classify objects on large-scale maps is described here. Based on a Fourier analysis technique applied to the boundary co-ordinates of an object expressed as complex numbers, Fourier descriptors are widely used in computer vision to describe and classify shapes.

The Fourier transform is a way of separating out the different frequencies that make up a signal. For example, applied to a musical sound, it can separate the high from the low notes. It can be applied to two-dimensional objects by imagining a point tracing out the boundary. The east and north co-ordinates of equidistant points on the boundary can be expressed as complex numbers, $x + jy$, where j is the square root of -1. Travelling around the boundary yields a sequence of complex numbers. The Fourier transform of this gives the Fourier descriptor values of that shape. These Fourier descriptors can be normalised to make them independent of translation, scale and rotation of the original shape. The first few descriptors give general characteristics of the shape, the remainder representing the small detail. Classification is performed by comparing descriptors of the unknown object with those of a set of standard shapes, finding the closest match. Objects of similar shapes will produce similar descriptors and, in particular, man-made objects, such as buildings and roads, with sharp corners will be distinctive from the curved shapes of more natural objects.

Most applications using Fourier descriptors deal with the classification of definite shapes, for example identifying a particular type of aircraft. To identify topographic objects, the technique needs to be extended to deal with general classes of shapes (buildings, fields, roads etc.). It is envisaged that Fourier descriptors will be only one of several techniques of object recognition, which will be combined to produce the optimal result.

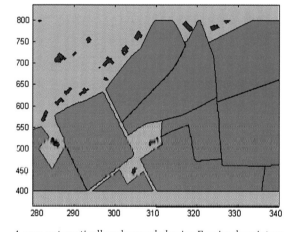

A map automatically colour-coded using Fourier descriptors to distinguish buildings and fields.

Contact: Dr Adam Winstanley,
Department of Computer Science,
NUI Maynooth;
E-mail: adam.winstanley@may.ie

Faculty of Science, NUI, Maynooth

The National University of Ireland, Maynooth, has a student population close to 5000, and approximately 1300 of these are in the Faculty of Science. The Faculty has over 100 full time academic staff and 22 Post-doctoral Research Fellows. There are more than 200 students registered for postgraduate degrees, including almost 100 for MSc or PhD degrees by research. Undergraduate students can choose from denominated entry degrees in Biotechnology or Computer Science and Software Engineering or over 20 single and double honours degree programmes offered by the Departments of Biology, Chemistry, Computer Science, Experimental Physics, Mathematical Physics and Mathematics.

The Faculty of Science at NUI Maynooth has undergone significant growth and development during the last few years. By the end of the decade, NUI Maynooth will have invested over £20million in developing its science facilities on the new campus. The second phase of a Science building programme was last year. The Departments of Chemistry and Experimental Physics moved into a new £10million building, which was opened by Her Excellency President Mary McAleese on 22 September 1998. Each of the Experimental Science departments can now boast state-of-the-art facilities for teaching and research, with Biology and Computer Science having moved into new buildings in 1993. Design work is underway for further development, including an extension to the Computer Science Department to cater for increased student numbers under the Skills Need Programme.

The new £10 million science building at NUIM, location of the Departments of Chemistry and Experimental Physics.

Prof. Kingston Mills, Dean, Faculty of Science, NUIM.

Research

Researchers in the Faculty of Science at NUI Maynooth have been successful in attracting major research grants from the European Union, The Wellcome Trust, The Health Research Board, Enterprise Ireland and other international and national funding bodies. The current committed funding from outside agencies for research in the Faculty is now close to £5million. NUI Maynooth was one of only three Universities in Ireland to be awarded a New Blood Lectureship under the new co-operative scheme operated by the Health Research Board and The Wellcome Trust. Dr Bernard Mahon was the recipient of this award, and took up his position in the Department of Biology on 1 October 1998.

NUI Maynooth has internationally acclaimed scientists in a variety of research areas from molecular biology to space technology. The Department of Biology has particular strengths in Immunology, Plant Science, and the study of nematodes for biological pest control. The Department of Experimental Physics is involved in the Planck Surveyor space mission *(see page 138)*. Dr Anthony Murphy is a principal investigator on this prestigious mission, whose aim is to provide precise details of fluctuations in the Microwave Background radiation, an experiment of great importance to Cosmological Theories. Some of the research areas in the different departments include:

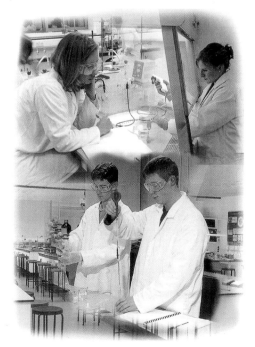

Biology
- Immunity to infectious diseases
- Mucosal immunology of the respiratory tract
- Molecular cell biology of programmed cell death
- Recombinant expression systems and diagnosis of infectious diseases
- Pathogenesis of *Candida* and other fungi
- Bacterial molecular genetics-stress responses in *Bacillus & Salmonella*
- Molecular genetics, physiology and behaviour of nematodes
- Development ecology and biological control of plant diseases and pests
- Plant molecular and development biology
- Bioinformatics – genomic evolutionary processes
- Ethical aspects of developments in biotechnology and medicine

Chemistry
- Group and graph theoretical applications
- The photochemistry of carbenes and catalysis
- Time resolved emission spectroscophy
- Bio-inorganic chemistry of organometallics
- Biomedical sensors for use in fundamental physiological research
- Advanced electrochemical analysis
- Heterocyclic and heteroaromatic chemistry
- Asymmetric synthesis

Computer Science
- Signal and image processing: computer vision, medical imaging, graphical analysis, and music processing
- Mathematics: formal methods and cryptography
  - Systems: embedded systems and dynamic systems analysis
  - Acousto-optical computing and neural networks

Experimental Physics
- Atomic and laser physics
- Astrophysics
- Space technology

Mathematics
- Analysis: functional, real, complex, harmonic, geometric
- Applied statistics and modelling
- Coding theory and cryptography
- Algebraic and differential geometry

Mathematical Physics
- Nonlinear physics
- Quantum field theory
- Non-perturbative gauge theories
- Quantum computing and quantum decoherence

Contact: Professor Kingston Mills, Dean of Science, National University of Ireland, Maynooth, Co. Kildare; Tel: +353-1-7083838; E-mail: kingston.mills@may.ie

NATIONAL UNIVERSITY OF IRELAND, MAYNOOTH — BRIAN DOLAN

Quarks, symmetry and cold electrons

Symmetry is all around us – witness the beautiful symmetry forms of beautiful and flowers. We can learn and understand much about Nature by studying her symmetries, but sometimes they are less than obvious. A new symmetry has recently been found in two very different natural phenomena, indicative of a curious connection between them. These phenomena are: Quantum chromodynamics (QCD) – the force that binds quarks, among the smallest known constituents of matter, inside protons and neutrons; and the strange behaviour of very cold electrons in semiconductors – the quantum Hall effect (QHE).

QCD is named after quantum electrodynamics, its close cousin. The term chromo (colour) is used in analogy with the three primary colours: as red, blue and green mix to white, so quarks can have one of three types of colour charge which add to zero. One can imagine hypothetical worlds where QCD would be different: the strength of the colour charge could be bigger than in our Universe, for example. A remarkable fact was recently discovered in some simplified mathematical models of QCD: the predicted physical properties of many of these hypothetical worlds are very similar – there is a symmetry relating them.

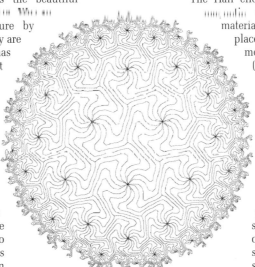

A Poincaré disc: think of points inside the disc as representing hypothetical worlds in QCD or plateaux in the QHE – any two points which can be related to each other by the symmetry of the picture have similar physical properties (copyright 1990-1999 by The Geometry Center, University of Minnesota; used by permission - see http://www.geom.umn.edu/admin/copyright.html).

The Hall effect occurs in semiconductors placed in a magnetic field. When a slab of semiconducting material, carrying a current along its length, is placed in a magnetic field, the field pushes the moving charged particles left or right (depending on the sign of their electric charge) building up a voltage across the slab's width. Normally this voltage is proportional to the magnetic field but, surprisingly, for pure, very thin samples in high magnetic fields and extremely low temperatures, the voltage increases in a series of steps or plateaux which are precise fractions of a basic unit. This is the QHE. Experiment shows that the physical properties of many of these plateaux are very similar – there is a symmetry relating them. Herein lies the connection between QCD and the QHE – they share the same symmetry! This symmetry is similar to that of a Poincaré disc, seen in the illustration.

One aim of my research is a deeper mathematical understanding of this connection – what can we learn about QCD from the QHE and vice versa?

More details and contact information at: www.thphys.may.ie/staff/bdolan

---

NATIONAL UNIVERSITY OF IRELAND MAYNOOTH — CATHERINE COMISKEY, GLORIA CRISPINO O'CONNELL, ELAINE HAND & ROBERT MILLER

Mathematics, modelling and medicine

A team of applied researchers working at the Mathematics Department at The National University of Ireland Maynooth, are looking at new ways in which mathematics can assist those working in some of the most topical areas of medicine. How can topics in applied calculus, statistics and numerical analysis illuminate real problems in diseases like HIV, AIDS and Hepatitis C in adults, measles and meningitis in children and, more recently, the heroin and ecstasy use in young adults? These are just some of the questions the team is addressing in collaboration with The Department of Health, The Garda Siochana, The Department of Education and the hospitals.

Many key questions on the spread of meningitis, say, need to be addressed for the health services. These include: What proportion of children carry meningococcal organisms? What level of vaccination is required to eradicate disease from meningitis and what is the optimum age we must vaccinate at in order to achieve this? A mathematical model can describe the transmission dynamics of the disease. This model used in conjunction with a series of computer simulations can assess the effects of changes in key epidemiological parameters on disease incidence in the varying age groups. Members of the research team have already completed several such studies in Ireland and internationally for measles, Haemophilus influenza or Hib, and HIV and AIDS.

Another two studies being conducted by the team are on the heroin and ecstasy epidemics facing Dublin. We ask the question: how many young people and, in particular, school attenders are using drugs? This study also assesses the effects of drug misuse on motivation, performance and early school leavers. To estimate the prevalence or number of heroin users in Dublin in 1997 we use the capture recapture statistical method* with anonymous identification data from three sources: the number of heroin-related hospital admissions; the number of heroin-related police arrests and, finally, the number of people in receipt of methadone treatment. In addition, this study also looks at how long it takes a drug user to progress to treatment.

The estimates produced by these studies will provide a valuable aid to the planning and provision of the necessary financial and health care services. Results will enable government authorities and those involved in the provision of the health care services to direct their planning and resources to those groups most vulnerable.

Opium poppy.

> * In the simplest capture recapture model, the first sample provides the individuals for tagging and is returned to the population, while the second sample provides the recaptures. Using the number of individuals caught in both samples (the recaptures) and the number caught in just one sample, it is possible to estimate the number not caught in either sample, thus providing an estimate of the total population size.

Contact:
Dr Catherine M. Comiskey,
Mathematics Department,
National University of Ireland Maynooth (NUIM),
Maynooth, Co. Kildare;
Tel: +353-1-708-3994;
Fax: +353-1-708-3914;
E-mail: cc@maths.may.ie

Probing the Early Universe with the PLANCK Surveyor

The Submillimetre-Wave Optics group in the Physics Department of NUI Maynooth is participating in an ambitious European Space Agency project known as the PLANCK Surveyor. The primary aim of the PLANCK mission is to make crucial measurements of a faint source of infra-red radiation known as the Cosmic Microwave Background which fills all of the Universe. This radiation field is the remnant afterglow of the explosive beginning of the Universe in the Big Bang. Astronomers observing this radiation today are effectively seeing the Universe at a very early stage in its history (about 30,000 years after the Big Bang, which happened over 10 billion years ago).

In 1965 Penzias and Wilson first detected the Cosmic Microwave Background at Bell Laboratories in New Jersey. Subsequently it became clear from careful measurements that the radiation is extremely uniform in all directions. Then, in 1992, a famous NASA satellite mission known as COBE revealed the existence of tiny variations or "fluctuations" in the temperature of the Cosmic Microwave Background across the sky. Imprinted on these "ripples at the edge of the Universe" is vital information about the origin and subsequent development of galaxies and large scale structures in the Cosmos.

Inspired by COBE, but carrying much more sophisticated ultra-sensitive detectors, the PLANCK Surveyor will actually be able to image and fully characterise these Microwave Background fluctuations. PLANCK is due to be launched in 2007 into a deep space orbit which will provide an environment suitably free of radio and infra-red interference to avoid degradation of the data from the mission. The results of the PLANCK project are set to have a revolutionary impact on cosmology, the study of the origin and evolution of the Universe as a whole. It will be possible to test the competing theories of the early development of the Universe and the origin of large scale cosmic structures such as galaxies and clusters of galaxies. Indeed it will be possible to determine the age of the Universe to within 1% and predict its ultimate fate.

Maynooth is involved in an international consortium of scientists to design and build the High Frequency Instrument on board the PLANCK Surveyor. The optical design of this instrument is being undertaken in collaboration with Queen Mary and Westfield College, University of London, the Institut d'Astrophysique Spatiale, Paris, the California Institute of Technology, and Stanford University, California. The formidable technical challenges implied in imaging the very faint Microwave Background temperature fluctuations require the development of new technologies and the very precise modelling of the system. This is an extremely exciting opportunity for NUI Maynooth, as it offers Irish participation in an extremely important cosmological experiment.

The PLANCK Surveyor will measure temperature fluctuations in the Cosmic Microwave Background.

Contact: Dr J. Anthony Murphy,
Experimental Physics Department,
NUI, Maynooth, Co. Kildare;
Tel: 01-708-3771; Fax: 01-708-3313;
E-mail: anthony.murphy@may.ie

Electrochemistry at NUI Maynooth

Electrochemistry is a very diverse and fast-growing field of research of great importance to modern man. In the Department of Chemistry at NUIM, there are two research groups studying the fundamental electrochemistry of corrosion, on the one hand, and bioelectrochemistry and neurochemistry, on the other.

Corrosion is one of nature's greatest nanotechnologies. Corrosion cells consist of local anodic and cathodic regions, which act as small reactors, feed on the energy in the local environment, and have the capacity to grow suddenly and consume even a massive structure, for example an oil rig. An understanding of what happens, during the very early stages of the corrosion event, is crucial to the development of any corrosion protection system, the fabrication of protective coatings and the design of advanced materials. Therefore, much of the recent research in the Corrosion Science Laboratory at NUIM is concerned with the early detection of corrosion events, through the development of local probe techniques. This new scanning technology is being used in conjunction with bulk electrochemical techniques, such as electrochemical impedance spectroscopy, in studies on pitting corrosion, electrodeposition, conducting polymers, protective coatings and photoelectrochemical reactions.

Electrochemical sensors are (small) devices, capable of selectively measuring the concentrations of specific chemicals in different media. Much of the recent work in bioelectrochemistry has been directed towards the development of electrochemical sensors for environmental, industrial and medical applications. By far the most common such device is the enzyme-based blood-glucose biosensor used by diabetic patients in their daily treatment regime. Research in the Bioelectrochemistry Laboratory at NUIM focuses on the development, characterisation and application of sensors and biosensors in biomedical research.

The laboratory is divided into the Sensors Development Unit and

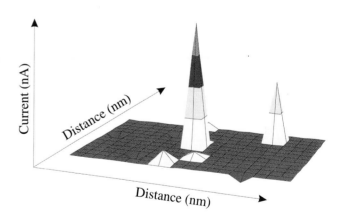

Scanning probe image of local breakdown events on a surface.

the Neurochemistry Research Unit. The former focuses on the development and characterisation of micro-electrochemical sensors for real-time monitoring of important biological chemicals such as glucose, glutamate, oxygen and ascorbic acid (vitamin C). In the Neurochemistry Unit these sensors are used in fundamental physiological research, such as investigating the chemical mechanisms controlling the operation of the living brain.

Contact: NUIM Chemistry Department;
Tel: 01-708-3677 (CB), 01-708-4633 (JL);
E-mail: cb.breslin@may.ie, john.lowry@may.ie;
URL: http://pwch23.may.ie

NATIONAL UNIVERSITY OF IRELAND, MAYNOOTH — SEAN DOYLE

Parvovirus B19 infection – an underestimated problem

Established in January 1998, the purpose of all research and development work carried out in the biotechnology laboratory at NUI Maynooth is twofold. Firstly it is our aim to carry out strategic immunological research, and secondly to translate this knowledge into valuable commercial realities. The group has two main areas of interest:

Microplate detection of viral nucleic acid.

1. Parvovirus B19.
Parvovirus B19 (B19) is a human pathogen which can cause a mild flu-like illness and joint pain. More importantly, foetal loss can occur if infection occurs during pregnancy, and it is conservatively estimated that upwards of 5,000 pregnancies are lost annually in the European Union due to B19 infection. The work of our group is primarily directed towards the development of sensitive viral and anti-viral antibody detection systems, and at understanding the fundamental immune mechanisms operating against this virus (in collaboration with Dr B.P. Mahon, NUI Maynooth).

A polymerase chain reaction (PCR) based assay for the detection of parvoviral DNA has been developed in collaboration with Biotrin International, and current efforts are directed towards the establishment of this test as a screening tool for testing both infected individuals and pooled blood products. Use of this robust and sensitive test system in clinical laboratories should aid in the successful treatment of pregnant women infected with B19 during pregnancy by identifying B19 both prior to and post-treatment, respectively.

Regarding immunity studies, we, along with others, have recently demonstrated that immunity against this virus may be compromised due to selective antibody loss in the post-infection recovery period (Kerr et al., *Journal of Medical Virology*, **57**, 179-185, 1999). In a Health Research Board funded project, sensitive *in vitro* cell culture systems, in association with highly purified recombinant B19 antigens, are being employed to explore the unique nature of the immune response to parvovirus B19. Insight into immune mechanisms against B19 should facilitate anti-viral treatment and vaccine development.

2. Immunodiagnostic assay development
Work in the biotechnology laboratory, in collaboration with a number of Irish biotechnology companies, is also directed towards immunodiagnostic assay development and the production of recombinant proteins for both immunodiagnostic and vaccine uses. To date, our laboratory has successfully completed one joint Enterprise Ireland/Tridelta Development Limited funded project to develop an automated assay system for the stress response protein, Haptoglobin.

Another Enterprise Ireland funded project has facilitated the establishment of a recombinant protein expression unit, and currently a number of model proteins (e.g. parvovirus B19 NS1 protein) are under investigation to establish the feasibility of using the baculovirus/insect cell expression system for the secretion of commercially relevant diagnostic antigens.

Contact: Sean Doyle; Tel: (01) 708-3858; E-mail: sean.doyle@may.ie

---

THE QUEEN'S UNIVERSITY OF BELFAST — S. RAGHUNATHAN

Research at The Queen's University of Belfast on reducing life cycle cost

The School of Aeronautical Engineering at Queen's University Belfast and Bombardier Aerospace Shorts have been conducting research to reduce life cycle cost of aircraft by trade off between manufacturing cost and fuel cost. In the aircraft industry, the specification of surface smoothness requirements, commonly referred to as manufacturing tolerances, arises out of the aerodynamic necessity to minimise drag, i.e. to reduce fuel burn for the sector of operations.

In the Nineties, for the majority of flight sectors, fuel burn constitutes nearly 10% to 15% of the Direct Operating Costs (DOC), while aircraft unit price contributes to nearly three to four times more.

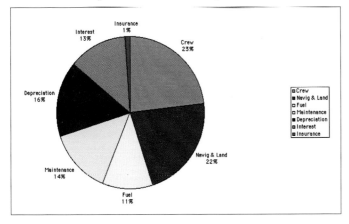

Breakdown of Direct Operating Cost.

Using AEA ground rules, a typical DOC breakdown of a 100 passenger subsonic jet aircraft on a flight over 1000 nautical miles is given in the *Figure*.

Aircraft unit price and parasite drag are related to the manufacturing tolerances.

The project offers an integrated methodology for industrial usage to arrive at a revised tolerance specification from the current practices in a concurrent engineering environment at the conceptual stages of design, as a customer driven design strategy to reduce aircraft DOC. The findings could lay the foundation for advanced research on Design for Manufacture. This project also offers the opportunity for national dissemination for standardisation of tolerance allocations for generic aircraft components.

> **Present study indicates that tolerance relaxation from the current specification can save up to 0.27% of the DOC, which is a significant saving in terms of cost for an airline.**

This research project was funded by the Engineering and Physical Sciences Research Council UK and Bombardier Aerospace Shorts Brothers plc, under an IMI managed programme.

*Contact: Professor S. Raghunathan,
School of Aeronautical Engineering,
The Queen's University of Belfast,
David Keir Building, Stranmillis Road,
Belfast BT9 5AG; Tel: 01232-335417.*

150 years of Engineering excellence at Queen's, Belfast

Engineering was one of the first subjects to be established at Queen's when courses started in 1849. One of the first full-time members of staff was James Thomson, brother of Lord Kelvin, who was the Professor of Civil Engineering from 1857-73. In his own right he made significant contributions to the fields of fluid mechanics and thermodynamics for which he was awarded an FRS.

Throughout the past 150 years Queen's has been successful in maintaining its position of eminence in engineering through the continued recruitment of outstanding students who have gone on to occupy key positions around the world.

Teaching
All eight Schools in the Faculty of Engineering have had their teaching assessed. Electrical and Electronic Engineering was judged to be joint first out of 75 universities within the UK. Civil Engineering was rated as joint second out of 47, and all others assessed in this way were in the top third. Current student enrolment is over 2,400.

Research
The Faculty's research performance is equally outstanding. It is comparable with Engineering at the top English and Scottish Universities. This success is due to the strength of the various research teams in the Faculty, which carry out internationally acclaimed research while contributing to the success of local industry. Queen's leads the world in digital sound technology (responsible for the sound in the film *Saving Private Ryan*), moulding of plastics, wave energy, and numerous other areas.

Within the Faculty, in addition to strong research groups within the individual Schools, leading edge research is carried out in the following Centres of Excellence for research:

The Queen's University of Belfast.

A measure of the overall international standing of Engineering at Queen's is the recent acceptance of the University into the Sterling Group (restricted to the top 20 UK universities with engineering faculties).

For further information, please contact those named or:
Professor A.E. Long, Dean, Faculty of Engineering,
Queen's University, Belfast, BT7 1NN;
Tel: 028 9033 5443; Fax: 028 9027 4536;
E-mail: a.long@qub.ac.uk; WWW: http://www.qub.ac.uk/feng/

Title:	Contact:
Polymer Processing Research Centre	Mr P. Grant (p.e.grant@qub.ac.uk)
International Combustion Engines Research Centre	Dr R. Douglas (r.douglas@qub.ac.uk)
Aerospace Centre	Professor S. Raghunathan (s.raghunathan@qub.ac.uk)
Digital Signal Processing Centre	Professor J.V. McCanny (j.mccanny@qub.ac.uk)
Industrial Process Control Centre	Professor G.W. Irwin (h.morrow@qub.ac.uk)
Electrical Power and Energy Systems Centre	Dr C. Tindall (ce.tindall@qub.ac.uk)
NI Semiconductor Research Centre	Professor H.S. Gamble (h.gamble@qub.ac.uk)
The Microwave & Millimetre Wave Centre	Professor V.F. Fusco (v.fusco@qub.ac.uk)
Centre for Supercomputing in Ireland	Professor R. Perrott (r.perrott@qub.ac.uk)
Centre for Image and Vision Systems	Professor F. Murtagh (f.murtagh@qub.ac.uk)
Environmental Engineering Research Centre	Dr R. Kalin (r.kalin@qub.ac.uk)
Structural Materials Research Centre	Professor P.A.M. Basheer (m.basheer@qub.ac.uk)
QUESTOR Centre for Clean Technologies	Professor W.A.J. Swindall (j.swindall@qub.ac.uk)

Transfer of silicon wafers using a robotic arm in a vacuum chamber at the School of Electrical and Electronic Engineering's clean room.

Research student Raquel Soares gets a mould set up before doing moulding on the Rotospeed Rotational Moulding Machine.

Image and vision systems

Research and R&D work in Image and Vision Systems (IVS) at Queen's continues to have a significant aerospace orientation. Staff members continue to have close links in image processing and data handling with the European Space Agency and NASA, in projects such as the highly-profiled Hubble Space Telescope observatory, and (in the past) the astrometry satellite, Hipparcos. Other expertise includes the development and assessment of vision-based automatic target detection and tracking systems for surveillance. Wavelet and multiresolution image processing software associated with the group is growing in usage worldwide, and includes such clients and users as the Jet Propulsion Laboratory and Caltech, the US Department of Agriculture, Unilever in Britain, the Ministry of Justice in the Netherlands, the French atomic energy authorities, CEA, and many others. In addition to aerospace applications, telecommunications is an important application area. Software and hardware aspects of video and large image compression and delivery technologies constitute a theme of considerable interest, for video-conferencing, telemedicine and other domains.

Figure 1: Footprint matching in forensic science.

Toolmarks and similar signatures present an interesting image database matching problem. Matching footprints *(Figure 1)* or other biometric characteristics is a forensic science objective of some importance. Fractal coding methods are being studied by us for matching against database exemplars. Among other forensic science – and image database – applications, the producers' symbols on Ecstasy tablets are being processed with wavelet transforms for background removal and feature detection.

High-resolution imaging, and high quality processing of the data products, are typical of such fields as astronomy and medical imaging. *Figure 2* (left) shows the galaxy cluster A2390 imaged with the ROSAT X-ray satellite, based on accumulating photons over a period of 8.5 hours. *Figure 2* (right) is a processed version, based on denoising in wavelet transform space. Very high quality image understanding and interpretation is also needed in medical imaging. Noise modelling is of critical importance for usage such as compression, in a telemedicine context, or feature detection and characterization, in a diagnostic context. The many modalities of medical imaging – X-ray, ultrasound, magnetic resonance, positron emission, and others – call for different models of the image formation process, and such modelling is needed for the best possible analysis.

Resolution scale is also of central importance in the Earth sciences. *Figure 3* (left) shows an ocean simulation model of the sea surface temperature off the Mauretanian coast, and *Figure 3* (right) shows data from a meteorological satellite. Unobserved parts of the images, caused by cloud cover, are shown in black. We have found that integrating multiple resolution scale information is necessary for information fusion from oceanographic model and observed satellite data. This work has involved extensive neural network modelling, with the ultimate objective of forecasting oceanographic upwelling.

Another forensic science application is the cleaning up of video data. Subpixel shifted images – in our case due to wind-occasioned movement of a video camera overviewing a football riot – allows for image coaddition, improving the signal to noise ratio. Video compression is a different application on video, and calls for real-time decompression. Such compute-intensive requirements can be met with parallel hardware and dedicated chips. Field Programmable Gate Arrays (FPGAs) offer a solution which has been fine-tuned by our group's research. Software environments have been created, around such hardware, to facilitate the task of implementing any given algorithmic application (such as video compression) on FPGA and other microprocessor hardware.

Many applications involve detecting and tracking independently moving objects against static or moving backgrounds. We are currently developing motion detection and tracking algorithms to process imagery from electro-optic sensors. Conventional computer vision based tracking algorithms have achieved only limited success to date, and require large computational budgets. Yet many biological vision systems routinely perform very similar tasks, often with quite limited sensors and neural processing budgets. Therefore IVS is looking to explore whether an integrated approach, in which unusual strategies employed by biological vision systems inform computer vision algorithms, may provide practical and robust solutions.

The expertise available in Image and Vision Systems at Queen's University is comprehensive, and targets integrated software and hardware solutions. With an expanding team of six faculty members and approximately ten graduate students and post-docs, and a local and worldwide network of closely collaborating colleagues, effective and comprehensive solutions can be provided for imaging and vision problems. We have just touched on some current work in our examples and illustrative figures. Past work has included imaging aspects of non-destructive testing by ground radar, printed circuit board inspection, woven fabric inspection, synthetic fibre quality assessment, and 1D signal processing – in astronomical spectral analysis, and financial and environmental modelling and forecasting. If a problem exists which has a component related to imagery, video or related digital information (and what problem does not?) then we would like to hear about it.

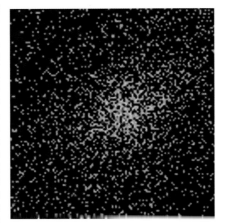 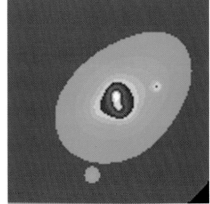

Figure 2: X-ray image of galaxy cluster (left), same image, interpolated and noise filtered using a wavelet transform method (right).

Contact: Professor Fionn Murtagh, Image and Vision Systems, School of Computer Science, The Queen's University of Belfast, Belfast BT7 1NN; Tel: 01232-274620; Fax: 01232-683890; E-mail: f.murtagh@qub.ac.uk

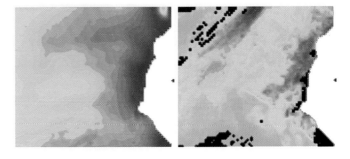

Figure 3: Ocean sea surface temperature off west Africa, from physical model (left) and observed data (right).

Queen's chips in with new designs

Digital Signal Processing (DSP) is key to many new innovations in the telecommunications, broadcasting and semiconductor industries. The growth of DSP components in consumer, communications, networking and computing devices has been spectacular. The DSP market has been growing at 40% per annum with a forecast of a $50 billion market over the next ten years. Market pressures demand constant innovation in terms of circuit performance, size and power consumption, as well as major reductions in design times, to ensure products come to market at the earliest opportunity.

DSiP Laboratories is a Centre of Excellence established in 1997. Its main function is to undertake research into new methods for the rapid design of advanced silicon integrated circuits for telecommunications, multi-media and broadcasting applications. Located in a purpose-built laboratory in the School of Electrical and Electronic Engineering, the research facility is equipped with the very latest high powered computer workstations and state-of-the-art software for the design and simulation of complex Digital Signal Processing (DSP) systems. The Centre extends Queen's research activity in this important technology and further enhances the University's international reputation for pioneering work in the field of DSP Intellectual Property (IP) and System-on-a-Chip (SOC) design.

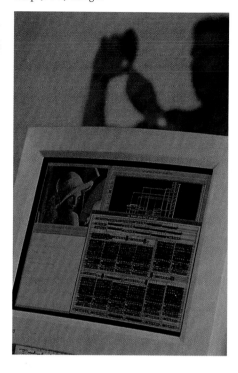

DSiP Laboratories bridges the gap between the fundamental research activities of the School of Electrical and Electronic Engineering and the requirements of the market place. Major research programmes focus on allowing complex DSP chips to be developed in a fraction of the time traditionally required. The Centre has an important role in encouraging and supporting the small but rapidly expanding local industry operating in this advanced technological area.

A key objective of DSiP Laboratories is to further expand joint research programmes by establishing links with other research organisations and high technology companies. Academic staff associated with the Centre have an impressive track record of collaborative research leading to successful commercial exploitation.

*For further information contact
Professor John McCanny, Director,
DSiP Laboratories, School of Electrical and Electronic Engineering, Ashby Building, Stranmillis Road, Belfast BT9 5AH;
Tel: +44(0)1232 335438;
Fax: +44(0)1232 663992;
E-mail: j.mccanny@qub.ac.uk;
http://www.ee.qub.ac.uk/dsp*

The UK Multiphoton and Electron Collisions HPC Consortium

Involving six academics from Queen's University Belfast (Professors P.G. Burke, K.T. Taylor, H.R.J. Walters and Drs J. McCann and M.P. Scott of Applied Mathematics and Theoretical Physics, and Dr N.S. Scott of Computer Science), as well as academics from Cambridge, Durham, London and Oxford Universities, the UK Multiphoton and Electron Collisions High Performance Computing Consortium has recently been awarded resources to the tune of more than £1M on the high-end supercomputers newly installed at the University of Manchester. Led from QUB, this Consortium has undertaken to investigate seven problems in multiphoton absorption by matter and six in electron collisions with atoms, molecules and their ions. The multiphoton work is crucial to the interpretation of experiments carried out at the Central Laser Facility of the UK Rutherford Appleton Laboratory and at other high-power laser facilities world-wide. The results of accurate electron collision calculations have application, for instance, in the understanding of astrophysical and laser fusion plasmas.

Typical of the multiphoton calculations is

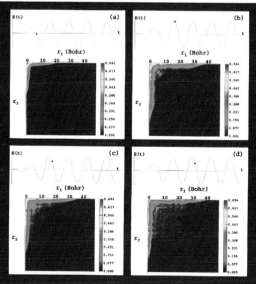

A helium atom exposed to a highly intense femto-second laser pulse.

that treating the helium atom exposed to a highly intense femto-second laser pulse. The primary interest here is in laser intensities sufficiently high to bring about ionization of both atomic electrons. The frames of the *Figure* display the overall electronic probability density plotted against the two electronic radial co-ordinates at four distinct instants (indicated by the location of the red dot) during the interaction of the atom (initially in its ground state) with a laser pulse with a wavelength of 194nm and of 8.0×10^{15} W/cm² peak intensity. In frame (a), the laser pulse has yet to achieve peak intensity and only one electron (either electron 1 or electron 2) has moved towards ionization. In frame (b), peak intensity has just been reached and the first indication of both electrons moving simultaneously towards ionization (probability density about the $r_1 = r_2$ line) is apparent. A very short time later – frame (c) – probability density bands appearing parallel to the axes are the first manifestation of a sequential ionization mechanism. Finally, by frame (d), a further cycle of the laser pulse has passed and both simultaneous and sequential ionization processes are fully developed.

Further details of the Consortium can be found at www.am.qub.ac.uk/mecpc

*Contact: Professor K.T. Taylor;
E-mail: k.taylor@qub.ac.uk*

Chasing solar eclipses at Queen's

A total eclipse of the Sun is an awe-inspiring event. Gradually, the moon edges further over the face of the Sun, and for just a few minutes the Sun's disk is completely covered, leaving only the ghostly light of the corona visible. We now know that the corona has a temperature of about two million K, which for the past 50 years has remained one of the great unanswered questions of solar physics. The energy source that keeps the Sun hot and shining is at its centre. Therefore, the temperature is expected to fall as you move away from this central "heater". The temperature of the corona, however, instead of dropping as you move away from the Sun, actually increases! There are currently two competing theories as to why the corona is so hot. One is that the energy is produced by many tiny flares in which magnetic energy is transformed into heat, while another assumes that hydromagnetic waves are responsible. The study of the mechanisms for transporting and depositing this energy is a worldwide area of research and one in which Queen's is actively involved.

In order to study the possibility that waves are responsible for heating the corona, we have designed and tested the Solar Eclipse Coronal Imaging System (SECIS). This project involves several groups across Europe, including Professor Francis Keenan and PhD student Peter Gallagher at Queen's, scientists at the Rutherford Appleton Laboratory and a group from the Astronomical Institute at the University of Wroclaw in Poland. The Principal Investigator for the experiment is Professor Ken Phillips from the Rutherford Appleton Laboratory.

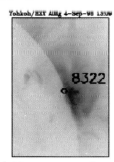

Comparison of a SECIS image taken at Sacramento Peak, New Mexico with X-ray data from the Yohkoh satellite (Japan/UK/USA) and an extreme ultraviolet image from the TRACE satellite (Lockheed-Martin/NASA).

SECIS was designed to search for short-period variations in the solar corona seen either during a total eclipse or with a coronagraph (a specially designed telescope). The charge-coupled device (CCD) cameras used in SECIS have the capability of imaging a selected portion of the corona at a rate of 50 frames a second, with the intensities in each pixel digitised in 12-bit levels. In the past, tests of SECIS have been carried out during the 1998 eclipse visible in Guadeloupe (French West Indies) and with a coronagraph at Sacramento Peak, New Mexico. The accompanying figure shows the first results from New Mexico.

We will be taking SECIS to Bulgaria for the eclipse of August 11, 1999. This will be the first time that a total solar eclipse has been visible from Europe since 1961, and offers us an ideal opportunity to observe the solar corona with SECIS. All we need now are clear skies!

Contact: E-mail: p.gallagher@qub.ac.uk; http://ast.star.rl.ac.uk/secis/

The QUILL Research Centre – rewriting the future of solvents

The QUILL (Queen's University Ionic Liquid Laboratories) Research Centre was launched on 20 April 1999 with 16 founder industry Members drawn from three continents. The structure of QUILL is based on the award winning QUESTOR Centre (Queen's University Environmental Science & Technology Research Centre). Both Centres have an environmental theme. The Chair of the Industry Advisory Board of QUILL is Dr David Moody of Zeneca, and the Co-Directors are Professor Ken Seddon (who is an Engineering & Physical Sciences Research Council and Royal Academy of Engineering Clean Technology Fellow) and Professor Jim Swindall OBE (who is also Director of QUESTOR).

So what are ionic liquids, and what have they to do with the environment? Volatile organic solvents that damage the atmosphere are the normal media for the industrial synthesis of organics (petrochemical and pharmaceutical), with a current world-wide usage of *ca.* £4,000,000,000 p.a. However, the Montreal Protocol ban on the use of many solvents has resulted in a compelling need to re-evaluate many chemical processes that have proved otherwise satisfactory for much of this century.

Ionic liquids offer a "green" alternative to molecular solvents. Because they have no effective vapour pressure, they can't escape into the environment and cause the problems that many molecular solvents do. And they're easily recycled. Thus, the principal aim of the QUILL programme will be to explore, develop and understand the role of ionic liquids as media for industrially-relevant chemistry, and to provide all the physical and chemical engineering data necessary in order to facilitate the design and operation of pilot plant. Ionic liquids possess, *inter alia*, the following desirable properties:

(1) they have a liquid range of 300°C, allowing tremendous kinetic control,
(2) they are outstandingly good solvents for a wide range of inorganic, organic and polymeric materials: high solubility implies small reactor volumes,
(3) they exhibit Brønsted, Lewis, and Franklin acidity, as well as superbasicity,
(4) they have no effective vapour pressure,
(5) they range from hydrophobic to hydrophilic, from water-sensitive to air-stable,
(6) they are thermally stable up to 200°C, and
(7) they are relatively cheap, and easy to prepare.

Unlike water and other hydroxylic solvents, they will dissolve a wide range of organic molecules: exploratory work in our own laboratories (carried out in collaboration with BP Chemicals and Unilever) has demonstrated that a wide range of catalysed organic reactions (including oligomerizations, polymerizations, alkylations, and acylations) occur in room-temperature ionic liquids, and that these are serious candidates for commercial processes. The reactions we have observed represent the tip of an iceberg – all the indications are that room-temperature ionic liquids are the basis of a new industrial technology. They are truly designer solvents.

QUB is building a new suite of laboratories for QUILL, including state-of-the-art handling facilities. A team of at least ten researchers will work on a four-year programme, which will include feasibility studies on the construction of at least two pilot plants.

Contact: E-mail: j.swindall@qub.ac.uk; WWW: http://quill.qub.ac.uk

Oxidative metabolites of aromatic hydrocarbons: chemistry between QUB and UCD

A partnership of microbiology and chemistry offers access to molecules not attainable by the chemist or microbiologist alone. This has been a guiding principle for the School of Chemistry at The Queen's University of Belfast working in association with microbiologists in Northern Ireland (QUESTOR Centre at Queen's), England (University of Warwick) and the USA (University of Iowa). Thus using wild type, mutant and recombinant strains of bacteria which are normally associated with the biodegradation of aromatic molecules in the environment (e.g. *Pseudomonas putida* and *Sphingomonas yanoikuyae*) it has been possible to oxidise aromatic rings to yield single enantiomers of both mono-*cis*-dihydrodiols (e.g. **1**) and bis-*cis*-dihydrodiols (e.g. **2**), and to oxidise dihydroarenes to yield enantiopure arene hydrates (e.g. **3**). Yields in excess of 10g/litre of single-enantiomer bioproducts from the culture medium are now being obtained using fermenters of up to 120 litre capacity and dioxygenase enzymes which catalyse the mono- or poly-hydroxylation of aromatic substrates *(Figure 1)*. Chemical modification of the *cis*-dihydrodiol and benzylic alcohol bioproducts has been widely used in the synthesis of both natural products and chiral auxiliaries. To date more than three hundred chiral *cis*-dihydrodiol metabolites have been isolated, and some have been used in the synthesis of cyclitols, alkaloids, modified sugars and substituted shikimic acids.

A focus of this article is a collaboration with the Chemistry Department of University College Dublin. The isolation and determination of the structure and stereochemistry of the diols and hydrates have been carried out at Queen's University using modern instrumentation, including X-ray crystallography *(Figure 2)*. Work at UCD has involved assessment of the stability and reactivity of the isolated metabolites, particularly with respect to acid-catalysed aromatisation. This information has allowed us to optimise the conditions for the detection of new transient intermediates, and the isolation of more stable metabolites. Problems currently under investigation include the role of a hydride shift (the so-called NIH shift, named after the US National Institutes of Health where it was discovered) in controlling reactivity differences between *cis*-dihydrodiols (bacterial arene metabolites) and *trans*-dihydrodiols (mammalian arene metabolites), the influence of heterocyclic oxygen atoms on the stability of carbocation intermediates in the aromatisation reactions, and evaluation of the resistance to hydrolysis of arene oxides.

Arene oxides are the initial products of epoxidation of carbon-carbon double bonds implicated in the eucaryotic (including mammalian) metabolism of benzene and other aromatic hydrocarbons. The carcinogenicity of benzene may derive from spontaneous electrocyclic rearrangement of benzene oxide (**4**) to form the oxepine (**5**) followed by further oxidation to yield a transient intermediate (**6**) which can ultimately form covalent bonds with DNA. These reactions compete with the more benign transformations involving enzyme-catalysed hydrolysis to yield *trans*-dihydrodiol (**7**), or nucleophilic attack with glutathione to yield *trans*-hydroxy thioether (**8**). An important factor identified in the collaboration as strongly influencing the course of this metabolism is the abnormally low chemical (and probably enzymatic) hydrolytic reactivity of benzene oxide.

It is well-established that acid-catalysed hydrolysis of an epoxide occurs 10^6-10^7 times more readily than dehydration of a structurally related alcohol. This reflects an enhanced reactivity towards carbon-oxygen bond-breaking in the molecularly strained three-membered oxide ring compared with the unstrained alcohol. Remarkably, benzene oxide proves to be *less* reactive towards aromatisation than the analogous alcohol, benzene hydrate (**9**). The origin of the exceptional thermodynamic and (or) kinetic stability of this simple molecule remains a mystery.

Figure 1. 50 litre fermenter at QUB (QUESTOR).

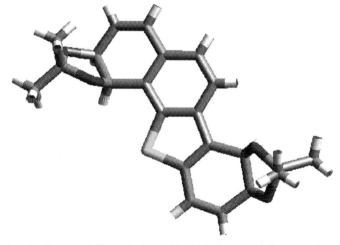

Figure 2. X-ray crystallography is used in the isolation and determination of the structure and stereochemistry of metabolites.

Contacts:
Professor Derek Boyd;
E-mail: dr.boyd@qub.ac.uk;
Dr John Malone;
E-mail: j.malone@qub.ac.uk;
Professor Rory More O'Ferrall;
E-mail:
rmof@macollamh.ucd.ie

TRINITY COLLEGE DUBLIN · EOIN P. O'NEILL

The TCD Enterprise Centre

Strategically located on an island site in the heart of Dublin, Trinity College is the most productive internationally recognised research centre in Ireland by many yardsticks. In it, through close links between Research and Innovation, and by making connections with the world outside, TCD has run a virtual science park for many years, sharing its campus with giants like Hitachi and Elan, nurturing SMEs like Nutriscan and Magnotic Solutions, and sometimes spinning off major enterprises like Iona Technologies. A constant need for more space in competition for more strictly academic purposes has curtailed the growth of technology transfer by the spin-off mechanism in recent years.

The IDA and the Department of Education have recently come to the rescue by enabling TCD to acquire the Pearse Street Enterprise Centre for development as a College-led Innovation and Enterprise Centre. About seven to ten minutes walk from the campus, and in the heart of Dublin's Docklands Area, this centre already houses three TCD spin-offs, the Dublin Business Innovation Centre, and several other companies which draw on TCD's research strengths.

It also has a Craft Centre, and room for substantial development of a new Innovation Centre. With nearly 10,000 square metres of built space gradually coming available, the strategy for Innovation at TCD can move forward dynamically in the next decade to populate the empty units and gradually to build up a reputation for knowledge based SME generation that will harness the College's output of research and new knowledge. The Centre will also be used to house any industrial R&D laboratories that can develop new products in collaboration with experts in TCD. Software companies and the IT industry generally, Biotechnology, and New Materials are expected to be technological sources for many new companies. But the Arts areas of College will also play a role. The most recently launched campus company, Eneclann Limited, harnesses the skills of two history postgraduates: their specialist area provides Internet services in genealogy, and electronic publishing. Indeed internationally traded services and virtual enterprises will probably emerge as strong contributors to the ethos of the Enterprise Centre.

The TCD Enterprise Centre.

Substantial opportunities for investment will arise from the clustering of expertise, young companies, and systems to link entrepreneurs, investors, international companies seeking new products and services, and academics with missions to apply their discoveries in commerce.

Those seeking to start-up new ventures or expand ventures that need TCD's research input should contact: innovation@tcd.ie (or call to see us at the Innovation Centre, the O'Reilly Institute, TCD).

TRINITY COLLEGE DUBLIN · DESMOND CORRIGAN

The School of Pharmacy at TCD

The School of Pharmacy, comprising four Departments and the Centre for Pharmacy Practice, offers courses leading to the B.Sc. (Pharm), as well as the Pharmaceutical Technicians Diploma and the graduate Diploma in Pharmaceutical Manufacturing Technology.

The Pharmaceutics Department also provides a taught M.Sc. in Pharmaceutical Technology. Its research focuses on the formulation of drugs into medicines and the design of drug delivery systems by investigating drug release mechanisms from formulations, microencapsulation methodologies, oral delivery of macromolecules, and the development of experimental and theoretical models to determine and predict bioavailability, particularly from the oral and transdermal routes. The Pharmaceutical Technology Centre provides specialist services in powder technology and thermal analysis to the pharmaceutical industry.

Pharmaceutical Chemistry offers a taught M.Sc. in Pharmaceutical Analysis building on one aspect of its research in pharmaceutical and medicinal chemistry. Research involves the design and synthesis of drugs, as well as the isolation, characterisation, analysis and properties of pharmaceuticals, augmented by studies of drug receptor interactions and drug metabolism. Current projects include work on novel antioestrogens of potential value in the treatment and prevention of breast cancer, impurity profiling of amphetamine drugs of abuse, stability studies for medicines registration purposes, hydrolysis kinetics of prodrugs, investigation of the mechanism of action of anti-inflammatory drugs, and the development of new carbapenem antibiotics for use against serious *Pseudomonas* infections.

The new Pharmaceutics Laboratory, School of Pharmacy, TCD.

Pharmacognosy deals with the phytochemical and biological evaluation of medicinal plants. Both native *(Taxus, Drosera, Ajuga)* and foreign plants *(Papaver, Dionaea, Leontopodium)* are being examined as potential sources of new anti-inflammatory, anti-cancer and anti-malarial drugs. The drug discovery programme also includes work on the chemical modification of the alkaloids, flavonoids, peptides, terpenes and naphthoquinones which have been found in the plants being studied. Biotechnology research has concentrated on the development of transformed root cultures of medicinal plants. Herbal medicines are also of interest, especially their evaluation and quality control. The Department forms part of Trinity's new Addiction Research Centre, and work continues on the methadone protocol and on analyses of Cannabis and Ecstasy.

Pharmacology research includes work on polyamines which activate the glutamate "NMDA" receptor which is implicated both in epilepsy and stroke. The Polyamine Research Group has recently shown that a novel polyamine analogue is the most potent polyamine antagonist available to date, and its activity as a prototype agent for the treatment of epilepsy or stroke is being investigated.

The Immunodulator Research Group is concerned with improving the treatment of auto-immune disease such as rheumatoid arthritis. A number of novel anti-inflammatory and immunosuppressant agents have been patented. Some of these compounds offer exciting possibilities for new approaches to immunosuppressant therapy.

Contact: Fax: 01-608-2810; E-mail: dcorrign@tcd.ie; www.tcd.ie/Pharmacy/

The TCD Department of Psychology

Psychology is a diverse science, concerned with problems such as repairing damaged brains on the one hand to social and cultural problems on the other. The Department of Psychology at TCD is organised into four major research groupings:

i. Developmental and Social
Developmental and Social: Research in child development and the psychology of women. The attached *Children's Research Centre (see article below)* researches early school leaving, juvenile crime, children in the care of the state and children's experience of drug problems *(Contact: Dr Sheila Greene)*.

ii. Neuroscience and Cognition
Thinking and reasoning research: Experimental and computational research on processes involved in thinking, reasoning and imagination. Basic issues in human rationality and imaginary thought are investigated, as is understanding thinking in applied settings, and disorders of thinking and emotion *(Contact: Dr Ruth Byrne)*.
Neuro-rehabilitation: The moulding of the brain by experience is an important question for the normal brain and for rehabilitation after brain damage. How experience changes the brain depends on the integration and competition between brain systems. We investigate attention, competition and integration in the brain, and develop methods

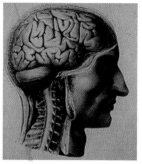

A basic problem for psychology is understanding the functions in the normal and damaged brain.

for fostering positive changes in the damaged brain *(Contact: Prof Ian Robertson)*.
Biology of learning and memory: Investigations centre on brain structures (such as the hippocampal formation) critically involved in the changes which occur during learning and memory.
Spatial function in humans after stroke: Investigations of patients who have had a stroke in the right-hand side of the brain and focus on locomotor and visually-guided tasks requiring feedback from body movements, the balance sense or sight *(Contact: Dr Shane O'Mara)*.

iii. Organisational Psychology and Ergonomics.
Aerospace Psychology Research Group: Investigates organisational safety systems and human factors in aviation, especially the role attitudes, culture and innovation play in safety systems *(Contact: Dr Nick McDonald)*.
Driver Behaviour Group: The modelling of human risk-taking in socio-technical systems (such as transport) and the design, development and assessment of interventions to enhance system safety *(Contact: Dr Ray Fuller)*.

iv. Health, Clinical and Counselling
Culture and Health: Investigations of how patients and clinicians, health service organisations, local communities and societies are influenced by, and represent, cultural constructions of illness and health.
Amputatees and Prosthetics: The Trinity Amputation and Prosthesis Experience Scales (TAPES) measures the experience of limb loss and prosthesis use. The TAPES predicts and differentiates stump and phantom-limb pain. Extensions of this work use virtual technologies to aid adjustment and acceptance of amputation and prosthesis *(Contact: Dr Mac MacLachlan)*.
Dyslexia Research Group: A particular interest is the development and evaluation of diagnostic assessments and remedial interventions for dyslexia *(Contact: Dr Ray Fuller)*.

Contact: Tel: +353-1-608-1886;
WWW: http://www.tcd.ie/Psychology/

Researching children's experiences: parental separation, homelessness, foster care, and prevention of early school leaving

Children are rarely consulted in research, even when it directly concerns them. The Children's Research Centre, Trinity College, aims to be inclusive of children's views and has a programme of empirical and evaluation research that focuses on children's own perspectives on their experiences and needs. Here we outline four studies, planned and on-going.

The Centre proposes to conduct a study on children's experience of **parental separation and divorce**. This study will explore the experiences of children in middle childhood and adolescence whose parents have separated. Key objectives will be to understand children's support needs at critical times of transition during and after the process of parental separation, based on their own experiences, identification of developmentally appropriate and child-centred structures and services to meet those needs.

A second empirical study will focus on the experience of **youth homelessness**. The Centre plans to conduct a follow-up study of a sample of young people who sought refuge in a homeless hostel service between the years 1976-1986. The participants, who are now adults, will be interviewed about their experiences as young homeless people, with the aim of identifying the factors that led to their becoming homeless, the services they received, and their progress to date. The findings will have implications for policies and practices of statutory and state-funded services for homeless youth.

The Centre is also embarking on a research project that will explore the needs and experiences of children who are placed in **foster care**. One element of this is an evaluation of the **Family Placement Initiative**, a pilot initiative in which, for the first time in Ireland, a voluntary agency is recruiting and supporting foster families. The aims are to understand children's subjective experiences as they make transitions from residential care into foster families, and to evaluate the delivery of such a service by a non-state agency.

The Centre is also conducting an evaluation of the **8-15 Early School Leavers Initiative**, which has been established by the

Department of Education and Science as a school-based response to preventing early school leaving. The main aim of the initiative is to pilot and examine the potential of different models of integrated area-based actions, both in-school and out-of-school, which prevent early school-leaving, and support the return to school of those who are no longer in the formal school system.

Through these and other projects, the Children's Research Centre is providing children in Ireland, particularly those who may be at risk, with opportunities to have their perspectives reflected in research that has clear implications for policy and practice.

Contacts: Barry Cullen, Dr Diane Hogan, & Louise Hurley.

Medicinal Chemistry at TCD

Plans are now taking shape for an exciting new degree in medicinal chemistry in TCD, and they complement research which is currently progressing in the Trinity Chemistry Department. Many distressing malfunctions of the human body are chemistry based, and a knowledge of what is going on at the molecular level helps scientists to devise suitable treatments.

One aspect of this is aimed towards the targeting, imaging and sensing of biologically important species. For example, the physical properties of elements such as Europium, Terbium, Gadolinium and Ytterbium (lanthanides) are being exploited in medicine. Synthetic macrocyclic complexes of these elements are being developed as luminescent emitting sensors for detecting ions and molecules, as contrast agents for magnetic resonance imaging (MRI), as probes for immunoassay and as catalysts or "ribozyme mimics" for the site-specific hydrolysis of mRNA. Such mimics open new avenues in the development of gene-specific therapeutic drugs.

In another direction, micro-cracks in bones can now be imaged using fluorescence. This can be seen in the *illustration*, where minute cracks show up as small elongated lines. This work is being developed further, using Computed Tomography for the early diagnosis of conditions such as osteoporosis, and is being carried out in collaboration with Dr Clive Lee in The Royal College of Surgeons in Ireland and Professor David Taylor in the Department of Mechanical Engineering at TCD.

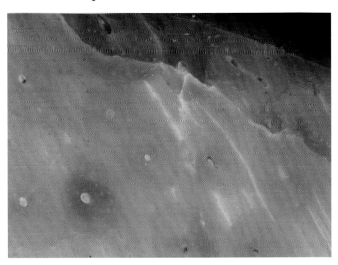
Micro-cracks in bones can be imaged using fluorescence: they show up as small elongated lines.

Another active research area has to do with chemicals in the body called pteridines. These were originally discovered in butterflies' wings, but they are now known to occur all over nature. Some of them are essential for the proper formation of DNA, which of course is what genes consist of. Other pteridines are required for the proper functioning of nerve impulses from and to the brain. How nature forms pteridines in the living cell is currently provoking some fascinating research, for without this process neither man nor animal can live. A series of enzymes controls the complicated transformation, and the properties of these enzymes are being examined in the Chemistry Department of TCD, in collaboration with research groups in Germany, Portugal, the UK, and Israel. This research is funded by the EU.

Contact: Dr Peter Boyle; E-mail: pboyle@tcd.ie;
Dr Thorri Gunnlaugsson; E-mail: gunnlaut@tcd.ie

Exercise ability with obesity

Obesity is excess fat content within adipose tissues stores, and is becoming more prevalent as we move towards the twenty-first century. Obesity results in adverse effects to daily health and longevity, and is associated with an increased risk of diabetes, heart disease, hypertension, gallbladder disease, arthritis, breathing problems, and some forms of cancer.

The aetiology of obesity displays a relative body fat component >20% for men and >30% for women, with body mass indices >31.1 and >32.3, respectively. [BMI = Weight(kg)/Height^2(m)]. Not only are there associated health risks with obesity, but also rising economic costs. In the United States, the 1996 National Health and Nutrition Examination Survey found the estimated economic costs of obesity and related diseases was in excess of $40 billion, while the public was spending a further $33 billion annually on weight-reduction products and services.

The pathogenesis of obesity is unknown. Both genetic and environmental factors are implicated to be involved in its development, including excess caloric intake, decreased physical activity, and metabolic and endocrine abnormalities. Unsuccessful treatment is commonplace with most attempts failing to result in a sustained reduction in obesity. Strictly controlled diets combined with exercise are considered the safest, and thus most regularly employed techniques to treat obesity. However, it has recently been shown in genetically obese mice (ob/ob) that a reduced skeletal muscle mass is a typical feature of obesity. This may limit exercise capability, and thus limit fat metabolism and the weight reduction achieved with exercise. Despite the knowledge that reduced skeletal muscle mass is associated with obesity, the functional capability of the remaining muscle is unknown. In the Department of Physiology at Trinity College, we have studied this question in the ob/ob mouse.

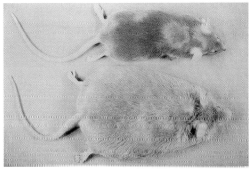
Two littermates. The genetically obese strain is used extensively as a research model of obesity.

The results suggest that skeletal muscle is functionally impaired in mice genetically predisposed to obesity, and point towards a deficiency in obese mice to utilise their skeletal muscle. Interestingly, the measured characteristics of obese skeletal muscle mimic those found in muscle from aged mice.

If similar characteristics occur in obese humans, they would limit the ability of obese individuals to undertake exercise. Given the similarity to aged skeletal muscle, exercise programs to treat obesity should perhaps be similar to those prescribed for the elderly. When combined with balanced low-calorie diets, optimal exercise would be of the low-impact aerobic type, comprising bouts of short duration, low intensity, and on a regular basis with two to three day rest intervals. This would ensure an elevated metabolism and subsequent reduction in adipose stores.

Although this would not be a cure for obesity, with time and hard work the benefits of diet and exercise should become apparent, and maintenance exercise should minimise potential relapse into the obese state.

Contact: Dr Stuart Warmington;
E-mail: warmings@tcd.ie

The Ocular Genetics Unit, Trinity College Dublin

Research into inherited eye disorders has been in operation in the Genetics Department, Trinity College Dublin, since 1985. During this period the research has grown and developed significantly, initially employing a single graduate student (JF) to its current status employing approximately twenty research scientists.

The focus of attention has been a group of inherited eye disorders involving progressive photoreceptor cell degeneration. Included in this group, amongst others, are Retinitis Pigmentosa (RP) and age related macular degeneration (ARMD). The accompanying figure is a picture of a human retina showing features typical of RP: the blackish pigment present in the retina is frequently observed in retinas from affected individuals. It is of note that ARMD affects approximately ten percent of people over sixty years of age: given the changing demographics of western populations, this group of patients represents a large and indeed growing untapped market in terms of therapeutics.

The research of the Unit has had a number of aspects, initially focusing on the elucidation of the underlying genetic pathogenesis of some inherited retinal degenerations. The team was the first to implicate a number of autosomal genes and one mitochondrial gene as causative of some inherited retinal degenerations. Recently, the focus of the team's research has shifted somewhat to the generation of appropriate animal models using transgenic technologies to study the disease pathology more closely, and to the exploration of novel therapeutic strategies. These strategies involve "replacement" for recessively inherited diseases where the disease pathology is due to the absence of a normal gene product. In contrast, therapeutic strategies for dominantly inherited diseases, where the pathology is due to the presence of a mutated gene and hence an abnormal gene product, involves suppression of the dominant disease gene.

With regard to the latter, during the course of the research, three novel and inventive platform technologies for dominant gene suppression have been developed. Notably, these technologies are not solely appropriate for inherited retinal degenerations but would indeed be appropriate for many of the one thousand plus dominantly inherited human disorders characterised.

In conclusion, the primary aim of the Unit currently is the development of novel therapeutic approaches for a group of debilitating inherited eye disorders.

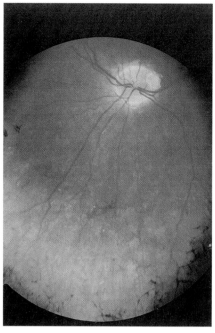

A human retina showing features typical of advanced Retinitis Pigmentosa (courtesy of Dr Paul Kenna).

Contact: Dr Jane Farrar;
E-mail: jane.farrar@tcd.ie

Inflammation and Cancer

Inflammation and stress response

Cellular stress occurs as a consequence of tissue injury, inflammation, infection and autoimmune responses, and represents a novel perspective on a wide variety of clinical conditions.

Pro-inflammatory Cytokines

Inflammatory Diseases include rheumatoid arthritis, ulcerative colitis, inflammatory bowel disease, psoriasis and other inflammatory skin diseases, multiple sclerosis, and septic shock. Investigations in Dr O'Neill's Laboratory focus on the pro-inflammatory cytokines interleukin 1 (IL1) and tumor necrosis factor (TNF), both of which have been recognised by the biotechnology industry as key mediators of inflammatory disease. The overall aim of the work is to identify targets for novel therapeutics and to exploit compounds with potential to block inflammatory mediators which play a determining role in the pathogenesis of chronic inflammatory disease.

Interleukin 1 and tumour necrosis factor signal transduction

Interleukin 1 and tumour necrosis factor are considered central mediators of the inflammatory response. Research is underway to uncover the signal transduction pathway activated by these cytokines and, particularly, the role of G proteins (both heterotrimeric and small G proteins), protein kinases (particularly p38 MAP kinase, MEK and JNK) and transcription. By utilising the yeast two hybrid screen, novel genes have been identified which code for proteins likely to be important in IL1 signalling. Recent work has focussed on the involvement of two recently-identified mediators of IL1 signal transduction in T cells, Traf6 and Irak.

NFkB transcription factor

NFkB has been identified as a key transcription factor - a DNA binding protein which regulates the expression of many genes which play a role in immunity and inflammation e.g. IL2, TNF, adhesion molecules and acute phase proteins. This project entails a full characterisation of the mechanism by which IL1 and TNF activate the NFkB system. In particular, the involvement of protein phosphorylation, proteases and reactive oxygen intermediates in the activation process are being examined.

NFkB and cancer

Recent studies carried out with Dr Dermot Kelleher of St James's Hospital, Dublin, have suggested that in Sezary's lymphoma (a type of cutaneous T cell lymphoma), aberrant NFkB leads to TNF production which drives proliferation in the cells. A blocking antibody to TNF decreases NFkB and is also anti-proliferative. This raises the possibility that anti-TNF therapy may be useful in this disease, and this study is currently being expanded to other cutaneous T cell lymphomas. Dr O'Neill's group has made the novel observation that anti-neoplastic agents, such as daunorubicin, are potent activators of NFkB. This may be important in drug resistance, and the possibility that limiting this response may improve drug efficacy is being investigated.

Contact: Dr Luke O'Neill, Dept of Biochemistry,
Biotechnology Institute, Trinity College Dublin;
Tel: +353-1-608-2439;
Fax: +353-1-608-2755;
E-mail: laoneill@tcd.ie

Atmospheric change studies at Oak Park

In 1987 the first studies investigating the impact of atmospheric pollutants on growth and yield of crops were initiated at Oak Park, Carlow, by the Botany Department, Trinity College Dublin and Teagasc, Oak Park. Initial experiments used open top chambers to quantify the effect of gaseous pollutants on the growth and yield of crop plants.

Ozone has been found to be the most prevalent pollutant in agricultural areas, although localised problems can be caused by high concentrations of oxides of nitrogen and/or oxides of sulphur. Although ozone is a natural constituent of the atmosphere, its concentration in the lower part of the atmosphere (troposphere) can be elevated by the action of sunlight on mixtures of primary pollutants (oxides of nitrogen, hydrocarbons). As a very reactive species, ozone can cause damage to plants and, when present in high enough concentrations, can reduce the yield of crop plants. This action is in contrast to the beneficial effects of ozone in the upper atmosphere (stratosphere) where it prevents harmful ultra violet radiation from the sun reaching the surface of the earth.

Studies at Oak Park initially concentrated on investigating the effect of ambient concentrations of ozone on the growth and yield of crop plants, including beans, sugar beet, radish, clover, spring wheat and spring barley. Although concentrations of ozone in the Irish atmosphere were not high enough to reduce yield, experiments conducted elsewhere have found that concentrations on the European mainland can reduce spring wheat yield by 10%. Time-concentration studies were carried out using the chambers at Oak Park in which equal amounts of ozone were added to treatments either as high concentrations over short time intervals or lower concentrations over longer time intervals. These studies illustrated the importance of peak concentrations, and the resulting information was used to formulate indices; descriptions of ozone exposure which are used as standards for crop protection.

The focus of the investigation using the open top chamber facility subsequently shifted to studying the interactive effects of carbon dioxide and ozone. Carbon dioxide is another natural constituent of the atmosphere and is of pivotal importance to the growth of plants as it is absorbed into leaves and used as the essential building block by which plants grow. Its concentration in the atmosphere is increasing by approximately 0.5% per annum, and the increased availability of this essential building block is expected to have a significant impact on the future growth of plants.

The European stress physiology and climate experiment (ESPACE) investigated the

An open top chamber, at Oak Park, Carlow.

interactive effects of carbon dioxide and ozone on wheat crops, and was funded by the EEC. The experiment was conducted in seven European countries: the Irish experiment was carried out at Oak Park between 1994 and 1997. It was found that elevated concentrations of carbon dioxide increased spring wheat yield, whereas elevated concentrations of ozone reduced yield, and that elevated concentrations of carbon dioxide partially compensated for the damaging effects of elevated ozone on grain yield.

The open top chamber facility is currently being used to study the interactive effects of carbon dioxide and ozone on potato crops. The project, Changing Climate and Potential Impacts on Potato Yield and Quality (CHIP), is a programme of co-ordinated experiments and mathematical modelling at sites across Europe which aims to establish the impact of future climatic change and associated stresses on potato growth, yield and quality.

In response to concerns about the impacts of climate change and loss of biodiversity on European grasslands, a collaborative programme of experiments and modelling activities (MEGARICH) was begun in 1998. Six experiments were established, in different locations throughout Europe, and the aim is to investigate the long-term response of this wide range of grasslands to climate change (elevated CO_2 and temperature) and the interactions with management regimes (nitrogen fertiliser, cutting frequency and grazing). The Irish experiment is located adjacent to the open top chamber site at Teagasc, Oak Park, Carlow. In 1998, 48 grassland cores (monoliths) measuring 0.4 m x 0.4 m x 0.4 m were extracted from a representative Irish permanent pasture in Co. Kildare and transported to the field site at Oak Park for experimentation. Monoliths were arranged within a miniFACE (free air CO_2 enrichment) exposure system which consists of a control unit and 12 exposure rings. The experiment is based on a randomised two-factorial block design with three replicates. The treatments are: elevated CO_2 (600 μmol mol^{-1}), ambient CO_2 (355 μmol mol^{-1}), high nitrogen (250 kg ha^{-1} yr^{-1}) and low nitrogen (50 kg ha^{-1} yr^{-1}). Treatments started in April 1999.

A major challenge of this work is to develop sustainable grassland management systems which may mitigate the effects of climate change (through increased carbon storage in soils) and which will simultaneously preserve and/or enhance biodiversity.

Contact: Botany Department, Trinity College Dublin;
E-mail:
jfinnan@oakpark.teagasc.ie,
clrbyrne@oakpark.teagasc.ie
& jonesm@tcd.ie;
Web Page: http://www.tcd.ie/Botany/megarich

The miniFACE exposure system.

New signal processing techniques for use in medicine

Electronic engineering research is at the heart of many of the current advances in medicine. In particular, Digital Signal Processing (DSP) - which is concerned with the development of algorithms to enhance, extract, predict and model the information content in raw data - is poised to revolutionize many routine, but data-intensive, areas of current medical practice. Researchers in the Department of Electronic and Electrical Engineering at TCD are contributing to these developments. Some of their research is now described.

Texture-Based Image Segmentation

The capture, processing and inspection of still images and video sequences are activities found in practically all modern hospital departments, from ultrasonic fœtal examinations, to MRI scanning of the brain, and endoscopic examinations of gastric tissue. In contexts such as mass screening, diagnosticians are being confronted with vast amounts of image data which need to be communicated between experts, processed, classified and possibly archived.

In many contexts, anatomical structures and regions of diagnostic significance (tumours, ulceration, etc.) are distinguishable by the expert as a subtle variation of the *texture* in the image. Dr Anthony Quinn and his students, Ed Clark and Elena Ranguelova, are collaborating with gastroenterology experts at St James's Hospital, Dublin, on the development of state-of-the-art Bayesian statistical algorithms to segment images into regions of distinct homogeneous texture *(see illustration)*. This technology will be used as the basis for a Computer-Aided Diagnosis (CADx) system to aid the diagnostician in identifying structures and pathologies in endoscopic image sequences.

The segmentation algorithms can also be used to *compress* medical images. An approximation to the original image - which preserves diagnostic cues - may be regenerated from the segmentation map, using samples of each identified texture *(see illustration)*. The map and texture samples can be encoded using far fewer bits than are necessary for the original image.

These methods are in the vanguard of efforts to devise smart approaches to the retrieval of specific images (e.g. those illustrating a particular disease) from large databases. In fact, any application involving textured images can benefit from these image analysis techniques.

ECG Signal Processing

Dr Martin Burke has applied wavelet analysis techniques to the human electrocardiogram (ECG) in order to characterize the variation in its constituent components with changing heart rate. This has yielded equations describing the durations of the various segments of the ECG as a function of cardiac cycle time. These equations can be used to synthesise an ECG signal which is a realistic reproduction of the *in vivo* signal, but which has controllable parameters such as QRS complex amplitude, rise-time, fall-time and the relative amplitudes of the P- and T-waves. The duration of each component will automatically track the selected heart rate in a non-linear fashion, reflecting its true behaviour. This will provide an invaluable - but cost-effective - tool for testing, calibrating and maintaining electrocardiographic equipment in hospitals and clinics, and for the design and improvement of new and existing instrumentation.

Work is also in progress on the design of very low power amplifiers for use in dry-electrode recording of the ECG signal. It is hoped to be able to record high quality signals without the need for pasted or adhesive electrodes. An amplifier with very high input impedance, a wide range of automatic gain control and a microwatt level of power consumption, is under development. It will operate from a low voltage battery supply and can be mounted on the electrodes. This will allow ECG signals of diagnostic quality to be obtained in situations where clinical preparation is impossible. Contexts include amateur athletics, cardiac rehabilitation programmes, and ambulatory monitoring of hospital outpatients.

Digital Signal Processing for the Human Auditory System

Dr Brian Foley and Derek O'Reilly are investigating the use of DSP to model the human auditory system, and its potential use in the alleviation of hearing impairments affecting this system. Before DSP can be effectively employed, the problem of developing engineering models of these impairments must be addressed. Furthermore, an effective means of tuning these models for each individual must be found. Currently, two types of impairment are being investigated.

The first is *recruitment*. This impairment causes the sufferer to be unable to hear the normal range of sounds that an unimpaired individual can hear. It is due to an abnormally large growth of loudness sensation. The net effect is that – while sounds *are* heard – there is an inability to comprehend the information that they carry.

The second is *frequency smearing*. This impairment results in the sufferer's ear being unable to respond accurately to the frequency domain resonances in natural speech. Once again, the individual can hear sounds, but is unable to pick out those natural cues that allow similar sounds - and hence similarly structured words - to be distinguished.

Both of these impairments cause the sufferers to have reduced auditory comprehension of the world around them. Their effect is particularly detrimental in young children, where the development of language skills and awareness of the environment are impeded.

Time-frequency methods are being employed in the development of compensating functions for the above-mentioned impairments. These can be integrated into an externally-worn DSP-based hearing aid.

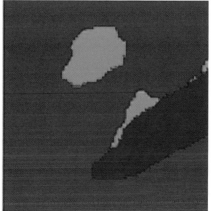

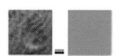   

The gastric tissue image (top left) is automatically segmented into three distinct textures (samples given on the second line). Their locations are indicated by the segmentation map (top right).

Contact:
Dr Anthony Quinn
Tel: +353-1-608-1863;
E-mail: aquinn@tcd.ie

High Performance Computing and Bioinformatics

In 1986 Trinity College (TCD) launched an initiative to bring High Performance Computing (HPC) to researchers, business and industry in Ireland. To date, strong partnerships have been established with Queen's University and with Hitachi Dublin Laboratory (HDL), and a number of technology transfer projects have been developed with companies in Ireland and in France. Over the last year, the Trinity initiative has focused on developing novel techniques to bring advanced computing to non-traditional users of supercomputing. Areas which are ripe for exploitation with these new techniques include the emerging field of Bioinformatics.

The Human Genome Project to identify all the estimated 60,000-100,000 genes in human DNA is now scheduled for completion in 2003.

Although molecular biologists have long known that a sea change will occur when the entire human genome is sequenced, the task of managing, analysing and designing appropriate research with the resultant information is very large. Yet it is a task that the molecular biologists and the Information Technology community must tackle together and solve. Now is the time to anticipate some of the needs for software, data management, training, and access to information. Recent work led by Dr Donal MacDonaill (Dept of Chemistry) and his colleagues shows that new approaches from the computational research side will enhance the ability of biologists to handle the new data. Interdisciplinary work emanating from teams which have strength in their own research specialisations will be a powerful aid to advancing science.

The development of advanced computing tools for molecular biologists will be a priority for the teams working in TCD. The programme will be in collaboration with industry specialists, and run by Dr James Sexton, TCD's Director of High Performance Computing, and Dr Luke O'Neill, the College's Director of the Biotechnology Institute. Other firms, particularly those from the pharmaceutical community, will be invited to participate and assist in the development of new computing tools.

In advanced computing work to date, experts at HDL and at TCD have elaborated some new tools for middleware, and have developed a strong competence in solving advanced modelling problems. Project plans at TCD call for a dedicated floor in the proposed IITAC (Institute for Advanced Computing) to house researchers in high performance computing from across College.

The liaison between industry and this new programme will be co-ordinated by the Irish Centre for the Transfer of Advanced Computing Technology, where the Marketing Manager, Ms Audrey Crosbie, is responsible for developing interest among niche SMEs and also larger companies in HPC projects and services. The goal of this liaison is to spin-out a company providing services to users of Bioinformatics tools.

Trinity's original initiatives in Bioinformatics, such as the Irish National Centre for Bioinformatics, will continue with their existing programmes and service to researchers in College and in Ireland.

Other projects in TCD in the Departments of Genetics, Microbiology, Clinical Medicine, Chemistry and Biochemistry will participate as resources permit. Each of these Departments has extensive programmes of research as users or developers of Bioinformatics, and these may be examined via the TCD Home Page (http://www.tcd.ie).

A major part of Bioinformatics involves the use of computer programmes to compare gene sequences and to use homologies between genes to search the growing database of gene sequences for related genes. This method gives clues as to the possible function of unknown genes and also indicates the regions in proteins important for their function. An alignment is shown for members of the Interleukin-1 receptor family, which are key players in the pathogenesis of inflammatory diseases. Important regions are identified (shown as boxes) which can then be used in efforts to understand how these proteins function. The information obtained can be used in the development of novel drugs with anti-inflammatory properties which target these regions.

Bioinformatics is as difficult for the layman to understand as the text on the Ogham stone, but its importance to the future is as great as the Ogham stone is to the past. This Ogham stone is from Coolnagort, Dungloe, Co. Kerry (photo courtesy of the Office of Public Works).

Contact: Dr Eoin O'Neill
(Research and Innovation Services);
E-mail: eponeill@tcd.ie;
or Dr James Sexton;
E-mail: james.sexton@tcd.ie

Mathematics in the Knowledge Economy

The Department of Pure and Applied Mathematics at Trinity has a long tradition of providing a broad training to its undergraduates and for carrying out fundamental research in a wide range of topics. This historical strength of the Department has placed it in a position where it can respond positively to the needs of Ireland's modern knowledge-based economy while maintaining an international presence in fundamental research.

Financial transactions, globalisation of economics, and the emergence of e-commerce mean that quantitative tools for analysing complex systems and for decision making are becoming more and more important. For instance, the security of e-commerce transactions is underpinned by sophisticated results from number theory, advanced optimisation techniques are regularly used for logistics and scheduling, and advanced stochastic methods are used in financial analysis and database mining. It is thus of utmost importance to have trained personnel who have the proper skills to respond to these changes. In general, these skills must include knowledge of mathematical modelling and computer simulation. Over the past few years, the Department has added many numerical modelling and computational elements to both its undergraduate and graduate programs.

Concurrent with the growth in the use of sophisticated mathematics for the world of commerce and industry, spectacular advances in mathematical research are occurring. The proof of Fermat's Last Theorem is an example. In fundamental theoretical physics, too, major advances in our understanding of the nature of space, time and black holes have been made in very recent times.

The Department of Pure and Applied Mathematics has fourteen permanent staff members, four funded post doctoral fellows, forty postgraduate students, and has research support from eighteen grants. It continues to attract excellent undergraduate students in mathematics and theoretical physics. Research programs in the Department include (among other items) string theory, analysis, non-commutative geometry, fluid simulation, and telecommunications.

One Departmental research area relates to String Theory and Black Holes. String Theory aspires to unify the four fundamental forces of nature, gravitation, electromagnetism, the weak force of radioactivity and the strong force present inside the nuclei of atoms. Black Holes are regions of the universe where the nature of space and time are profoundly modified. Stephen Hawking proved in the

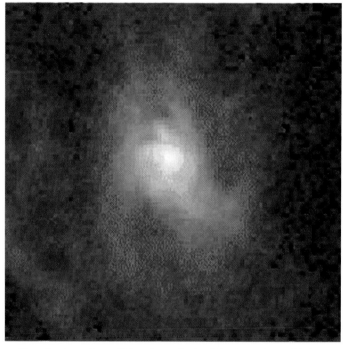

The galactic nucleus of M87. An analysis of the dynamics visible here suggests the presence of a black hole. This picture was created with support from the Space Telescope Science Institute and is reproduced with permission from AURA/STScI.

1970s that a Black Hole has many properties normally associated with complex systems, temperature, heat radiation and entropy (a measure of its complexity). A proper, universally accepted, understanding of these properties was, however, lacking until very recently. In 1996, using ideas from String Theory, it was shown how Black Holes could be constructed and their entropy calculated. Researchers in the Department, in collaboration with colleagues from University College Dublin, have clarified a number of conceptual problems by considering a mathematical model known as the BTZ Black Hole. A particularly interesting result established by the group (which appeared in the prestigious *Physical Review Letters* journal) shows how the BTZ Black Hole has a "holographic property" – i.e., all the properties of the three dimensional system are contained in its two dimensional boundary: just as a two dimensional hologram contains three dimensional images within it.

In pure mathematics, topics of current interest in the Department include completely positive and completely bounded operators. These relate to modern "non-commutative geometry", a subject where traditional geometrical ideas first appeared in algebraic terms and were then extended. Other pure mathematical topics include more traditional differential geometry, operator algebras and wavelets. Wavelets are a modern rival to Fourier analysis (long established as a fundamental tool for scientists and engineers) and may become established as decidedly preferable in certain application areas such as signal processing, pattern recognition in images or data filtering.

Another recent research activity in the Department is the Navier-Stokes Project. This project is concerned with accurate and robust computational methods for fluid flow problems having non-smooth solutions. The mathematical equation which is used to describe the flow of fluids is known as the Navier-Stokes equation. Solving this equation under different conditions is an important problem. The class of problems with non-smooth solutions of the Navier-Stokes equation is rapidly increasing, and many of the grand challenges in modern science and technology give rise to mathematical problems in this class. The methods being developed by the group, in collaboration with researchers from Dublin City University, the University of Limerick and Kent State University, Ohio, USA, are novel. The project has benefited from the expertise of Professor Shiskin of the Russian Academy of Sciences, Ekaterinburg. If successful, the project could have a tremendous impact on diverse fields from weather forecasting to chemical engineering.

Contact: Professor Siddhartha Sen,
School of Mathematics, TCD;
Tel: +353-1-608-1949;
E-mail: sen@maths.tcd.ie

Origin of the proton's spin

The spin of a proton plays a critical role in a variety of Magnetic Resonance studies such as the MR Imaging (MRI) of body tissues. Like the spin of its neutron partner, it contributes to the spin of the many atomic nuclei suitable for MRI applications and to the study of chemical compounds using Nuclear Magnetic Resonance (NMR). The inert gases helium-3 and xenon-129, for example, show considerable promise for the rapid MR Imaging of the lung when the gases are suitably spin polarised.

A profound combination of quantum and relativity theories reveals that spin is an intrinsic property of a particle, taking values related to a particular quantum unit. The component quarks of a proton, for example, have intrinsic spin, as do the gluons which bind those quarks together. A recent surprise is that less than a third of a proton's spin may be attributed to the spin of its charged component quarks. Electron accelerators at Stanford, Hamburg and CERN have deduced this result by probing the structure of the proton with photons which interact only with those quarks. The gluons in a proton are neutral and invisible to photons. Thus the major contribution to a proton's spin seems to come from its gluons. To probe the gluon structure one must use a beam of ions. A Relativistic Heavy Ion Collider (RHIC) at Brookhaven near New York will accelerate polarised protons to unprecedented energy from the year 2000. One aim will be to measure the spin contributions from a proton's gluon, quark and anti-quark components.

Aerial view of Brookhaven National Laboratory, Long Island, New York, showing the Relativistic Heavy Ion Collider ring.

Calibrating the polarisation of a beam with energies higher than 30 times the rest energy of a proton poses a difficult challenge. One method involves scattering polarised particles at the small angles where strong and electromagnetic forces are of similar strength. This is equal to the challenge provided the spin dependence of the strong interaction is sufficiently well understood. Bounds on such spin dependence have been derived from mathematical principles based upon causality, probability and invariance under the reversal of matter with antimatter, of time, and of space. A collaborative project, part-funded by Enterprise Ireland, including researchers from Trinity, CERN and Brookhaven is successfully addressing these issues.

Measurement of the gluon polarisation and the quark and anti-quark polarisation by flavour – up, down and strange – will then be possible using the enormous detectors at RHIC with their considerable facility for data analysis using high performance computing. A calibrated proton beam whose polarisation is known to sufficient accuracy will assist in understanding the source of a proton's spin.

Contact: Dr N.H. Buttimore, School of Mathematics, TCD; E-mail: nhb@maths.tcd.ie

Lattice QCD

Quantum Chromodynamics (QCD) is the theory which describes how nuclear matter is constructed. Its constituents are matter fields called quarks and force carriers called gluons. Quarks play a role identical to protons and electrons, and gluons play a role identical to photons in normal chemistry. Thus, studies of QCD are the nuclear matter equivalent of normal chemistry.

Unfortunately, the analogue with normal chemistry ends at that point. The experimental approach to QCD is hugely expensive, and involves building very large particle colliders to study, for example, what happens when protons and anti-protons are collided at high energies. One innovation in the field is the realisation that modern computer power can be used to analyse interactions at low energies and reduce the need for such accelerators.

The numerical formulation of QCD dates from the late 70s and is know as Lattice QCD. In this formulation, computer simulation replaces the theorist's pen and paper calculations, and the goal is to study experimentally accessible but theoretically difficult-to-calculate properties of nuclear matter such as the mass of the proton and neutron, and the decay properties of pions.

Lattice QCD has its own particular problems however and is enormously expensive computationally. The current understanding of the field is that a full solution will require a computer with a performance of 10 trillion operations per second. Research directions include development of advanced computer architectures and algorithms, software optimisation, low energy physics calculations to certify hardware and software systems, and application of the methodology in "blue skies" areas such as the identification of new types of matter.

The School of Mathematics has been collaborating for some years with the Hitachi Dublin Laboratory, and with UKQCD (a consortium of UK theoretical physics departments including Edinburgh, Southampton, Glasgow, Swansea, Liverpool, Cambridge, and Oxford) and participates in all these research areas.

A recent result is shown in the plot which demonstrates the performance of a new algorithm to evaluate vector-matrix-vector inner products of the form $(\psi, \log(M) \psi)$ where the matrix M is of size $10^6 \times 10^6$. The horizontal axis counts the number of matrix times vector iterations. The vertical axis

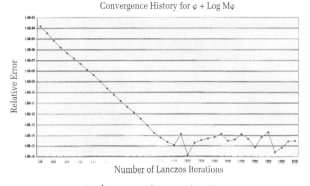

Performance of a new algorithm to calculate vector-matrix-vector inner products.

shows the absolute error in the estimation of the inner product. After about 1000 iterations, this new algorithm is seen to converge. We have separately proven that the converged answer is correct to machine precision. The basis of this new algorithm is the Lanczos matrix tridiagonalisation procedure, and one surprising spin-off from this work is an improved understanding of how finite precision roundoff affects this fundamental method in matrix analysis.

Contact:
Dr James Sexton,
School of Mathematics, TCD;
E-mail: sexton@maths.tcd.ie

TCD Department of Physics

The TCD Physics Department has a long and distinguished history of teaching and research. Richard Helsham, appointed in 1724, was the original Erasmus Smith's Professor of Natural and Experimental Philosophy (i.e. Mechanics and Physics). He was the first to lay out Newton's methods in a form suitable for the undergraduate, so that his lectures in Natural Philosophy were in use for a hundred years in the College. More recent holders of the Chair include G.F. Fitzgerald, famous for the Lorentz-Fitzgerald contraction, and E.T.S. Walton, the only Irish recipient of a Nobel Prize in Science. This tradition is maintained today through a very successful research program and respected degree courses.

Departmental research has spawned two campus companies, "Magnetic Solutions" and "Horcom", and the Programmes in Advanced Technology (PATs) "Optronics Ireland" and "Materials Ireland". We have approximately 60 graduate students working for the degree of M.Sc. (by research) or Ph.D. We recruit many of our own graduates into these programs, as well as graduates from many other countries, both inside the EU and beyond. Each year the Department raises well over £1 million in external research support, and publishes about 100 peer-reviewed papers.

Our degree course in Physics includes an Astrophysics option in the third and fourth years. There is a separate degree course in Computational Physics. We also run a Theoretical Physics degree course jointly with Mathematics, and a Science of Materials degree course jointly with Chemistry. Our undergraduates run the College Physical Society and organise a series of events each term.

All of our courses are accredited by the Institute of Physics. This is our professional association, and both students and staff may join. It spans Ireland and Britain, and is a member of the European Physical Society (EPS). The current President of the EPS, Denis Weaire, is our present Erasmus Smith's Professor, and the first Irish physicist to hold the presidency.

The aim of all our courses, as with our substantial research programmes, is to achieve a flagship level of excellence. Many of our students go on to win college scholarships and prizes, and ultimately high honours in their degrees. In 1996, the graduating class in Theoretical Physics all obtained first class honours. In both 1995 and 1997, students from our Department beat all other contestants from Ireland and Britain to win the Institute of Physics undergraduate lecture competition: the 1997 winner was 18 years old and still in his first year here. In 1994 and 1995, the Institute's Irish Branch Physics project prize was won by students from the Science of Materials course.

The Department is located in the Physics building at the East End of College, with additional research laboratory space in other areas, including the O'Reilly Institute. Construction will shortly begin of a large Institute of Advanced Materials Science adjoining the main Physics building, in which the Physics Department will occupy most of the space.

Because our degrees cover both experiment and a substantial element of theory, they are a general preparation for a wide range of careers. These include school and university teaching, government services such as meteorology, and industrial positions in research, development and management. The present fast-changing technologies require graduates who are versatile and flexible, and capable of applying themselves to very diverse fields. A physics degree with its broad base is an exceptionally good qualification for this new world. Indeed, a recent survey declared that "the demand for physicists on the European job market looks set to rise sharply".

> Recently there have been several major breakthroughs at the forefront of science by staff in the Physics Department. Professor Denis Weaire and his research student, Robert Phelan, have come up with a complex cellular "shape" which when packed together is the most efficient way to fill all space, thereby beating for the first time the shape suggested by the great Lord Kelvin a hundred years ago. Professor Michael Coey has put magnetic atoms together in such a way as to produce state-of-the-art magnetic materials, a basis for the new Campus Company, "Magnetic Solutions". Professor John Hegarty, with Hitachi's Trinity Lab, used nanostructures obeying strictly the laws of quantum mechanics to perform the brain-like function of pattern recognition. Professors Vincent McBrierty and Werner Blau patented a totally new laser process for making contact lenses, arising from a knowledge of how light and matter interact. These have been acknowledged as significant milestones in advancing our knowledge of and control over the world we live in.

An ultra-high vacuum, low-temperature scanning tunneling microscope in Dr Igor Shvets' group.

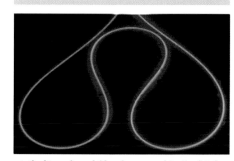

A thulium doped fibre laser used in Prof. John Hegarty's group for gas sensing.

The Physics Department Internet home page is at: http://www.tcd.ie/Physics/; The postal address is: Department of Physics, Trinity College, Dublin 2; for further information about the Department, contact Ms Michelle Duffy.

TRINITY COLLEGE DUBLIN — DENIS WEAIRE, SARA MCMURRY & STEFAN HUTZLER

Physicists take a look at foam

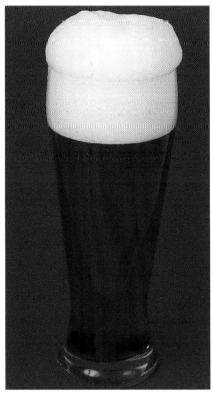

Interest in foam is not confined to poets and drinkers.

The head on a glass of beer presents many interesting effects to the discerning eye of the physicist. The drainage of liquid from between the bubbles forces them into elegant polyhedral shapes. The diffusion of gas changes their sizes. Eventually the rupture of thin films causes their collapse, in a metaphor for instability invoked by poets throughout the ages.

Interest in foam is not confined to poets and drinkers. Throughout the chemical industry, designing and controlling foams is important to processes and products, both liquid and solid.

No wonder then that the Foams Research Group of TCD has a collaborative link with the Shell Laboratories in Amsterdam and Leuven. One of their joint endeavours lies in the pursuit of better methods of monitoring foam density, by measuring local resistance or capacitance. The TCD team has helped to demonstrate the effectiveness of an instrument developed by Shell for this purpose. It has been applied to analyse the drainage of liquid in a foam column. In particular, it has been used to examine what happens when additional liquid is added to a foam (more or less as when you top up your glass of beer). It travels downwards in what mathematicians call a solitary wave. This means that it has a constant profile.

The solitary wave in drainage was discovered at TCD several years ago in basic research funded by Forbairt. Its precise profile was measured for the first time with the new instrument. It was found to agree with a mathematical theory which is also a joint Shell/TCD accomplishment, and has been recognised as a notable one by the Shell Laboratory.

The resistance profile monitor, as it has come to be called, will be developed further. There is even a suggestion that it may be used in research in space, where foam behaves very differently, in the absence of gravity.

Contacts:
Prof. Denis Weaire, dweaire@tcd.ie;
Dr Sara McMurry, smcmurry@tcd.ie;
Dr Stefan Hutzler, shutzler@tcd.ie

TRINITY COLLEGE DUBLIN — EOIN O'NEILL – FROM PAPERS OF MICHAEL COEY

Attractive materials

Permanent magnetism is a technology that we encounter in many phases of life: from the aids to navigation in the humble compass (China AD 1088), to the modern earphone in the Walkman. More important but less obvious uses of magnets in computers, motors, electrical generators and magnetic recording devices have spurred research into new and improved magnetic materials. Among the important qualities sought are that a permanent magnet will be strong (have a high energy product), will hold its magnetisation against opposing magnetic fields (coercivity), and will retain its properties when heated. The improvement in new materials during this century reflects research at every level, from metallurgy to theoretical physics (see *Figure*).

Leadership on an international scale in the search for new magnetic materials, the understanding of why materials behave as they do, and the application of new materials in magnetic devices have been features of Professor J.M.D. Coey's research in Trinity College for a decade and a half. Awarded the Charles Cree Medal of the Institute of Physics (London) in 1996, Mike Coey's research has attracted financial support and industrial and academic interest from around the globe (80% of his research funds are external investment in Ireland); he has led a European wide consortium (over 70 laboratories) created to re-establish the technological competitiveness of Europe's research on and technology competence in permanent magnetism relative to Japan and the USA. Currently he leads a European network on spin electronics.

With the support of Denis Weaire and Paul Coughlan, he set up a campus company, *Magnetic Solutions*, in 1994 to develop and manufacture instruments using new magnetic materials. These compact devices provide uniform, variable magnetic fields without cumbersome electrical power and water-cooling requirements. They can supersede electromagnets for many practical purposes.

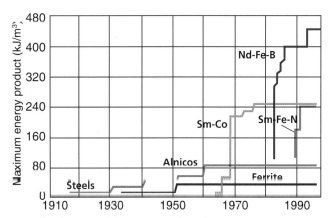

Development of the energy product of permanent magnets during the Twentieth Century.

The importance of new magnet technology grows as recording media and reading heads are miniaturised, and as complex processes for laying down thin films of materials are magnetically controlled. As devices using magnetism approach the nanotechnology level, the need for physicists with an understanding of the complex behaviour of magnetic materials and heterostructures will grow rapidly. An integration of magnetics with electronics (spin electronics) is a likely development, according to Coey. If we are to sustain in Ireland a modern IT industry with its own ability to master and advance new technologies, then we will need to expand and support the activities of Mike Coey's research group, since he and his colleagues will train the solid-state engineers of future industries.

Contact: *J.M.D. Coey, Physics Department,*
Trinity College, Dublin 2; Tel: +353-1-608-1470/2019;
Fax: +353-1-671-1759; E-mail: jcoey@tcd.ie

TRINITY COLLEGE DUBLIN — ALAN SHAW, BRENDAN ROYCROFT & JOHN HEGARTY

Lasers the size of a photon

Lasing is a property of light that has been known and studied since the 1960s. The first lasers (LASER stands for **L**ight **A**mplification by **S**timulated **E**mission of **R**adiation) were large inefficient versions that fitted on a bench top, but now they are taken for granted in CD players, laser pointers, eye surgery, and an ever expanding range of applications. Recent work at Trinity College Dublin has observed lasing in devices that are only three millionths of a metre in diameter, not much larger than the wavelength of light itself.

The devices are made from semiconductor material (as are the silicon chips in all personal computers), which emit a particle of light when a high energy electron falls suddenly into a low energy state. The emitted light can then actively stimulate another high energy electron to fall in energy and emit even more light – and this chain reaction is used to make a laser.

In our structures, the light is also trapped in a tiny cylinder. Because the cylinder is only slightly larger than the wavelength of the light emitted, the light can only fit inside it in a few pre-specified ways, either by bouncing around the perimeter (denoted by quantum number m), or by travelling backwards and forwards across the diameter (quantum number n). Two such modes are identified in the *Figure*, an (m=1, n=11) mode and an (m=2, n=11) mode. Such modes are also defined over a very narrow range of wavelengths, making them sources of very pure light.

Another exciting property of these structures is the possibility that the rate at which electrons fall in energy can be increased, simply due to the small size of the cylinders. The implication is that, not only can devices be made faster by making them smaller, but an extra effect due to quantum mechanics can be brought into play that will speed them up even further. It will be interesting, then, to see how much control physics will allow humans over its fundamental particles!

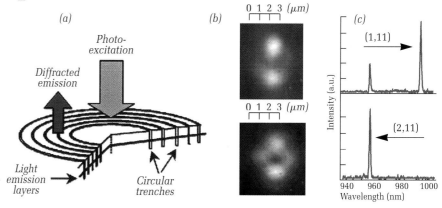

(a) Schematic of the structure. (b) Images of two of the lasing modes, (1,11) on top, (2,11) below. (c) Spectrum of the emitted laser light.

Contact: Dr Brendan Roycroft;
E-mail: brycroft@tcd.ie

TRINITY COLLEGE DUBLIN — BARRY LEHANE, ERIC FARRELL & TREVOR ORR

Geotechnical research at TCD

The Geotechnical Research Group at TCD has grown significantly in recent years in response to both the greater demands of industry and the expansion of the Civil Engineering Department.

The development of analytical models to describe the mechanical characteristics of geomaterials, which are often considerably more complex than other engineering materials, is central to the research effort at TCD. Numerical work is carried out in parallel with extensive laboratory testing of soil elements using state-of-the-art equipment. Significant progress has been made in developing realistic models for the glacial soil underlying Dublin, the soft estuarine clay beneath Belfast, and the peat that covers much of the Irish midlands. These models are checked and refined through their implementation in finite element analyses of monitored and instrumented structures. Ground conditions at the sites of these structures are evaluated using the TCD ground investigation unit.

Pile foundations have been used since Egyptian times, but it is only now that attempts are being made to model the physical processes involved. One such attempt is underway at the Soil Mechanics laboratory, where a specially developed transparent soil is being used, together with a high resolution TV camera, to measure the displacement fields induced by driven model pile installation. A 2.1m diameter pressurised chamber is used in a separate series of experiments to investigate the plugging mechanisms associated with driving of offshore open-ended piles. Full scale field experiments with instrumented piles are currently being used to investigate interaction effects in groups of piles and the response to combined axial and lateral loading.

Instrumented pile group under tension loading.

Many other research areas are being actively pursued: these include studies of a range of soil-structure interaction and hydro-geological problems, the stabilisation of peat, tunnelling and cyclic loading of sand.

Contact: Dr Barry Lehane,
Department of Civil, Structural
& Environmental Engineering,
Trinity College, Dublin;
E-mail: blehane@tcd.ie

Trinity College Dublin — John Monaghan

Research projects in Mechanical & Manufacturing Engineering at TCD

The Department of Mechanical & Manufacturing Engineering at TCD has established an international reputation for the scope and quality of its research activities in Mechanical and Manufacturing Engineering.

The Fluids & Vibration Group, which includes **John Fitzpatrick**, **Henry Rice**, **Darina Murray**, **Craig Meskell** and **Michael Carley**, has been conducting research for many years on modelling and analysis of **Flow/Structure Interactions** and **Heat Transfer Processes**. Work includes *Fluid Mechanics, Combustion and Heat Transfer, Acoustics & Vibrations, Signal Processing and Software Development*. Practical applications have included the identification of aeroacoustics sources using array focusing techniques, noise generation by air flows and by propellers/fans, flow induced noise and vibrations, non-linear vibroacoustic modelling of aerospace components, analysis and control of unsteady combustion, compact heat exchanger design and optimisation of fluidised bed systems.

Jim McGovern is involved in research projects on **Applied Thermodynamics** and he has an international reputation for his work on *the Efficient use of Energy* and *the Simulation of Energy Transforming Processes and Plant*.

Research in the area of **Materials Engineering** is undertaken by **David Taylor**, **Andrew Torrance** and **John Monaghan**. David works with many companies within the European motor industry on problems related to *the Fatigue and Fracture of Automotive Components*. Andrew has an international reputation as a Tribologist and is involved in industrial and EU projects, such as, *Lubrication and Wear of Vehicle Components* and *the Grinding of Hard Materials*. John is currently involved with Irish and international companies on *Materials Processing* research projects, with particular emphasis on components for the international automotive and aerospace industries.

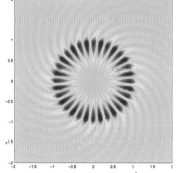

The noise field of a propeller calculated in the plane of the propeller disc.

The **Bioengineering / Biomechanics** group is recognised as being among the best in Europe. **Patrick Prendergast**, **David Taylor**, **Garry Lyons**, **Kevin O'Kelly**, and **Clive Lee** undertake work in this area. Patrick Prendergast has established an international reputation for his work in *Biomechanics*. A current project involves an investigation of *the Stresses within Various Prosthetic Joints*. **Garry Lyons** is currently investigating *the Mechanism of Jaw-Lash*, while **Kevin O'Kelly** is investigating *the Fracture Mechanics and the Constitutive Modelling of Cancellous Bone*. David Taylor has a well-established reputation for his work on *the Fracture and Repair Behaviour of Bone*.

The Department also incorporates the **Materials Ireland (Programmes in Advanced Technology) – Materials Processing Research Centre**, under the direction of **John Monaghan**. The Centre is involved in a range of industrial and longer-term R&D projects for companies within Europe and the USA. These projects include *Non-traditional Machining of High Strength Materials* and *the Use of Elasto-Plastic FEA to Investigate the Stress Conditions of Riveted Aircraft Joints*.

Further information on the activities of the Department can be obtained at our website: http://www.mme.tcd.ie, or from jmonghan@tcd.ie

---

University College Cork — Bill Lane

Silicon sensors and sensor systems at NMRC

The inexorable increase of computing power is creating both the demand and the opportunity for sensors that interface information systems to the physical world. To satisfy this demand, technology developed in the microelectronics industry is being used to fabricate miniaturised sensors permitting integration of the sensors with microelectronics processing systems on a common substrate in a single production environment. It is anticipated that, during the next decade, this technology will generate an order of magnitude increase in the number of sensors deployed.

The diversity of the sensing possibilities is reflected by the variety of work ongoing at the National Microelectronics Research Centre (NMRC) at University College Cork. This work can largely be divided into the areas of biosensors and surface micromachined mechanical and optical sensors. To assess the suitability of these sensors to real-world applications, NMRC has also developed expertise in instrumentation and sensor signal processing.

Silicon-based biosensors
DNA sensors, based on complementary binding of target sequences to immobilised probes, are being developed for Polymerase Chain Reaction (PCR) product detection and identification. The potential applications for this type of sensor are vast, including screening for genetic disorders, forensic examination, pathogen identification and gene expression monitoring. The sensors developed at the NMRC are more sensitive than traditional ethidium bromide staining detection, being capable of measuring picomoles of DNA. They have advantages over existing and many of the newly emerging detection methods in that they are rapid, direct, one-step and do not require labels or mediators. A novel miniaturised system for DNA sequencing, including PCR amplification and on-chip detection, is also under investigation. While single sensors have currently been fabricated, the approach taken is easily expanded to an array format: this work is currently underway. Current clinical targets include para-Tuberculosis screening.

Surface micromachined sensors on silicon
Surface micromachining involves the fabrication of micromechanical structures from deposited thin films. NMRC has developed low cost surface micromachining processes to fabricate microsensors on the upper surface on an integrated circuit wafer. Applications demonstrated to date include a 40*40 infrared detector array and a wide dynamic range absolute pressure sensor. The infrared array has resolution of <1°C and offers a hundred-fold cost saving over existing technology. This opens up many new commercial opportunities in security and inspection applications. The pressure sensor has been demonstrated over a pressure range from 12-110 PSI and targets automotive applications such as manifold and tyre pressure sensing.

The developed sensor processes are suitable for production of multiple innovative structures and are now being used to manufacture microbalance, ultrasonic, flow and vacuum gauge devices. Due to their small size and IC compatibility, these processes and their derivative sensors offer numerous opportunities for new solutions to sensing problems.

An array of micromachined infrared detectors. The detectors are used in a 256x256 array for scene imaging applications.

Contact: Bill Lane, NMRC, Lee Maltings, Prospect Row, Cork; E-mail: blane@nmrc.ucc.ie

Programming approaches subsumed by new model of computing

The Centre for Unified Computing (CUC), located in the Computer Science Department of University College Cork, was established in 1997 with a grant from the National Software Directorate, Enterprise Ireland. The core technology of the centre is the Condensed Graphs model of computing formulated by the centre's director, Dr John P. Morrison. The centre aims to secure Ireland's position as the leader in this emerging technology, consolidating existing expertise and nurturing new talent. As a centre of excellence, it will liaise with industry through seminars and technology transfer. Currently it serves as a training ground for postgraduate students interested in research into parallel computing.

Due to the proliferation of computer networks and the production of extremely fast and comparatively inexpensive personal computers, the hardware to enable parallel and distributed computing is now readily available. This trend coincides with the ever-increasing demand for processing power to drive more and more sophisticated applications. Ironically, the traditional approach to programming, being inherently sequential in nature, makes this processing power difficult to exploit.

Internal view of the Centre's 4-node SMP machine.

Nowadays, many sophisticated computer applications are so impressive that they seem to require superhuman intelligence on the part of the computer. In reality, the computer is a machine capable only of executing simple instructions (albeit at superhuman speed) and according to a "computing model".

There is a number of these models and they are distinguished from each other by the order in which they execute instructions. In the traditional "imperative model", this order is specified by the programmer who informs the computer to "do this, then do that"; in the "data-driven model", the criterion is "if it can be done, do it"; and, in the "demand-driven model", "do it only if the result is needed".

Because the order of executing instructions in the imperative model is prescribed by the programmer, the onus is on him or her to construct parallel programs from collections of interacting parts. This is a non-trivial task and the resulting program may not execute on other machines. In contrast, data-driven and demand-driven models are inherently parallel. Sequencing the execution of instructions is performed automatically without burdening the programmer – although moving programs from one machine to another may still be a problem. These two models have other drawbacks: data-driven computations can give rise to too much parallelism, whereas demand-driven computations perform only the minimum amount of work necessary to produce a result. Consequently, the computation time for the demand-driven model may be longer.

The Condensed Graphs model contains each of the imperative, data-driven and demand-driven models as a special case. Moreover, it manages to eliminate, side-step or mitigate many of the problems associated with each.

A number of projects to illustrate the power of the Condensed Graphs model are currently underway at the CUC. These include the construction of a computing engine to assemble and to distribute tasks for execution on an intranet or the Internet. The engine, known as WebCom, incorporates the technology of the World Wide Web to use machines which otherwise might be idle or underutilized. Surveys have shown that desktop workstations are grossly underutilized – often performing at less than 20% of their potential capacity. This waste of computing resources is significantly compounded in companies where tens or perhaps hundreds of computers are networked. Initial WebCom tests demonstrate its ability to exploit this underutilization by executing distributed applications on the spare computing capacity of the network.

David Power working on the WebCom system.

Thus making the most of an asset that would otherwise be inaccessible.

For WebCom to work well, the size of the tasks it manages have to be sufficiently large to warrant the cost of sending them to less busy machines. In general, exposing many sufficiently large tasks is problem-dependent. So WebCom is not suitable for every application.

When the task sizes are too small to merit distribution in the WebCom system, they may be executed efficiently on a cluster of computers with optimized communication protocols. The CUC has a 16-node Beowulf cluster of 350MHz Pentium II processors to which the Condensed Graphs model is being adapted. This cluster is only the second in a sequence of implementation platforms being investigated. The third is a 4-node 550MHz Pentium III Symmetric Multi Processor (SMP) machine with a Gigabyte of RAM. Due to the low communication costs associated with this machine, it is capable of efficiently running parallel programs whose individual tasks are small in size. At the far end of the implementation spectrum, the possibility of constructing dedicated Condensed Graphs processors is being investigated.

One exciting prospect is to be able to dynamically transform programs expressed as Condensed Graphs so that they can optimally exploit the architecture on which they run. This architecture typically changes when a program is moved from one machine to another. The contention is that by performing dynamic transformations, programs can be tuned to maintain optimum execution performance.

Contact: Dr John P. Morrison,
Centre for Unified Computing, Computer Science Department, University College Cork;
Tel: 021-902914; Fax: 021-274390;
E-Mail: j.morrison@cs.ucc.ie;
URL: http://www.cuc.ucc.ie/

Institute for Non-Linear Science at UCC

For many positive and practical reasons, the world of science and engineering has been organised and taught in an increasingly specialised and compartmentalised fashion since the early 1900s. While this approach has paid many dividends, such as focused research and training, as well as the development of a professional ethos and identity within individual disciplines, it is clear that something essential has been lost: the inter-disciplinary approach to science and technology; the culture of the broadly-based designer, inventor or troubleshooter; the ability to attack problems across a broad front and to devise solutions which satisfy several diverse criteria simultaneously.

Another difficulty has arisen because, again for perfectly rational historical reasons, most scientists and engineers are taught and trained on linear problems which are simplified and sanitised versions of reality. In linear approximations the outputs vary with the inputs as a straight line, so that a change in one is perfectly predictable, being proportional to a change in the other. Uncertainty and noise make the solution fuzzy but do not change its essential character.

Although some good basic education and training can be done in this manner, and some real problems are genuinely linear, it is entirely inadequate to address the world generally from a purely linear viewpoint. Real science, engineering and applications are messy and complicated. Problems have outcomes capable of rapid and dramatic change even when the inputs are changed only slightly. In scientific terms they are *nonlinear*, in that the outputs do not vary with the inputs as a straight line: changes in the outputs are not proportional to changes in the inputs, and unexpected outcomes are frequently observed. When we add other common complications such as feedback (whereby the outputs influence the inputs), control delays and randomness, the result is often highly complex and tangled.

There is clearly a need for programmes of scientific research, education and training directed at problems which lie at the boundaries between traditional disciplines, which are complex in nature, and which involve realistic non-linear approximations or assumptions. Spurred on by revolutions in the science of nonlinear dynamics and chaos, and by successes in the related field of nonlinear control, several inter-disciplinary centres of excellence in nonlinear science have been set up around the world, especially in the USA where funding agencies encourage large multi-investigator research contracts.

We have recently established an Institute for Nonlinear Science (INS) at UCC to develop this new philosophy, and to proceed further towards the practical applications of inter-disciplinary non-linear science and engineering than anyone else. INS has the unique dual mission of developing synergistic research in the fundamentals of the physical, mathematical and computational sciences, while addressing real world applied scientific and engineering problems. We intend to establish new efforts in information science, complex systems and biological physics, as well as building bridges between existing centres of excellence, for example in microelectronics, food science, optoelectronics, telecommunications, unified computing, and cellular physiology. In addition to driving innovative inter-disciplinary research, the INS will sponsor exchange visits, postgraduate education and training, summer schools, seminar series, technology transfer, and remote learning. Summer or part time fellowships for teachers will be founded, and work experience for students arranged.

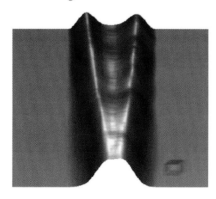

Figure 1: Measured spontaneous light emission profile in tapered laser, obtained by imaging through the substrate.

Scientific collaborations have already begun with several centres of excellence in the US, UK, Russia, Germany and Scandinavia, and other agreements are in process. There are strong connections with high technology Irish and multi-national companies: these have already paid dividends in the form of research funds and joint projects, as well as equipment donations, which have included several laser systems and a Cray vector supercomputer.

One of the projects being undertaken is the formation and selection of patterns in the light emitted from lasers with very large apertures. This problem is both fascinating from a fundamental point of view – providing an example of strong dynamic competition between nonlinearity, dispersion and randomness – and very important for applications of lasers in many areas including telecommunications/information technology, medicine, manufacturing, metrology and space science. This work is being carried out in a collaboration between the Department of Physics and the National Microelectronics Research Centre, as well as Optronics Ireland.

Figure 1 shows experimentally obtained images of the light field in a tapered semiconductor laser operating well above threshold: this structure has been optimised for a single stable field pattern (i.e. a single transverse mode) essential for focusing to small spots, propagating cleanly over long distances, or coupling into optical fibres. *Figure 2* shows the importance of optimising and controlling brightness, depicting calculations based on many-body semiconductor physics and nonlinear optical propagation. The optimisation consists of producing a gradual tapering off of the injected current in the transverse direction. *Figure 2* illustrates the time dependence of the spatial light output patterns of lasers with and without this optimisation: in a normal broad area laser, the beam flickers and jumps like a flame, and is useless for most precision applications, whereas when optimised it is stable and well pointed. The development of dynamically stable, high power semiconductor lasers is an enormous problem involving focused efforts between physicists, engineers, materials scientists, mathematicians and computer scientists.

The INS currently has a Director (Professor J.G. McInerney), two Associate Directors (Professors A.V. Pokrovskii and P.J. McKenna), three full-time research staff, approximately twenty part-time research affiliates, and several research students. Apart from large aperture lasers, other problems addressed include chaotic communication and cryptography, interaction of highly intense energy pulses with matter, ion transport dynamics in cellular physiology, excitation of singlet oxygen for cancer therapy, laser spectroscopy of biomolecules, wind wave patterns in oceanography, parallel computation and reliability of computer models, nonlinear control systems, dynamic memory systems for data storage, multi-fractal analysis of rainfall and river runoff, nonlinear behaviour of electrical power systems.

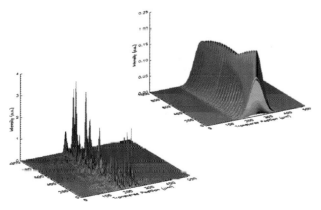

Figure 2: Calculated near field dynamics of broad area laser (a) standard top hat profile, (b) linearly smoothed profile.

For further information please see
http://www.ucc.ie/ucc/depts/physics/ins/index.html

University College Cork — Mary O'Brien, Ken Higgs & Ivor MacCarthy

Probing volcanic dust from the past

A new research project at the Geology Department in UCC is focussing on the significance of volcanic dust deposits or tuffs which are preserved in Devonian and Carboniferous sediments in south Munster.

This area was the site of a major sedimentary basin during the Devonian and Carboniferous Periods (370 to 320 million years ago) and accumulated a thick (8.5km) succession of sediments. Analysis of this succession has revealed that there was a major environmental change associated with the transition from the Devonian into the Carboniferous which took place about 363 million years ago. This involved dramatic changes in both plant and animal populations comparable to mass extinctions which have been recognised elsewhere in the geological record.

Detailed analysis of the sedimentary succession on either side of this important boundary has revealed the presence of thin volcanic tuffs, which appear to have been the product of major volcanic eruptions. These would have thrown clouds of volcanic dust into the atmosphere, which subsequently settled out from suspension and became incorporated into the sedimentary record. The fine grained texture of the tuffs indicates that their source volcanoes may have been located a considerable distance from their depositional site in south Munster.

The present study aims to identify the occurrence of tuffs throughout the sedimentary basin and to study the environmental conditions under which they were deposited.

Tuff level (pale layer) of late Devonian age interbedded with marine sediments at Whitebay, County Cork.

This is employing sedimentological and palynological (i.e. fossil spores and pollen) techniques. A petrographical (rock study) analysis of the tuff layers is also being undertaken using a Scanning Electron Microscope. The study is also employing geochemical analytical techniques and X-Ray diffraction analysis to identify the range of compounds and elements present. The resulting data will provide information for comparing and contrasting the various tuff levels and identifying the type and location of volcanic activity which may have been responsible for the deposits. This will permit comparison and possible correlation with similar deposits preserved in other sedimentary basins.

The study focuses on a number of interesting questions e.g.:

(1) Similar tuff deposits have been preserved in sediments of comparable age in Canada, Germany and Australia. Could these tuff occurrences have been derived from the same volcanic source?

(2) Can the study throw light on the atmospheric circulation patterns in the past? The four localities where tuffs have been identified lay in an equatorial position during the Devonian and Carboniferous *(see Figure)*. This would have been the optimum siting for the most efficient global distribution of dust from a single volcanic source.

(3) The coincidence of the tuff bands with a time of mass extinction is particularly interesting. Could the volcanic eruptions which supplied the tuffs have played a part in bringing about these extinctions either by acting as a polluting agent or by affecting global atmospheric temperatures due to high concentrations of dust in the atmosphere? This would have resulted in a Greenhouse Effect which would have had an impact on floral and faunal populations.

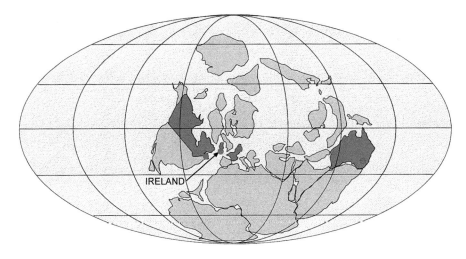

Simplified palaeogeographical map showing the distribution of the continents during the Late Devonian and the locations of some tuff deposits shown in red.

Contact: Mary O'Brien,
Department of Geology,
University College, Cork;
E-mail: khiggs@ucc.ie; imaccarthy@ucc.ie

UNIVERSITY COLLEGE CORK — MICHAEL JOHN O'MAHONY, ALEX BLACQUE, COLM HURLEY & JOHN REAVY

The Caledonian granites of Ireland and Scotland: where, why and how?

On initial inspection of a geological map of Ireland and Scotland, several large red-coloured areas are apparent. These features are granites. The main granite bodies are shown on the map. Granites are coarse-grained intrusive igneous rocks which contain the minerals quartz, feldspar, biotite, hornblende and occasionally muscovite, and are the products of crustal scale tectonic activity, typically melting during continental collision and orogenesis (mountain building). All the granites on the map are of the same age, around 400 million years, and were produced by the closure of the Iapetus Ocean and subsequent continental collision during the Caledonian Orogeny.

The presence of these large granitic bodies within the earth's crust raises several questions. What is the exact source material of the granite melts? How is the melt extracted from the source material, transported through the crust, and assembled to produce the granite bodies we see exposed on the earth's surface today? The Granite Research Group at the Geology Department, UCC, is endeavouring to answer some of these questions.

Insights into the source material can be obtained by investigating the mineralogy and geochemistry of the granite bodies. A particularly important aspect is: why do the granites occur in certain areas of the crust and how did they get there? The ongoing work at UCC has indicated that the granite bodies are sited along deep crustal fractures and

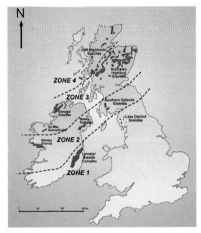

The position of the Caledonian granites of the British Isles and tectonic zones. Zone 1 - SE Ireland; Zone 2 - Galway-Southern Uplands; Zone 3 - NW Ireland-Grampian Highlands; Zone 4 - NW Highlands.

faults and, furthermore, that intrusion is fundamentally controlled by these structures. This insight is achieved through detailed mapping and analysis of the structural and metamorphic features from both granite and surrounding host rock.

Compilation of this data from the various granites across the Caledonian Orogen provides a better understanding of orogenic processes. The granites act as "tectonic fingerprints" and have allowed us to divide the orogen into four sectors which have distinctive tectonic histories *(see map)*. In Zone 1, the source of the granite is melting of the local sedimentary rocks, and intrusion is controlled by deep crustal fractures. In the other zones, the source rocks are deep crustal rocks with mantle involvement. The differences in intrusion styles within these zones is a reflection of changes in orientation of deep crustal faults during continental collision. This rationale can also be applied to other orogens worldwide, and may be useful in mineral and hydrocarbon exploration.

Contact: Dr John Reavy, Geology Department, University College Cork; Tel: 021-902886; E-mail: j.reavy@ucc.ie

---

UNIVERSITY COLLEGE CORK — RICHARD FITZGERALD, THOMAS CROSS & MAIRE MULCAHY

Optimising the aquaculture production of turbot and halibut

The primary objective of this three-year, multi-disciplinary, project (EU FAIR Contract No. PL 97-3382), involving the Aquaculture Development Centre (ADC), UCC, the University of Bergen, Norway, and the University of Nijmegen, Holland, is to create a basis for the diversification of European aquaculture production into the higher-valued species, e.g. turbot and halibut. By examining the physiological adaptations of different turbot and halibut strains to environmental factors (like temperature, salinity), one can define optimal rearing conditions, thereby maximising growth rates, improving yield and enhancing survival.

The primary tasks of the work programme include: defining the impact of environmental factors and their interactions on the basic production characteristics (growth and feeding efficiency, stress resistance and immunocompetence) of turbot and halibut from the north Atlantic, including strains from Norway, Iceland, Greenland and Canada. The primary experimental work is being undertaken at the aquatic facilities in the Bergen High Technology Centre (EU Large Scale Facility) with laboratory analyses being carried out in the home laboratories of the partners. This empirical work will then be translated into a schema of best available strategies for implementation in industrial situations (fish farms).

The involvement of the ADC is in three key areas:

Immunocompetence
To characterise baseline immune parameters of turbot and halibut, of different strains, held under different environmental conditions, and to determine their resistance to defined bacterial disease challenge.

Baseline immune values have been established for halibut and turbot, including total blood cell count, total white cell count, haematocrit, differential cell count, as well as the protein level and the lysozyme, antiproteinase, and iron-withholding activities of serum. Macrophage phagocytic activity and intracellular superoxide production were also measured.

The halibut disease challenge indicates that the different strains exhibit differential resistance to specific *Vibrio* infection. The Norwegian group survived best. This result may be correlated, at least in part, to the greater non-specific immunocompetence, in particular the macrophage activity, of the population.

LiCor image of one of the Cork loci, Smax-03, used in the study of wild populations and reared strains of turbot.

Molecular genetics
To determine the population structure of both turbot and halibut at the macro-geographic level across their natural oceanic distribution, and to examine the genetic differences between reared strains.

To undertake these investigations, novel microsatellite DNA loci have been developed for turbot *(see illustration)* and halibut, for the first time worldwide, and are currently being used to screen for within, and between, population and strain variability.

Industrial Application
To exploit the results obtained in the project under industrial conditions, it is intended to monitor the current practices and rearing strategies used on commercial farms. Already, an in-depth audit of current best practice has been carried out on two farms, one in Iceland and one in Ireland.

Later in the project, in the light of project findings, optimal production strategies will be devised and implemented on selected farms.

Contact: Aquaculture Development Centre, Department of Zoology & Animal Ecology, UCC; E-mail: nadc@ucc.ie

Atmospheric chemistry at UCC

Ozone "Holes", Photochemical Smogs and the Greenhouse Effect have entered the vocabulary of everybody who reads newspapers or watches TV science programmes. Each of these phenomena can have significantly adverse effects on our environment, and a huge international effort is underway in order to understand them.

A new research group has recently been assembled in the Department of Chemistry at UCC and is investigating the chemistry associated with some of these processes. It is lead by Professor John Sodeau, whose major research interests are in laboratory studies of heterogeneous chemical reactions that occur in the troposphere and stratosphere. Both Raman and infrared spectroscopies are used to probe the mechanisms of reactions occurring in Polar Stratospheric Clouds (PSCs) that lead to the release of ozone-depleting halogen compounds. Research is also directed towards understanding the chemistry associated with sea-salt particles which is thought to be responsible for the "sudden" ozone depletion episodes observed in the Arctic troposphere. The latest experiments involve infrared studies of reactive aerosols.

Two newly appointed College Lecturers, Dr John Wenger and Dr Albert Ruth, also have research interests in atmospheric chemistry. Dr John Wenger's research activities are centred on the gas-phase reactions of volatile organic compounds (VOCs) that are released into the atmosphere. Kinetic and mechanistic studies are carried out at UCC and also at the outdoor European Photoreactor in Valencia, Spain. The information gained from these studies can then be used to ascertain the environmental impact of alternative solvents and fuel additives. Dr Albert Ruth is a chemical physicist whose main interests are in the application of ultra-sensitive laser-based techniques, such as cavity ring-down absorption spectroscopy, to atmospheric sensing. His ultimate goal is the development of instruments, capable of monitoring both trace gas pollutants in real-time and reactive intermediates such as the NO_3 radical, which plays an important role in night-time chemistry.

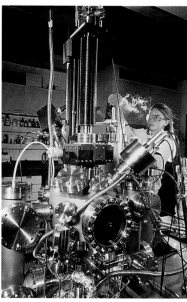

Apparatus for investigating chemistry in polar stratospheric cloud mimics.

Laboratory research in atmospheric chemistry must be performed in order to improve our understanding of the reasons for the occurrence of elevated pollutant levels. We also recognise that the field is data-limited, primarily because of the challenge of measuring the key chemical constituents in the global environment. A further ambition of the team is therefore to develop analytical methods, which gather so-called "Air Truth" data in order to either validate or replace emissions inventories.

A variety of research positions are currently available in the group for scientists interested in the study of atmospheric chemistry.

Contact: Atmospheric Chemistry Group, Department of Chemistry, UCC, Cork; Tel: 021-902680; Fax: 021-903014; E-mail: j.sodeau@ucc.ie; URL: http://www.ucc.ie/ucc/depts/chem/

NMR spectroscopy for chemistry research and teaching

Nuclear magnetic resonance (NMR) spectroscopy - based on the interaction of matter (chemical compounds) with electromagnetic energy (radio frequency range) in the presence of a magnetic field, is widely employed by chemists for the determination of molecular structure, the study of molecular properties, and chemical reaction dynamics. A related technique, magnetic resonance imaging (MRI) is used in clinical medicine.

A modern NMR spectrometer operates in the pulsed Fourier transform mode and consists of a superconducting magnet system, a spectrometer console, and a work-station for instrument control, data acquisition, processing and storage.

The Chemistry Department at UCC has implemented a major initiative in NMR spectroscopy. This involves two specialised NMR laboratories to support a broad spectrum of teaching and research activities at undergraduate and postgraduate level. New, state-of-the-art NMR equipment has been purchased from the AVANCE series manufactured by Bruker, Karlsruhe, Germany - a world leader in the field. Each instrument has unique facilities accessed through dedicated Silicon Graphics O2 computers and

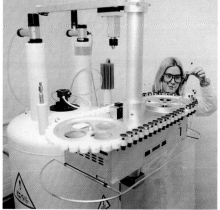

Flora McSweeney preparing to measure an NMR spectrum on the Bruker AVANCE 300 spectrometer.

is networked to the UCC Intel Chemistry Computer Laboratory.

The AVANCE 300 is available on open access to research students and has an auto-tune multinuclear probe. This will facilitate measurement of high quality spectra on a wide range of organic, inorganic and organometallic compounds. An auto sampler with a 60-position carousel will permit optimum use of instrument time and rapid sample processing.

In contrast, the AVANCE 500, equipped with a shielded magnet and a field-gradient probe, is a flagship instrument possessing outstanding capabilities for the measurement of high resolution NMR spectra. The instrument can perform an array of advanced multi-dimensional NMR experiments on complex molecular structures.

UCC now has excellent NMR facilities for chemistry teaching and research in the following areas: conformational analysis of organic molecules, heterocyclic chemistry, enantioselective synthesis of bio-active compounds, pharmaceutical chemistry, palladium complexes related to anti-cancer compounds, catalysis and inorganic materials. Collaborative, NMR linked projects with the biological and food sciences and the Irish pharmaceutical industry are envisaged.

Purchase of the new equipment involved costs in the region of IR£430,000 and was supported by funds from the Higher Education Authority, the Department of Education and Science, and UCC.

For further information, contact: Dr D.G. McCarthy, Chemistry Department, UCC; Tel: 021-902449; Fax: 021-274097; E-mail: stch8038@ucc.ie.

UCC's Department of Anatomy

John Fraher & Kieran McDermott, University College Cork

New Neuroscience Degree

The Department of Anatomy in UCC has recently introduced a BSc (Neuroscience) degree, a four year honours interdisciplinary course, the first of its kind in Ireland. The course covers the structural organisation, structure-function relationships, and the development of the human body as a whole and, in particular, of the nervous system. Neuroscience is a rapidly expanding field and this course is intended to provide graduates with a thorough basic training in this area of increasing importance. Neuroscience is studied at the molecular, cellular and systems level. In the second year, foundation modules in anatomy and physiology deal with cells and tissues, the principles of human structural organisation and the relationship of structure to function. In the third year, the detailed study of the human nervous system starts, and this is placed in the general context of the structure and development of the human body as a whole. Other modules concentrate on physiological principles, molecular biology, and the biochemistry and pharmacology of the nervous system. The final year entails detailed study of the nervous system, focusing on its internal architecture, microscopic structure, development and ageing. Modules spanning neuroendocrine control mechanisms, neuroimmunology and psychology are also taken. A substantial research project is undertaken and is supported by a module in statistics and morphometric methodology.

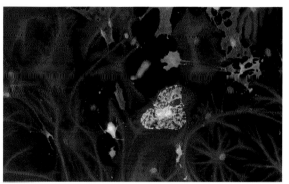

Immunofluorescently stained astrocytes (red) and oligodendrocytes (green) in cell culture.

Departmental overview

The Department has five full-time academic staff, and several part-time/temporary academic staff. Courses in anatomy are provided to medical, dental, nursing and science students. Anatomy teaching to medical, dental and nursing undergraduates emphasises the continuum of basic and clinical science. Areas such as surface, clinical, radiological and functional anatomy, and the genesis of congenital abnormalities, are stressed.

Research interests within the Department are primarily in the field of neuroscience. Ongoing research focuses on:

1. Central Nervous System regeneration studies to develop treatment for paraplegia and stroke;
2. neural cell lineage and differentiation;
3. growth factors and nerve cell transplantation in models of Parkinson's disease;
4. models of motor neurone disease and diabetic neuropathy.

There are six research laboratories, and excellent state-of-the-art facilities are available for microscopy (light, electron and confocal), cell culture, immunocytochemistry and morphometric analysis.

For further information contact:
Professor John Fraher; Tel: 021-902246;
E-mail: j.fraher@ucc.ie;
or Dr Kieran McDermott; Tel: 021-902247:
E-mail kmcd@ucc.ie

Regulating cell survival and lifespan

Rosemary O'Connor, University College Cork

Understanding what keeps us alive and what causes us to age is one of the most fundamental and fascinating problems in biology. All of the cells in the body depend on survival signals from neighbouring cells or nearby tissues to stay alive and to protect them from damage that could lead to premature death. Defects in the proteins and the signalling pathways that regulate cell survival can lead to many diseases. For example, tumour cells have enhanced survival proteins, and this is one reason why a cell that turns cancerous survives to form a tumour. Neurodegenerative diseases like Alzheimers disease are associated with too few survival signals, where nerve cells do not survive in response to stress, and this leads to eventual dementia.

The insulin like growth factor 1 receptor (IGF-1 receptor) is one of the major proteins that mediates the survival of human cells. It is activated by the hormones IGF-I and IGF-II, or by high concentrations of insulin, and it is absolutely essential for normal human development and growth. When the gene for the IGF-1 receptor is inactivated in mice, they cannot form normally developed embryos, and they die. The IGF-1 receptor is more abundant on tumour cells than on normal cells, and high levels of IGF-I in the blood have been associated with a greater risk of breast or prostate cancer.

Researchers in the Cell Biology Laboratory in the Department of Biochemistry at UCC are internationally recognised leaders in the study of IGF-1 receptor-mediated cell survival. They have identified regions of the IGF-1 receptor that are essential for its cell survival activity, and they are using molecular biology, genomic, and proteomic technologies to identify previously unknown signalling pathways that regulate the activities of the IGF-1 receptor in normal and diseased cells. The Cell Biology Laboratory has several collaborations with academic institutions and biotechnology/pharmaceutical companies in North America and Europe. One aim of this research is to develop new methods for isolating drugs for the treatment of cancer and inflammatory diseases.

An interesting twist to the function of the IGF-1 receptor comes from recently published studies of a gene called *daf-2*, which is the most primitive version of the IGF-1 receptor found in animals. *Daf-2* regulates the survival and lifespan of the small soil worm called *Caenorhabditis elegans*. It is a common pattern in biology that proteins with important functions are maintained and used in the same way by all animals. Therefore, the tantalising prospect is that the IGF-1 receptor and its interacting proteins might regulate the lifespan of humans.

Contact: Rosemary O'Connor;
E-mail: roc@ucc.ie;
www.ucc.ie/ucc/depts/biochemistry/cellbiology

Sensing membrane stress

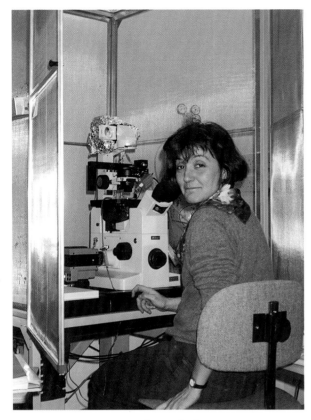

Dr Valerie Urbach

Animal cells are subject to mechanical stress which poses a threat to rupture cells. This type of stress is experienced by the cell wall (membrane) when red blood cells squeeze through small blood vessels or when kidney cells swell upon exposure to dilute solutions. It is thought that sheer stress on the cell membrane can trigger a protective response by the cell to preserve its structure and volume so as to avoid rupture.

Mechanical stress on the membrane can also be used by some cells as a detector of sensation, such as sound waves or pressure. The translation of a mechanical stimulus to a biological response is called mechano-transduction, and much research is currently focused on the cell signals which are responsible for this type of translation. Mechano-transduction in kidney cells is receiving much attention because of its role in sensing changes in blood pressure and producing appropriate response in the kidney's ability to filter and reabsorb salt and water.

Recent research at the Cellular Physiology Research Unit at UCC has uncovered the mechano-transduction signal in kidney cells which senses changes in membrane stress produced by an expansion in cell volume[1].

When kidney cells swell, a response is rapidly initiated which causes the loss of ions (salt) and water from the cell in an attempt to avoid rupture of the cell membrane. It is thought that the membrane stress caused by the cell swelling activates a cell signal which triggers the release of salt and water from the cell.

To test this hypothesis we used a combination of microscopy techniques to measure cell volume, and electrical techniques to measure single ion channel activity in the cell membrane. The experiments showed that, when the membrane was stretched, a pore (channel) in the membrane was opened which allowed calcium ions to flow into the cell *(see illustration)*. The rise in calcium levels in the cell could be measured using fluorescence microscopy while, at the same time, the opening of the ion pore could be detected using a tiny glass electrode pressed against the cell membrane. As soon as the mechanical stress was removed, the ion pore became shut or closed *(see illustration)*.

These studies showed that the rise in calcium levels in the cell was the signal which caused the cell to release salt and water. If the calcium signal was prevented by removing all the calcium from outside the cell, the membrane ruptured when the cell was swollen.

Because the cell uses the rapid opening and closure of ion channels in its membrane to sense mechanical stress and the "state of its own volume", it effectively is digitizing a physical force into a biological response. Thus the cell's biological response is finely tuned to the degree of membrane stress.

Contact: E-mail: v.urbach@ucc.ie

Reference
1. Urbach, V., Leguen, I., O'Kelly, I., & Harvey, B.J., Mechano-sensitive calcium entry and mobilization in renal A6 cells, *Journal of Membrane Biology*, **168**, pp. 29-37, 1999.

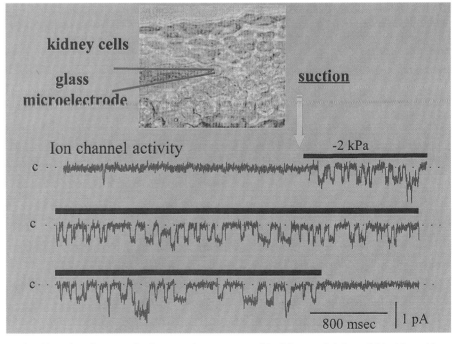

The effect of application of sub-atmospheric pressure (black bars - 2 kiloPascals) inside a glass micro-electrode attached to the membrane of a kidney cell. The negative pressure causes the membrane to stretch and opens an ion channel (downward deflections in the electrical current recording). The ion channel is normally closed in the absence of membrane stress and rapidly turns off (closes) when membrane stretch is removed. The dashed line "c" is the closed state of the channel when no electrical current flows. The channel opens briefly as indicated by the time calibration bar in milliseconds. When open, the channel allows calcium ions to cross the membrane into the cell, and this produces a very small electrical current measured in picoAmperes (10^{-12} Amps).

UNIVERSITY COLLEGE CORK, UNIVERSITY COLLEGE DUBLIN & CHINESE UNIVERSITY OF HONG KONG   BRIAN HARVEY, ALAN BAIRD & WING-HUNG KO

Chinese takeaway

Nature is still competitive with synthetic chemists in providing potent biologically active substances. These include, on the one hand, the most poisonous substances known and, on the other, plant-derived therapeutic drugs such as those which have revolutionised successful transplant surgery.

One example of a natural pharmacologically useful substance is an alkaloid called berberine. Berberine is obtained in its pure form from a variety of plants, including the shrub *Berberis* which is extremely common in Irish gardens. Berberine has been employed for millenia in oriental medicine as a herbal remedy for the treatment of diarrhoea, although its mechanism of action has not been established[1].

Our research in this area is an object lesson in collaboration between specialised laboratories. Studies on berberine began when we were carrying out experiments on mast-cell contribution to food allergic disease. We noted that mast-cell activation in the intestine produced symptoms including diarrhoea. Since berberine was already known to bind to mast cells, we postulated that this was the mechanism underlying the effectiveness of the drug in preventing mast cell-dependent diarrhoea. Extensive biochemical and physiological experiments proved this was not the case. Following a reductionist approach over three years, during which quite a large number of scientists-in-training were exposed to the problem, the answer to the question of the mechanism of action of berberine was discovered[2].

The cellular basis of secretory diarrhoea is transepithelial electrolyte secretion into the lumen of the gut. The consequent osmotic gradient for water changes the intestine from an absorptive to a secretory organ. Our results show that that the anti-secretory activity of berberine in human colon is due to blockade of potassium ion channels that are responsible for driving chloride secretion and the underlying water movement (see Figure). Berberine also exerts additional therapeutic benefits by inhibiting contractions of the gut and decreasing inflammation[3].

Almost 40% of all new approved drugs over the last ten years were natural products or derived from natural products. In fact, 70-80% of antibacterial and anti-cancer drugs were derived from natural products. Only a tiny fraction of the world's biodiversity has been tested for biological activity, so it can be assumed that nature will continue to offer leads for novel therapeutic agents. Ways to improve communication between practitioners of molecular drug discovery and natural products researchers should ensure that the chemical diversity that exists in nature will continue to fuel modern drug discovery. To that end, we have already established a strategic alliance between researchers in Physiology Departments at UCC and the Chinese University of Hong Kong, and the Pharmacology Department at UCD.

References
1. Tang, W. & Eisenbrand, G., in: *Chinese Drugs of Plant Origin*, Springer-Verlag Press, London, pp. 361-371, 1992.
2. Taylor, C.T., Winter, D.C., Skelly, M.M., O'Donoghue D.P., O'Sullivan, G.C., Harvey, B.J., & Baird, A.W., Berberine inhibits ion transport in human colonic epithelia, *European Journal of Pharmacology*, **368**, pp. 111-118, 1999.
3. Winter D.C., Taylor C.T., Skelly M.M., O'Donoghue D.P., O'Sullivan G.C., Baird A.W., & Harvey, B.J., The mechanism of the anti-secretory effect of berberine in human colon, *Gastroenterology (in press)*.

E-mail: harvey@ucc.ie

The mechanism of action of berberine to reduce secretory diarrhoea. Berberine inhibits the three main contributing factors to diarrhoea by reducing the secretion of salt and water from the blood side into the lumen of the gut, and by decreasing inflammation and inhibiting muscle contraction of the gut wall.

UNIVERSITY COLLEGE CORK   FERGUS SHANAHAN

Immunophysiology of inflammatory bowel disease

The inflammatory bowel disease (IBD) research unit is funded primarily by the Health Research Board, and involves several investigators across the Departments of Medicine, Physiology, Microbiology and Surgery within University College Cork. Since it was set up almost four years ago, the major achievements have been:

- the mapping of neuropeptide receptors such as the substance P receptor within the human gut in health and disease;
- the discovery of the "Fas counterattack" as a novel method by which tumour and other cells may evade the immune system;
- the role of oncogenes in the epithelial life cycle;
- the unravelling of fundamental mechanisms of electrolyte transport in the context of inflammatory mediators.

There have been in excess of 70 full publications from the collaborators within the unit (Profs Shanahan, Harvey, O'Sullivan, Collins and Dr John Morgan) and Prof. Shanahan was recently invited to deliver the state-of-the-art lecture on IBD at the American Gastroenterological Association annual meeting based on the work from the unit. This work has also stimulated new research initiatives in the area of the genetics of IBD, dysplasia and cancer in colitis, and the role of functional foods and probiotic bacteria in its treatment.

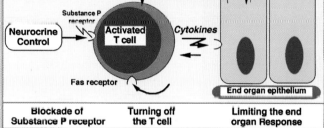

Taming the angry T cell in inflammatory bowel disease. Persistent activation of mucosal T lymphocytes plays a pivotal role in IBD. Receptors for the neuro-peptide, substance P, are present on mucosal lympocytes and are greatly unregulated in IBD.

Contact:
Fergus Shanahan, M.D.,
Department of Medicine,
Cork University Hospital, Cork;
Tel: +353-21-922221 or 901226;
Fax: +353-21-345300 or 343722;
E-mail: fshanahan@ucc.ie

UNIVERSITY COLLEGE CORK

Food Science and Technology for the 3rd Millennium

For over 70 years, the Faculty of Food Science and Technology has been educating students for careers in the food, drinks and healthcare industries. Thousands of UCC Food Graduates are part of a sophisticated workforce with companies and research institutes in Ireland and around the world.

In partnership with Government, EU, State Agencies and industry, UCC continues to develop its food-related teaching, research and continuing education and training programmes. In addition to some general information about the Faculty of Food Science and Technology, some recent initiatives within the Faculty are also highlighted here.

Faculty of Food Science and Technology Complex, UCC.

UNDERGRADUATE EDUCATION

UCC has a strong commitment to excellence in teaching, research and continuing education in the food area and offers five undergraduate degree programmes, each of four years duration:
- BSc Food Business
- BSc Food Science
- BSc Food Technology
- BSc Nutritional Sciences
- BE Food Process Engineering

More than 600 undergraduates in total are involved in these degree programmes at UCC. Teaching is through lectures, tutorials and practical work in the laboratories and in the Food Processing Hall, with strong emphasis also on communications, information technology, language skills and work placement. Undergraduate work placement is a key feature of the degree programmes, where students are placed in industry for six months from April to September in the third year of their studies. During the final year of each degree programme, a relevant research project is undertaken by the students on an individual or team basis, depending on the course structure.

Students in the BSc Food Business and BSc Food Technology degree programmes undertake comprehensive product development projects which encompass all stages from idea generation to full commercialisation.

Working from their own ideas or from ideas from companies associated with the projects, students examine all aspects of product development, market research, prototype development, recipe optimisation, production technology, packaging, determinations of safety, and sensory evaluation. They also prepare full business plans as part of these final year projects.

An Exhibition Showcase is organised each Spring to give staff, students and industry an opportunity to sample the innovative food products prepared by the students. The products are assessed and prizes awarded, sponsored this year by the Limerick County Enterprise Board. The winning entry for 1999 was "Putog", an innovative low-fat black pudding which uses fish instead of pork as the main ingredient. It was selected on the basis of product concept, innovation and marketing plan. The other new food products in the showcase were from the dairy, meat, beverages and bakery sectors.

Prof. Joe Buckley (right) with Third Year BSc Food students with their prize-winning new product "Putog", a fish-based black pudding, at the 1999 UCC New Food Product Exhibition. (Photo courtesy of Tony O'Gourman, Irish Farmers' Journal.)

VIBRANT RESEARCH PROGRAMME – POSTGRADUATE OPPORTUNITIES

The Faculty is associated with three research centres, the National Food Biotechnology Centre, the Centre for Co-operative Studies, and AMT Ireland – UCC Centre. The Faculty also has a major involvement in national and EU research programmes and, through its strong links with international universities and research centres, can provide opportunities for postgraduate students to study abroad.

The extensive interaction of the Faculty with the food industry also provides opportunities for postgraduate students to work on industry-related projects. The Faculty offers an exciting environment for postgraduate studies:
- PhD and MSc research programmes in:
  Food Chemistry, Food Economics, Food Engineering, Food Technology, Microbiology, Nutrition.
- MSc Degrees combining coursework and research in:
  Food Business, Food Chemistry, Food Microbiology, Food Technology, Nutrition, Co-operative Organisation/Food Marketing/Rural Development.
- Higher Diplomas by coursework in:
  Food Science, Food Technology, Nutrition, Co-operative Organisation/Food Marketing/Rural Development.

The current research themes include such areas as:
- Food Ingredients
- Dairy, Meat & Fish Products
- Prepared Consumer Foods
- Brewing & Beverage Technology
- Food Packaging
- Sensory Science
- Food Fermentations
- Food Biotechnology
- Food Quality, Safety & Toxicology
- Molecular Genetics & Physiology of Microorganisms
- Nutritional Sciences
- Diet & Health
- Functional Foods & Disease Prevention
- Functional Foods/Probiotics
- Food Processing
- Minimal Processing Technology
- Food Business
- Co-operative Organisation, Food Marketing & Rural Development

NEW EXTENDED FOOD FACILITIES

The growing research opportunities in the Faculty have been made possible through the extension of the Faculty's food research facilities which were officially opened on Friday, April 16th 1999, by Mr Joe Walsh, TD, Minister for Agriculture and Food. The event marked a further milestone in the continued expansion of UCC and the increased capacity to undertake leading edge research.

The expansion of the food activities has mirrored the growth of UCC itself and has been supported by both the Government and the food industry. This support has included the provision of Structural Funds for capital projects for food research in 1993 and for Food Marketing in 1997, and the funding of two PAT (Programme for Advanced

Technology) centres, namely the National Food Biotechnology Centre (BioResearch Ireland) and, more recently, the AMT (Advanced Manufacturing Technology) Ireland – UCC Centre.

NON-COMMISSIONED FOOD RESEARCH PROGRAMME

UCC participates in the Non-Commissioned Food Research Programme (NCFRP) which is financed from EU and National funds and administered by the Department of Agriculture and Food. The Programme has provided a major stimulus to food research in the Universities and in the Teagasc Food Centres with a total investment of £41 million in the period 1994 to 1999. This investment has provided a critical mass of physical resources and human capital to underpin the future development of Ireland's key indigenous industry – The Food Sector. The Programme includes a major Research Dissemination Project which helps to ensure that the results of the research are communicated to the food industry.

UCC has received funding of approximately £13 million from the NCFRP to support research in a wide range of themes, such as Food Safety, Nutrition and Health, Dairy and Food Ingredients, Cheese, Processed Meats, Consumer Foods, Bakery and Cereal Science, Brewing, Food Process Engineering, Food Marketing, and Food Policy. The Programme provided state-of-the-art research equipment, as well as new pilot-scale facilities for bakery and cereal science, processed meats and brewing. These facilities have enabled UCC to greatly diversify its teaching and research activities in tandem with the changes in the Irish and global food industries. The pilot-scale brewery, for example, enables the Faculty to interface with the economically important drinks sector and emphasises fermentation and separation technologies that find wide application in industry.

The NCFRP has been an excellent catalyst for collaborative research between the institutions in Ireland, and UCC acknowledges its interaction with Teagasc, UCD and TCD in areas such as meat research, food marketing, food policy, nutrition, starter cultures and cheese research.

The UCC/Food Industry Partnership Board has provided matching funds for the expansion of the research facilities and for a number of key developments within the research and teaching programmes. The Board is chaired by Dr Declan Scott and has representatives from the Agri-Food Sector. This Partnership between Government, UCC and the food industry has been a vital element of the strategy to develop the food facilities at UCC and to establish a critical mass of human resources dedicated to food.

HUMAN RESOURCES

The academic staff complement in the Faculty of Food Science and Technology has been increased to 40 by the creation of two new Professorships – one in "Food Microbiology" and the second in "Food and Health". The new Chair in "Food and Health" reflects the increasingly close ties between UCC's Food and Medical Faculties, with the objective of addressing key aspects of the role of diet in healthy living. The Faculty has a food research income of approximately £4 million per annum, which supports 80 research staff and 180 postgraduate students, who represent a key human capital resource for the food sector.

FOOD COMMUNICATIONS INFORMATION SERVICE

UCC is very aware of the huge public interest in all matters to do with food and has, therefore, established a Food Communications Information Service to liaise with food interest groups, including consumers, regulatory agencies and the food industry, and to provide information on a wide range of food topics. The service uses a specific

Mr Joe Walsh, TD, Minister for Agriculture and Food at the Official Opening of the Extended Food Facilities at UCC in April 1999.

Ms Glenda Marmion, Research Brewer UCC, Mr Joe Walsh, TD, Minister for Agriculture and Food, and Prof. Gerry Wrixon, President of UCC, in the pilot-scale research brewery at UCC

web site (www.ucc.ie/fcis) and a newsletter to disseminate information.

CONTINUING EDUCATION AND TRAINING

The Food Industry Training Unit (FITU) was established in 1993 within the Faculty of Food Science and Technology. It services the part-time training, continuing education and professional development requirements of people working in or associated with the food and drinks industry. The activities of the FITU include:

- Certificate and Diploma in Food Science and Technology
- Management Development Programmes
- Short courses on specialised topics
- In-company training

The FITU works in partnership with the Faculty of Food Science and Technology, the Faculty of Commerce, the Centre for Adult and Continuing Education, and the National Food Biotechnology Centre (BioResearch Ireland) in UCC. In addition, the FITU has close links with a number of external agencies including FÁS, Enterprise Ireland, and the NCFRP's Dissemination Project.

29TH ANNUAL FOOD SCIENCE AND TECHNOLOGY RESEARCH CONFERENCE

University College Cork will host the 29th Annual Food Science and Technology Research Conference from the 15 to 17 September 1999. Last year the event was attended by over 200 delegates from research institutes all over Ireland, and the number of presentations – 146 in total – exceeded that of all previous Conferences. It is anticipated that this year's Conference will generate even more interest. This is a reflection of the major increase in food research in Ireland that has been stimulated by the Non-Commissioned Food Research Programme (1994-1999).

ON THE WEB

The Web Site for the Faculty of Food Science and Technology is available at http://www.ucc.ie/acad/faculties/foodfac. This web site provides the Internet user with information on the many activities of the Faculty including:

- Education
- Departments
- Academic Staff
- Research Opportunities
- Upcoming Events
- Food Industry Training Unit
- Centre for Co-operative Studies
- AMT Ireland – UCC Centre
- National Food Biotechnology Centre (BioResearch Ireland)
- Non-Commissioned Food Research Programme
- Food Communications Information Service

FOOD SCIENCE AND TECHNOLOGY AT UCC FOR THE 3RD MILLENNIUM

This brief overview of the activities of the Faculty of Food Science and Technology reflects its commitment to teaching, research and continuing education. As we approach the 3rd Millennium, the food, drinks and healthcare industry will require a dedicated workforce to meet exciting new challenges that arise. UCC aims to meet these needs by continuing to be a leading provider of graduates for the food industry into the 21st century.

Contact: Faculty of Food Science and Technology, University College Cork; Tel: 021-902007; Fax: 021-276398; E-mail: dean.food@ucc.ie; Web site: http://www.ucc.ie/acad/faculties/foodfac

Oral health, fluoride toothpaste and fluorosis: information based planning for Europe

Fluoride has been a major public health success in preventing dental caries, a disease which, although now present at a much lower level than previously, is still almost ubiquitous. It is generally accepted that a major factor in the improvement in oral health has been the fluoridation of water supplies and the widespread use of fluoride toothpastes.

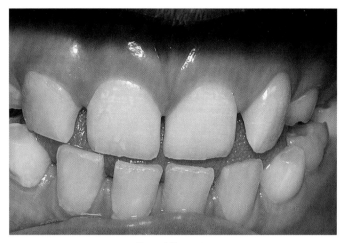

Dental fluorosis.

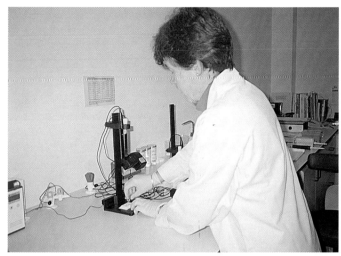

Laboratory work – measuring amount of fluoride child spits out after brushing.

Over the last five years, there have been suggestions that ingestion of fluoride from toothpaste by children has given rise to an increase in dental fluorosis. Fluorosis is characterised by the presence of paper white patches or fine white lines on the teeth. In its mildest form the markings are visible only when the teeth have been air-dried. As it becomes more pronounced, i.e. with higher levels of fluoride ingestion, the marks become increasingly visible. Only fluoride ingested at the time of mineralisation of the enamel of the teeth can have this effect. For anterior teeth the period to avoid excess fluoride ingestion is between 1.5 and 3 years of age.

In some countries industry has reacted by introducing low fluoride "paediatric" toothpaste. There is no substantial evidence to show that this latter toothpaste is effective in preventing dental caries. Hence, an acceleration of this recent trend may lead to a worsening of oral health in Europe.

This is a classic public health dilemma in which, due to misuse by a tiny minority of the population of a very effective measure, the benefits of this measure may be reduced or withdrawn from the population as a whole. Central to this debate are two measurement issues, namely the measurement of fluoride ingestion by young children through use of fluoride toothpaste, and the measurement of fluorosis. Consensus on standardised procedures and methods to measure dental fluorosis and fluoride ingestion is essential to facilitate the design of studies to determine the appropriate level of fluoride in toothpaste in the EU.

In this project, standardised procedures and methods were developed for measuring fluoride ingestion by young children from toothpastes and for measurement of dental fluorosis. These methods were utilised to measure fluoride ingestion from toothpaste in the participating countries and to measure enamel fluorosis in the deciduous and permanent incisors of children. In each country a sample of 1.5 to 3.5 year old children was selected, the tooth brushing practices were recorded, and the amount of toothpaste ingested was measured. 24-hour urine samples were collected and the amount of fluoride in the urine was measured. In addition, a standardised photographic method was used for recording fluorosis in the permanent incisors of a sample of older children.

The methods developed in this project will be used in future studies of fluoride ingestion and in epidemiological studies of dental fluorosis

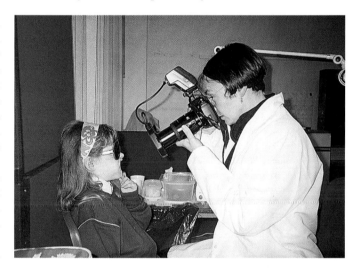

Photographing teeth.

in Europe. They will ensure comparability of results across Europe. Furthermore, the results of this project will inform discussions pertaining to the future development of fluoride toothpastes in the EU.

Dr Judith Cochran was the principal scientist working on the project over its three year lifetime: she has recently been conferred with her PhD in Dentistry. The laboratory work for all the countries was conducted by Ms Eileen MacSweeney, a laboratory technician at the Dental School in Cork.

The project was led by Professor Denis O'Mullane, University Dental School, Cork, and the participating countries were Greece, The Netherlands, England, Finland, Portugal, Iceland, and Ireland.

The project was funded by the EU under the BIOMED 2 programme.

Contact: Dr Helen Whelton;
E-mail: h.whelton@ucc.ie

UNIVERSITY COLLEGE CORK — SASKIA VAN RUTH & PATRICK MORRISSEY

Instrumental flavour research

Aroma, taste, texture and mouthfeel account for the major stimuli that make up flavour. Whereas taste involves four major stimuli, i.e. sweet, sour, salty and bitter, aroma can be composed of thousands of different sensations. Therefore, it is not surprising that a major part of flavour research has dealt with aroma. Sensory analysis can measure the overall aroma quality of a food. However, it can only provide limited information on individual compounds, as humans can identify by name only three or four individual aroma compounds in a mixture. Individual compounds can be responsible for off-flavours and, in certain foods, individual (character impact) compounds need special attention, as they largely determine the aroma of the food, e.g. 2-methoxy-3-isobutylpyrazine in bell peppers.

An instrumental approach to aroma characterisation comprises isolation of volatile compounds at concentrations and in the proportions as present in the human mouth. Analytical techniques have mainly focused on the total volatile content of a food and the profile of volatile compounds present in the air above a food (headspace). Total volatile contents have been determined by using extraction and concentration procedures before analysis. It is now realised that these extraction procedures can cause formation of artefacts, due to the high temperatures and long extraction times required. It has been difficult to relate this volatile content to the profile expressed when foods are eaten. An alternative is to measure the aroma compounds present in the headspace of a food. In classic headspace analysis, volatiles are removed from the food without any attempt to simulate mouth conditions during eating. This profile is probably more closely related to the odour perceived as food approaches the mouth before eating.

Artificial mouth system.

Recently, an artificial mouth system *(see illustration)* was developed by Dr Saskia van Ruth, presently working at UCC. In this artificial mouth, the volume and temperature of the mouth and air flow are taken into account, sample size is one tablespoon, artificial saliva is added and the food is masticated. Volatile compounds are removed by nitrogen gas, trapped on an absorbent and analysed by gas chromatography. Aroma release in this apparatus was not significantly different from release in the mouth of consumers. The apparatus has successfully been used to determine the aroma composition of several vegetables, oils, emulsions, dressings, chocolate and cheeses. The data correlated very well with sensory analysis.

Representative isolation of compounds is an inevitable step in instrumental flavour research when compounds contributing to (off-) flavours have to be determined or when relationships between instrumental and sensory data are established. The artificial mouth is an important tool in this type of research.

Contact: Dr Saskia M. van Ruth, Department of Food Science & Technology, Division of Nutritional Sciences, UCC; Tel: +353 (0)21-902496; Fax: +353 (0)21-270244; E-mail: vanruth@tinet.ie

---

UNIVERSITY COLLEGE DUBLIN — GERALDINE WARREN

AMT Ireland

AMT Ireland is the leading Irish provider of advanced manufacturing management and technology-based consulting services. Our team of over twenty experienced manufacturing professionals combine specialist expertise and practical delivery capability. We deliver National coverage.

As members of the European Foundation for Quality Management, our approach to the identification and realisation of competitiveness improvement opportunities draws on the vision implicit in that organisation's Business Excellence Model. Solutions and programmes typically integrate human, technical and business considerations. Our consulting strategy combines hard expertise with proven improvement delivery methodologies – programmes are results driven and focus on the generation of real-world benefit for the client. Client companies can achieve reduced costs, improved quality, improved product competitiveness and better customer service.

AMT Ireland fees are frequently offset by financial support obtained directly from the relevant state agencies and funding programmes.

SERVICES

Manufacturing Excellence
Manufacturing Excellence is a structured method for achieving world class performance in manufacturing. AMT Ireland facilitates client companies in the achievement of measurable competitiveness improvement using the Manufacturing Excellence methodology.

New Product Introduction
AMT Ireland facilitates the installation of concurrent product development and new product introduction using a team-based Business Process Re-engineering methodology. AMT Ireland also supports the introduction of a variety of supporting innovation tools, technologies and techniques.

Manufacturing Engineering
Line, cell and workstation design and improvement. Manufacturing process planning, visualisation, modelling, improvement and simulation using the Ergomas suite of integrated software tools. Facilitation of engineering-based manufacturing cost reduction and process improvement programmes. AMT Ireland's portfolio of applied manufacturing engineering services supports the achievement of reduced costs and improved process and operator performance in client companies.

Training
AMT Ireland delivers a comprehensive range of management and technical training in related topics. Standard courses may be customised to client requirements and delivered in-house if required. Public courses and conferences are also offered from time to time.

For further information:
Tel: 01-706-1881;
E-mail: amtirl@ucd.ie;
Web Site: www.ucd.ie/~amt

University Industry Programme – support for innovation and technology transfer at UCD

In the late 1980s the University Industry Programme was established to assist UCD and academic staff in the promotion of innovation, technology transfer, continuing professional education and other forms of interaction with the industry and business community. Over the decade considerable progress has been made, and the commitment to commercialisation and co-operation has increased. One manifestation of this is the level of campus companies activity. About 50 companies have been established and two-thirds of recent start-ups have been in the software, information technology and multimedia areas. Some of these companies have expanded rapidly and have attracted investment from venture capitalists in Ireland and overseas.

Campus companies are knowledge-intensive, highly flexible and innovative, and they represent ideal partners for the university in promoting innovation, not only in the exploitation of new technologies for new products and processes, but also in the development and communication of new scientific concepts. The examples outlined below, which are drawn from the Computer Science and Chemistry Departments at UCD, show how campus companies can contribute to the education and research mission of the University.

WBT Systems – Delivering Educational Programmes on the Internet

This campus company grew out of an EU-assisted research programme, undertaken under the direction of Henry McLoughlin from the UCD Computer Science Department. The main objective of the programme was to explore how computer-based technology could be used to support and enhance open and distance learning. In 1994 the research team succeeded in implementing a prototype which proved to be very effective in supporting the learning process, and the advent of the World Wide Web provided a technology platform upon which to build the system.

In 1995, WBT Systems was formed to market TopClass, a new system for the management and delivery of educational programmes on the Internet. The company employs over 50 people and has attracted over $10m in investment. Its head office is in Boston, although most of the development is still carried out in Dublin. Some 4,000 TopClass licences are now in use in over 40 countries.

Changing Worlds – Personalising the Internet for the Needs of Individual Users

As a result of the unprecedented growth of the Internet, it is becoming more and more difficult to find the right information at the right time. Changing Worlds Ltd is a new

The University Industry Centre at UCD.

campus company that has a practical and effective solution to this problem.

Changing Worlds Ltd brings a technology to the market place that is capable of personalising the Internet for the needs of individual users. By using advanced artificial intelligence and user profiling techniques (developed as part of an ongoing research programme headed by Dr Barry Smyth in the Department of Computer Science), it is possible to learn automatically about the needs and preferences of individual users and to present each user with information that has been selected and presented to match their profile. In short, each person can view a version of the Internet that has been carefully personalised to reflect his or her individual tastes *(see also page 178)*.

The PTV system, the first Changing Worlds product, is an intelligent Internet service that provides users with personalised, electronic television guides, each guide listing only those programmes that are likely to be of interest to the relevant target user. PTV currently attracts 6000 regular users and is about to be launched as a co-branded service on Ireland.com (the portal site run by *The Irish Times*).

Cell Media – Simplifying Complex Scientific and Technical Information

Cell Media Ltd is a spin-off company from the Chemistry Department, which is creating new ways to communicate information, mainly scientific, technical and medical. Its approach blends software, computer art, publishing and other related skills to produce unique new ways to share information either by CD, or via the Internet.

Cell Media's first product, "The Dynamic Cell", was created for Springer Verlag, the international publisher. "The Dynamic Cell" offers a new way of demonstrating how the interior of a biological cell works. It was created by a multidisciplinary team of programmers, artists and scientists. The groundbreaking product has attracted wide recognition, not only in the publishing industry, but also in the pharmaceutical, medical and biotech sectors. The Cell Media approach, which helps to simplify the communication of complex scientific and technical information, has obvious marketing applications in these sectors.

Nanomat – Developing New Products from Nanotechnology

Nanomat Ltd, a spin-off company from the UCD Chemistry Department, has been established to identify innovative nano-materials research, to develop marketable technology and licence it to industry for commercial exploitation.

Nanomaterials are materials possessing defined structure on the scale of one thousandth of one millionth of a metre. Components, which incorporate novel nanomaterials technologies, are expected to have improved manufacturing costs, operational efficiencies and functional performance. Nanomaterials are, therefore, expected to influence future manufacturing techniques.

Currently the company is developing products based on technology licensed from UCD and from EPFL in Switzerland. These products include a light-sensitive rearview mirror, and a shelf-edge label which displays commodity prices electronically.

Intelligent Materials – Developing New Concepts in Biomaterials

Intelligent Biomaterials Ltd, which originated in the UCD Chemistry Department, is a new scientific company which is developing the concept of biologically compatible materials that possess intelligence. It is closely allied to the Centre for Colloidal Science and Biomaterials, which has been developing the academic science aspects of this new field. The company is pioneering a new concept that it believes will gain wide acceptance in the medical device industry. The approach is quite new, and much work has still to be done to transmute these scientific developments to products in an industry that values prudence and safety beyond all other features.

Projects with a number of the main medical device companies are in progress, and some are close to completion. These represent proof of concept and proof of ability to transform the science to technology to product. The results to date are encouraging.

Contact: Dr Pat Frain, Director, University Industry Programme, UCD; Tel: +353-1-706-1676; Fax: +353-1-283-8189; URL: http://www.ucd.ie/uip

Conway Institute of Biomolecular and Biomedical Research

UCD has many internationally recognised biomolecular and biomedical scientists, including many world leaders in their fields. Much of their current work is carried out within traditional faculty boundaries, most notably in the Faculties of Science, Medicine, Veterinary Medicine, Agriculture and Engineering. This combination of faculties on one campus is unique in Ireland. In addition, UCD also houses the National Virus Reference Laboratory and the National Institute of Technology Management.

Biomolecular and biomedical research is facilitated greatly when a critical mass of high-quality researchers come together in a dynamic, multidisciplinary, collaborative research environment where scientific innovation is underpinned by ready access to a core of high technology equipment and technical expertise. It is now proposed to create new organisational and operational structures in the biosciences which transcend traditional faculty boundaries. This new reality will be embodied in the Conway Institute of Biomolecular and Biomedical Research, named after Professor E.J. Conway, FRS, one of Ireland's most distinguished scientists and the first Professor of Biochemistry and Pharmacology at UCD (1933-1963). The new Institute will be a focal point for multidisciplinary biosciences investigation on the Belfield campus. The Conway Institute will adopt a holistic approach that draws on the breadth and depth of talent among UCD's Faculties of Medicine, Veterinary Medicine, Science, Agriculture and Engineering. The goals of the Institute will be realised through the establishment of a critical mass of researchers investigating biomolecules through activities which include:

- Synthesis and analysis of structure;
- Function at the molecular, cellular and integrated organism level;
- Molecular mechanisms of disease.

The organisational structures of the Conway Institute will include procedures for development and exploitation of intellectual property and fast-tracking of discoveries to bio-industry. The Conway Institute will be managed by a Director reporting to a Board of Management. The scientific focus and productivity of the institute will be monitored on a biennial basis by an external advisory board.

The Mission
The Conway Institute will promote knowledge, health and economic advancement through excellence in the biomolecular and biomedical sciences.

The Vision
The overall vision is the creation of an International Centre of Excellence that will:
- Enrich the academic milieu at UCD;

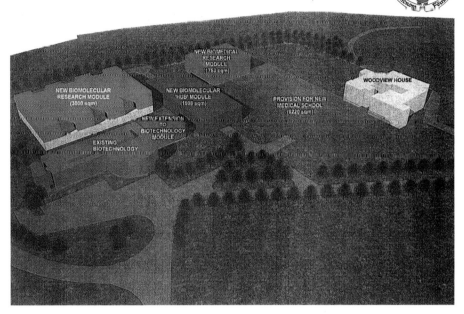

Location of the Conway Institute of Biomolecular and Biomedical Research at UCD.

- Serve as a national resource for researchers in biomolecular and biomedical science;
- Yield new knowledge that will advance the treatment of disease;
- Fuel the emergent Irish biotechnology industry;
- Increase the attractiveness of Ireland as a base for pharmaceutical research and development programmes.

The Objectives
The objectives of the Conway Institute will be:
- To advance knowledge in the biomolecular and biomedical sciences by fostering dynamic interactions among innovative researchers within a multidisciplinary collaborative environment that is supported by "state-of-the-art" equipment and technical expertise;
- To increase understanding of the pathogenesis of disease, identify new therapeutic targets, and improve the treatment of common human and animal diseases;
- To consolidate and strengthen fundamental research at UCD in the core disciplines at the biomolecular and biomedical research interface through the establishment of a critical mass of top-quality researchers and technologies;
- To provide a first-class educational environment, in which teaching is delivered by research active academic staff, to equip its graduates for careers in the expanding knowledge-based bio-industry and healthcare sectors;
- To offer partnership through collaborative research, strategic alliance and consultancy to the chemical, pharmaceutical, medical and healthcare professions and related industries in order to embed these industries in Ireland;
- To help meet the needs of the Irish bioindustry for the 21st century, based on the foresight exercise carried out by The Irish Council for Science, Technology and Innovation;
- To consolidate and enhance participation in collaborative, international scientific research projects such as the European Community Fifth Framework Programme for Research and Technological Development (1998-2001), in particular, the thematic programme "Quality of Life and Management of Living Resources";
- To build bridges to the community at large by continuing education courses and expansion of the "Merville Lay Seminar Programme" which provides a forum for postgraduate students to present their research projects to a non-scientific public audience.
- To enhance the international standing of UCD as a centre of scientific excellence and to be an asset for the wider Irish academic and industrial community.

For further details and information on the Conway Institute of Biomolecular and Biomedical Research contact:
Professor Michael P. Ryan;
Tel: 01-706-1558; Fax: 01-269-2749;
E-mail: michael.p.ryan@ucd.ie
Professor Paul C. Engel;
Tel: 01-706-1525; Fax: 01-283-7211;
E-mail: pengel@macollamh.ucd.ie, or
Professor Hugh R. Brady;
Tel: 01-803-2190; Fax: 01-830-8404;
E-mail: hrbrady@mater.ie

Broad spectrum of activities at UCD's Department of Zoology

Classically, Zoology is thought of in the context of David Attenbrough and treks into the wild to study rare and endangered species. However, modern Zoology is much more, dealing with all aspects of animals from genetics and cell biology to ecology, animal behaviour and diversity. The Department of Zoology in UCD prides itself on maintaining a diverse curriculum, allowing students to choose from a wide range of disciplines, including marine and fresh water biology, immunology, ecology, population genetics, developmental biology and zoonotic and other diseases. The policy of a broad curriculum extends into the Department's research activities, where post-graduates undertake research that crosses many traditional disciplines. The central goal of this research is to gain a greater understanding of how animals develop and interact with their environment both as individuals and as populations.

For example, for more than a decade Dr Tom Hayden has undertaken extensive studies of the deer population in the Phoenix Park in Dublin. Using this large herd, he has attempted to understand the inter-relationship between the life history, behaviour and their mating success, with a view to establishing the most appropriate management of the herd.

In contrast, Professor John Bracken has for many years been interested in understanding why a river makes a good fishery, and has studied the effects of water quality and animal diversity on many river systems in Ireland.

"Mad Cow Disease" or BSE reached epidemic proportions in the UK in the 1980s and 90s with severe consequences for the beef industry in the UK and Ireland. Dr Mark Rogers has been studying the agent that causes this disease for more than a decade, and is involved in the development of diagnostic tests to allow more rapid detection of infected animals and their removal from the food chain.

A very active population genetics unit has been operating in the Department for some time now under the supervision of Professor E. Duke. The genetic techniques used here have rapidly become indispensable in population genetic studies. DNA fingerprinting, RAPD-PCR and microsatellite DNA analyses can all be used to investigate possible inbreeding within isolated populations and to identify specific genetic markers for these populations. Examples of populations being examined are deer, horses, herring, hares and the harbour porpoise.

All aspects of ecology are covered in the research programmes of the Department. Dr Tom Bolger heads a research group on terrestrial/soil ecology. They are interested in the responses of forest ecosystems to acid deposition and global change. The structure and function of the animal communities in the forest are investigated. Dr Bret Danilowicz supervises the group working on marine biology/ecology. This group investigates the fundamental organisation of marine communities, involving influences of behaviour, ecology and oceanography.

Professor Bracken and Dr Declan Murray are in charge of the freshwater ecology section, which carries out extensive studies on all aspects of animal biology in Irish lakes and rivers. Dr Michael Ryan's interests are in chemical aspects of ecology, based primarily on chemically mediated plant/insect coevolution. Investigations into the effects and mode of action of naturally occurring substances in plants, which act as insecticides or as insect repellents or attractants, is being extended to tropical rain forests.

Dr Kay Nolan leads a group working on various aspects of cell biology. Recent funding from the Higher Education Authority has resulted in the Department being one of the best equipped in the country for the teaching of cell biology.

Dr Paddy Joyce works on immuno-parasitology. This involves the analysis and characterization of parasitic antigens leading to the development of immunodiagnostic tests for human and veterinary medicine.

These examples give only a flavour of the work undertaken in the Zoology Department, maintaining strength through diversity.

Fallow deer buck calling in the Phoenix Park.

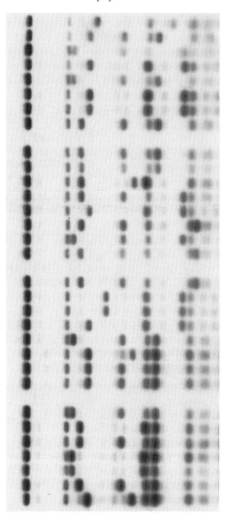
DNA fingerprinting used to study the epidemiology of bovine TB.

Contact: Professor E. Duke
or Professor J. Bracken,
Department of Zoology,
University College Dublin, Belfield, Dublin 4;
Tel: +353-1-706-2265; Fax: +353-1-706-1152;
E-mail: eamonn.duke@ucd.ie

Medicinal chemistry and carbohydrates at UCD

Most drugs are small molecules that work by interfering with a biological process. The bioactive agent has groups that allow it to bind to an enzyme or another receptor and have a positive (agonist) or negative effect (antagonist, inhibitor) which might be useful in treating an illness. For example, the antibiotic penicillin kills infectious bacteria by inhibiting a transpeptidase that is crucial for bacterial cell wall development.

Carbohydrates, like glucose and starch, are well known for their importance in diet. They also play critical roles in biological systems, and thus researchers are interested in how they function, as this information provides a basis for the discovery of new drugs. A class of biologically active carbohydrate polymers are glycosaminoglycans (GAGs) which can bind to receptors and influence their function. A well known example of a GAG is heparin, which helps to deactivate thrombin and thus is widely used as an anticoagulant.

As part of our research programme, we use organic synthesis to prepare novel analogues of GAGs which may be useful in the treatment of a variety of illnesses such as heart disease, cancer and inflammation. We are interested in discovery of inhibitors of sulfotransferase enzymes as they have potential in regulating processes mediated by GAGs.

Combinatorial chemistry is being applied to preparing large numbers of novel structurally diverse agents: this will increase the possibility of success. Biological studies are carried out by Dr Kathy O'Boyle (UCD, GAG-protein interactions), Dr Jerzy Osipiuk (University of Gdansk, sulfotransferases) and Dr Martin Walsh (Rome, protein crystallography).

In addition, we are interested in structure-based drug design using 3D structures of receptors. It is possible to predict novel agents that bind to active sites using computer software called GROWMOL. We have prepared some of the structures designed by computer modelling and will shortly evaluate their potential as agents for AIDS and heart disease.

Access to GROWMOL and biological assays for this project are available through collaboration with Professor Amos B. Smith at the University of Pennsylvania.

We acknowledge the collaborators mentioned above, and Professor Richard Taylor (University of York). We appreciate financial support from Enterprise Ireland, the Health Research Board/Wellcome Trust, Pfizer and UCD.

The X-ray structure of a GAG analogue bound to two Fibroblast Growth Factors (FGF). FGFs are proteins that promote cell growth. The structure implies that dimers of FGF induced by GAGs are biologically active and could provide a basis for discovering drugs useful in cancer, wound healing or heart disease.

Contact: Dr Paul V. Murphy,
Department of Chemistry,
University College Dublin,
Belfield, Dublin 4;
Tel: +353-1-706-2504;
Fax: +353-1-706-2127;
E-mail: Paul.V.Murphy@ucd.ie

Genetic modification of wheat and food safety

Crop performance is determined by two primary factors: the genetic make-up of the variety and the environment in which it is grown. Improvement in the genetic component comes when a plant breeder alters genetic composition (via introgression of novel genes, etc.) in a manner that leads to enhanced expression of important crop characteristics. This is the focus of the wheat breeding research programme in the Department of Crop Science, Horticulture and Forestry, UCD. It is located at Lyons Research Farm, and relies, almost totally, on classical recombination breeding techniques, although contemporary molecular techniques are intermittently deployed for specific objectives.

Crop characteristics that are targeted because of relevance to wheat growers are those which affect the output potential of the crop (level and stability of grain yield per hectare) and those that minimise yield losses or reduce costs of production (disease, lodging and shedding resistance). Grain characteristics targeted in the interests of processors include milling quality (endosperm texture) and breadmaking quality (grain protein content, protein quality and α–amylase activity). Traditionally, the needs and concerns of the consumers of crop products have been accorded little weight when formulating breeding objectives. However, with growing awareness and concern about food safety among consumers, aspects of composition and contamination of crop products are coming under increasing scrutiny.

Under this heading, our wheat breeding programme has embarked on a significant project focussed on the synthesis of new wheat varieties that are genetically resistant to fungal pathogens (*Fusarium spp.* and *Microdochium nivale*) that attack wheat and cause the disease known as Fusarium Head Blight. These pathogens are potentially very serious for Irish wheat producers and the consumers of wheat-based products. They have the capacity to cause significant reductions in the yield of grain per hectare and they can also produce an array of highly dangerous myctoxins (nivalenol, deoxynivalenol, etc.) in infected grains. Mycotoxin-infected grain and food items produced from it represent a serious threat to human health.

In the initial phase of the project, wheat germplasm originating from Europe, North America, Mexico and China was evaluated for the presence or absence of genes that confer effective resistance to Head Blight under field conditions. This has led to the identification of three wheat lines that are far more resistant to Head Blight than any commercial wheat variety available to Irish growers. In the current phase of the project, these novel resistance genes are being introgressed into agronomically adapted wheats.

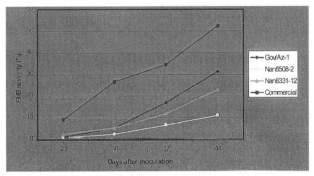

Mean Fusarium Head Blight severity on three resistant selections and on commercial varieties 23, 30, 37 and 44 days after innoculation (1998 data).

Contact: Dr Edward Walsh;
E-mail: edward.walsh@ucd.ie

Nutrition and fertility: a dilemma for the modern dairy cow

The association between the amount of food eaten and fertility is well known. In humans, both excessively low and high food intake can result in disruptions in normal menstrual cycles. Despite substantial differences in the digestive system between humans and farm animals such as cattle and sheep, similar effects of extremes of dietary intake on fertility are evident. Of particular interest in our research programme is fertility in the modern high-yielding dairy cow. Genetic selection, and the use of imported semen from high genetic merit bulls in particular, has resulted in a dramatic increase in the ability of many cows to produce milk. An increase in food intake is required to compensate for the higher level of production. Thus, in a practical context, many high-yielding cows require extremely large quantities of food each day. When these cows are offered a diet consisting predominantly of grass, they may fail to consume enough food. It is then inevitable that they begin to mobilise body reserves to meet their nutritional requirements for milk production. This discrepancy between food intake and the requirements to satisfy demands of production can have disruptive effects on the reproductive system. For a normal pregnancy to develop, co-ordinated signals must be transmitted to control several hormones secreted from the hypothalamus, anterior pituitary gland, follicles growing on the ovary, and the embryo developing in the uterus.

Balancing the conflicting food intake requirements for milk production and reproduction is a difficult task for modern high-yielding dairy cows.

The aim of this research programme is to study the effects of food intake on the hormonal regulation of reproductive function of farm animals, and in dairy cows, in particular.

Results to date show that reduced fertility evident in high-yielding dairy cows is not due to disruptions in secretion of reproductive hormones from the pituitary gland. Nor is it due to disruptions in the expression of behavioural oestrus. The main cause of reduced fertility in these cows appears to be due to disruptions in the development of oocytes, the female gametes that develop in the follicle before ovulation. This results in a developmentally-compromised embryo after fertilisation, that has a much lower chance of survival and establishment of pregnancy. Production of inferior quality oocytes appears to be related to genetic merit, milk yield and time since parturition.

The physiological causes for the production of developmentally compromised oocytes and possible management strategies that may be employed to reduce this problem at farm level are currently under investigation.

Contact: Diarmuid O'Callaghan, Department of Animal Husbandry and Production, Faculty of Veterinary Medicine, Ballsbridge, Dublin 4; Fax: 01-660-0883; E-mail: docall@ucd.ie

Endocrine disrupting compounds and reproductive problems

A number of scientific publications have suggested a decline in human semen quality and an increase in testicular abnormalities over the past 50 years. While most of the attention has been focused on the male, a number of other publications have identified abnormalities in the female, including earlier onsets of puberty and menopause and increases in the number of cases of breast and uterine cancer.

A number of hypotheses have been proposed to explain these reproductive anomalies. A variety of chemical compounds have been identified in the environment that have an endocrine disrupting effect *in-vivo* and *in-vitro*. These chemicals could act as oestrogenic, anti-oestrogenic, androgenic or anti-androgenic mimics, and include the organochlorine pesticides, PCBs, phthalates, phytoestrogens and alkylphenols. Many of these compounds bioaccumulate in the lipid of living organisms. It is long known that the development of the reproductive axis during fetal life is highly susceptible to the detrimental effects of exposure to foreign compounds, especially endocrine disrupting compounds.

In the Faculty of Veterinary Medicine, UCD, we are particularly interested in the alkylphenols, such as octylphenol, which is an oestrogenic compound *in-vivo* and *in-vitro*. The sheep is used as an experimental model as the timing of the development of the hypothalamic – pituitary – gonadal axis in the fetal sheep is similar to the fetal human, thus making it a suitable model for both humans and farm animals. We have identified that exposure of the fetal sheep to octylphenol results in a reduction in testis size and the total number of Sertoli cells in the testis. This is extremely important as the Sertoli cell is the cell which supports the developing sperm cell. The total number of Sertoli cells is laid down during fetal and early post-natal life. Hence any factor which prevents normal Sertoli cell development during fetal life will result in reduced sperm capacity in the adult. We have also identified that animals treated with octylphenol during fetal life or post-natal life have abnormal sexual behaviour. In the female, we have identified that exposure to octylphenol can advance the onset of puberty and cause abnormal uterine development.

Can exposure to endocrine disrupting compounds influence the reproductive potential of humans and farm animals?

The issue of EDCs in the environment is of serious concern. The recent ban on the dumping of UK sewage sludge into the North Sea means that sewage sludge will have to be spread on farm land as fertiliser. There is then the two-fold problem of the impact that such compounds would have on farm animals, and how much would be passed onto humans in milk and meat. This problem has formed the basis of a grant recently funded by the EU.

Contact: tsweeney@vetmed.ucd.ie

Some beneficial effects of Yucca plant extracts in sheep and other domestic animals

The Yucca plant *Yucca shidigera* grows in the south western deserts of the United States and in the Baha California region of Mexico. It dominates the desert landscape, achieving a height of twelve to fifteen feet, growing at a rate of three feet per year. Numerous factors limit the growth of the plants – most notably the availability of moisture. To sustain a high growth rate it is also essential that adequate supplies of nitrogen are available. The successful colonisation of the desert environment by the Yucca plant has been due to its ability to bind ammonia at high concentrations and to retain this essential nutrient in a non-volatile, non-toxic form.

Commercial products, such as De-Odorase (manufactured by Alltech Inc.), are prepared by crushing Yucca logs and then concentrating and refining the juice. The most common form available in the USA is sold as a multi-combination product consisting of Yucca plant extract (30% solids) on calcium silicate (as an inert support), combined with dried microencapsulated bacteria and enzymes which stimulate bacterial growth. De-Odorase, as sold in Europe, does not contain these additional components. The standard formulation (supplied by Alltech Ireland Limited, Dunboyne, Co. Meath) is a simple refined concentrated powder consisting of 60% solids dried onto a finely powdered inert gypsum carrier. This form is usually recommended for addition at levels from 120–240 g/tonne to feedstuffs during various stages of growth in beef, swine, poultry, and also to control odours in cats and dogs.

Dietary supplements incorporating Yucca plant extracts have a long history of safe use as a food material for both humans and livestock. Yucca extracts were established as a suitable forage for cattle in the United States over seventy years ago, and in 1965 it was approved for use in human food without restriction. Extracts or preparations from the Yucca plant are used in the soft drinks industry as flavouring and foaming agents, in the cosmetic industry for their surfactant properties, and in the animal feed industry to promote digestion and improve production performance. The commercial extract, De-Odorase, is used in the livestock industry to reduce ammonia odours in lagoons, ponds, manure and animal confinement buildings. Improving the atmosphere in which animals such as chickens and pigs are housed has been found to enhance livestock weight gains.

Curious ruminants!

The extract contains two active ingredients: a glycocomponent fraction which binds ammonia, and a saponin fraction which does not bind ammonia but has surfactant properties. Certain glycosides called saponins have been known to be present in Yucca plant extracts for over fifty years. The term saponin itself derives from the Latin word for soap, as these substances confer a foaming character to the extract. Saponins act as biological detergents and when agitated in water form a soapy lather. They form oil-in-water emulsions and may act as protective colloids.

Saponins can act as powerful haemolytics, dissolving red blood cell membranes if injected directly into the blood stream. Saponins also poison fish by entering through their gills, and Yucca extracts were used for this purpose by the aborigines of South America. However, due to their chemical structure, saponins are non-toxic when ingested orally. They are not absorbed from the digestive tract. Yet, their surfactant properties may reduce the surface tension around cell membranes and in this way may aid nutrient absorption.

The Yucca plant.

For the past five years, in the Department of Veterinary Physiology and Biochemistry, we have been looking at the effects of De-Odorase on ruminal digestion of various feedstuffs by sheep. The ruminant forestomach acts like a fermentation vat where food is digested by a large number of different rumen microbes.

It has been suggested that the beneficial effects of De-Odorase are based on its ability to bind ammonia. However, the amount of ammonia bound is small, so the observed enhancement of livestock performance, particularly in ruminants, cannot be explained simply in terms of direct binding.

De-Odorase may act by selectively promoting bacterial growth. This could prove beneficial because the nitrogenous products of protein fermentation may be re-cycled into useful protein, thereby reducing the ammonia, which could otherwise be lost to the diet. Another mechanism whereby De-Odorase could affect rumen fermentation is by reducing rumen protozoal levels. The susceptibility of rumen protozoa to saponins may be explained by the presence of cholesterol in their membranes. Reduction of ciliate protozoa would reduce predatation of bacteria and could similarly improve the protein nutrition of the ruminant by increasing the availability of microbial protein.

Yucca extracts increase organic matter digestion and, with mixed roughage diets, they also increase plant fibre degradation. This effect could be due to the wetting action of the saponins, which act as surfactants to soften fibrous material. Supplementation of feedstuffs with De-Odorase also generally increases the rate of ruminal volatile fatty acid production, mainly acetic, propionic and butyric acids. Concomitant effects on ruminal pH are slight due to the strong buffering action of ruminal fluid, but are inversely related to any volatile fatty acid changes.

It may be concluded that the overall effects of De-Odorase are to reduce ammonia in a wide variety of situations and to increase ruminal fermentation and efficiency in ruminants such as sheep and cattle.

For further information contact:
Dr Philip Ryan,
Department of Veterinary Physiology
& Biochemistry, Veterinary College,
Ballsbridge, Dublin 4;
Tel: 01-668-7988;
E-mail: philip.ryan@ucd.ie

UNIVERSITY COLLEGE DUBLIN — THERESE KINSELLA

Cardiovascular disease – the missing link

Cardiovascular disease is the major cause of premature death in Ireland and the western world. But what are the main factors that lead to cardiovascular disease, and are these factors inter-related? Where, through pharmaceutical intervention, one tries to regulate a given factor that can lead to cardiovascular disease, will one inadvertently influence other factors? The cardiovascular group at the Department of Biochemistry, UCD, is investigating this complex question at both the molecular and cellular levels and through clinical collaborations.

A number of clinical and epidemiological studies have identified clear links between the development of cardiovascular disease and either (a) elevated levels of blood cholesterol, arising due to dietary excesses or elevated biosynthesis, or (b) imbalances in the synthesis of hormone metabolites of the lipid arachidonic acid, namely the prostanoids thromboxane A_2 (TXA_2) or prostaglandin I_2 (PGI_2). TXA_2 and PGI_2 play key yet opposing roles in the regulation of blood flow and vascular hemostasis. TXA_2 mainly acts as a platelet aggregator, promoting blood clotting, and as constrictor of vascular smooth muscle causing a narrowing of the blood vessel. In contrast, PGI_2 inhibits platelet aggregation and leads to vaso-relaxation. These opposing actions of TXA_2 and PGI_2 are mediated through activation of their specific cell surface "prostanoid" receptors, leading to downstream activation of specific intracellular responses, and are essential to the normal control of vascular hemostasis.

In this switch control mechanism, activation of the PGI_2 receptor can inhibit the function of the TXA_2 receptor. On the other hand, inhibition of the PGI_2 receptor can, in turn, lead to loss of control of the TXA_2 receptor. Different (iso)forms of the TXA_2 receptor have subtle functional differences which may in turn have differential effects on PGI_2 receptor signalling. Thus, we have complex "cross-talk" between TXA_2 and PGI_2 receptors. To further complicate this "cross-talk", my group has recently discovered that metabolites of the cholesterol biosynthetic pathway are actually required for the normal function of the PGI_2 receptor to signal correctly, thereby identifying new links between cholesterol and PGI_2 within the cardiovascular system. Therapeutic drugs belonging to the Statin family function to lower blood cholesterol levels by inhibiting its synthesis in the body. We have discovered that inhibition of cholesterol synthesis with the Statins indirectly inhibits normal signalling by the PGI_2 receptor which may, in turn, affect the regulation of vascular hemostasis.

We are currently investigating the molecular links between inhibition of cholesterol biosynthesis with interference in intracellular signalling by the TXA_2 and PGI_2 receptors, and hope to identify the resultant knock-on effects on platelet responses and vascular hemostasis.

Contact: Therese.Kinsella@ucd.ie

CITY ANALYSTS LIMITED — MAURA QUINN, MARY HYLAND, MIRIAM BYRNE & ALAN SHATTOCK

Water quality: Trouble in the pipe-line?

The Irish public are increasingly concerned about what goes into their mouths – fears about BSE and GMOs have focused consumer attention on quality issues relating to the food chain. However, the safety of the water we drink is the most fundamental health issue facing Ireland today, since this life-giving element can also act as a deadly carrier of human disease or toxins. The establishment of rigorous EU water quality regulations, new pathogen detection technologies, and increased consumer demands, have combined to change the face of water quality management.

Consolidating conventional microbiological and chemical analysis with newer technologies, City Analysts Limited was formed in 1997 at the state-of-the-art biotechnology facility at UCD, and is a dedicated "one-stop-shop" for chemical analyses and for the detection of bacteria, fungi, viruses and parasites in food, water and effluents. It is headed up by specialists with extensive experience in the university, public, food and biotechnology sectors, bringing both focus and perspective to the stated mission of

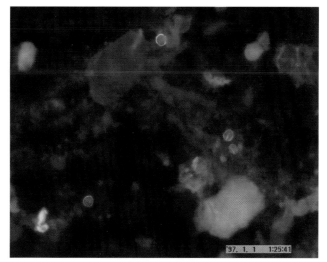

Cryptosporidium *oocysts stained green using fluorescein isothiocyanate.*

providing a gold standard environmental quality index for clients. The great need to take an active technology lead in environmental analysis is a strong focus for City Analysts, complemented by an expanding programme of R&D aimed at developing new diagnostic systems for emerging threats to human health.

However, it was the pioneering work on the development of methods to detect complex parasites such as *Cryptosporidium* ("Crypto") and *Giardia* which put this young company "on the map" and brought it to the attention of the general public. "Crypto" is found in animal faeces and will be familiar to many as an opportunistic infection of AIDS patients, causing a severe life-threatening diarrhoea. Its ability to withstand conventional water treatment processes and enter homes via tap water poses a serious threat to the general public, with the young, aged and immuno-compromised at special risk. Incredibly, before the establishment of City Analysts, analysis for this parasite was sent abroad. The situation is now reversed, with City Analysts performing the analysis for international clients, proving that an Irish water analysis company really can make a big splash in Europe!

For further details contact:
Tel: +353-(0)1-706-2819;
Fax: +353-(0)1-706-1145;
E-mail: miriamb@ollamh.ucd.ie;
Web: http://www.ucd.ie/~analysts/

The imperative to green design

Since the industrial revolution, but particularly in the present century, the twin phenomena of more widely-diffused wealth and relatively cheap energy have resulted in widespread increases in energy use. The cost of maintaining a high-efficiency artificial light source is one thousandth that which a tallow candle represented 100 years ago. Such reductions in proportional cost, and greater affordability, apply not only to energy but to materials produced or transported using energy, which includes all building materials. As a result, the cost of building and operating buildings has fallen many times over, and for some decades it was unnecessary to consider every design issue from an energy-cost viewpoint.

The 1973 and 1979 oil crises prompted governments to seek secure sources of energy and reduce dependency on imported fuel. But reduction of dependency on oil is not the main imperative to green design.

It is now impossible to ignore the global environmental crisis, whether this be the destruction of the ozone layer by chlorofluorocarbons, the loss of wildlife habitat and diversity through pollution, desertification and deforestation, or the increasing levels of carbon dioxide caused by emissions from building heating and other inputs. It is primarily for environmental reasons that the European Union, national governments and private citizens invoke higher standards of building design.

Green design has other advantages. The continuing financial savings which energy-efficient design may achieve can be of real importance in daily life. Winter heating costs can consume a significant portion of family income, and the extra floor area afforded by a simply-constructed sunspace is welcome in many crowded households for spatial as well as economic reasons.

The other reason to promote green design is that of architectural quality. Buildings with more natural and fewer artificial inputs are very often better. Daylit buildings are, in general, more pleasant than artificially lit ones; natural ventilation, if clean air is available from a quiet external environment, is more acceptable than mechanical; the fewer heat emitters, the better; and so on. Mies van der Rohe said that "Less is more"; today, a better way of putting it may be, as Alexandros Tombazis says: "Less is beautiful". Classic design elegance is found in the complete, simple solution.

The recent publication *A Green Vitruvius*, coordinated by the Energy Research Group, is intended as a single-point general reference for those wishing to design and realise sustainable buildings.

Since 1975 the School of Architecture UCD's Energy Research Group has undertaken research, development, consultancy, education and dissemination activities on energy utilisation in buildings and sustainable architectural design. For most of this period the Group's activities have had a strongly European dimension. The Group's "clients" have included EC DGsXII, XIII, XVI, and XVII, as well as Irish public and private agencies and professional practices.

Further information: WWW Internet: http://erg.ucd.ie

Centre for Water Resources Research

The Centre for Water Resources Research (CWRR) engages in R&D activities through sponsored and contract research for public and private sector clients, and through academic research projects associated with postgraduate degrees. It co-ordinates and develops the resources of the Civil Engineering Department of UCD to provide an integrated and multidisciplinary team with extensive practical experience and academic expertise in the areas of Hydrology, including surface water and groundwater and their interaction; Hydraulics, including pipe network analysis, waterhammer modelling and river hydraulics; Water Quality Modelling, including estuary modelling; Water and Wastewater Treatment Science and Technology; Water Pollution Control; Multi-Criteria Decision Support; and Environmental Impact Assessment.

The Dargle catchment.

CWRR has initiated and managed projects at national and international level and has participated in a large number of national and international world-wide research projects ranging from the Aral Sea to Lake Nyasa.

CWRR is developing 2-D models of runoff from agricultural areas which support erosion, and chemical and biological wash-off estimates for use in catchment management tools and for assessing impacts on river, estuary and marine/beach water quality. CWRR (M. Bruen, P. O'Connor) is a partner, with the Departments of Biochemistry (B. Masterson, co-ordinator) and Geography (M. Thorp), in an INTERREG project investigating the contribution of river catchment inputs to bathing water quality. The project focuses on the Dargle catchment (*see Figure*). It is part of a co-operation with the Centre for Research into Environment and Health, which will study catchments in Wales.

CWRR was the project manager for the completed TELFLOOD international project which links meteorological precipitation forecasts to hydrological models to provide useful flash flood forecasts in steep catchments, such as Dodder and Dargle in Ireland, the Reno in Italy, and Pepparforsen in Sweden.

In additional to its expertise in individual specialisations, CWRR has the capability to integrate the latest computer technology, software, technical modelling and decision theory into practical and useful decision support tools, e.g. for catchment and water quality management, for flood warning, and for treatment process monitoring and control. One example is a Decision Support System for the Dodder catchment (supported by Dublin Corporation) which integrates a variety of runoff generation, flood routing and water quality modules.

For further information see our web-site at
www.ucd.ie/~civileng/cwrr.html or E-mail michael.bruen@ucd.ie

UNIVERSITY COLLEGE DUBLIN NICK KUSHMERICK, BARRY SMYTH, GREG O'HARE, MARK KEANE & JOE CARTHY

Advanced internet applications at the Smart Media Institute, UCD

The Smart Media Institute (SMI) is a Research Centre established in the Department of Computer Science at UCD to carry out basic research into the next generation of Internet applications. This research involves using advanced techniques like machine learning, case-based reasoning and conceptual indexing to develop fundamental solutions to problems which will be faced by a wide range of future applications. This year, Enterprise Ireland has funded several new projects at the Institute.

One project, headed by Nick Kushermick, uses machine learning techniques to perform information extraction, integration and retrieval. The Internet delivers mountains of information to the ordinary user's desktop. In the future, users will require intellgent tools to sort, filter and re-present this information in a palatable form. For example, many commercial Internet sites display advertisements. Several systems have been developed for filtering out advertisements, but all require users to continually create and repair hand-crafted rules for detecting advertisements. In contrast, we have developed an AdEater system that uses machine learning techniques to automatically generate and maintain such rules. Users show AdEater example advertisements and the system generates rules that discriminate advertisements from regular images. Experiments demonstrate that AdEater's learned rules are highly accurate, even when generated from relatively few examples.

There are several projects, involving Barry Smyth, Greg O'Hare and Mark Keane, that use case-based reasoning (CBR) in advanced Internet applications. CBR is a problem solving technique that relies on a corpus of past experience in the solution of future problems. New problems are solved by retrieving and adapting the solutions to similar problems that have been solved in the past and stored as *cases* in a *case-base*. CBR has been used in *intelligent* Web-site and e-commerce applications – Web sites that are capable of automatically learning about their users, and adapting their information content to match an individual's preferences and needs. For instance, the PTV system (http://ptv.ucd.ie) is a personalised, web-based television listings service that learns about the viewing preferences of its users in order to deliver

personalised electronic TV guides. The true potential of PTV becomes clear with the advent of Digital television – with hundreds of channels and thousands of programmes to choose from on any given day, the traditional television guide will be rendered useless, whereas PTV can deliver highly customised guides, listing only those programmes that a given individual is likely to watch. The technology underlying the PTV system can be easily adapted to personalise any Web site and, with this in mind, a new campus company, Changing Worlds Ltd *(http://www.changingworlds.com)*, has been established to bring this groundbreaking work to the market-place.

A third area, led by Joe Carthy, concerns the area of information retrieval, and deals with topic detection and tracking (TDT) in electronic texts. TDT techniques detect the occurrence of a new event such as a plane crash, a murder, a jury trial result, or a political scandal in a stream of news stories from multiple sources, and the tracking of a known event. This project aims to improve topic detection using *conceptual indexing* based on the WordNet computational lexicon, to identify the concepts underlying the terms that occur in texts, and use this for more advanced retrieval techniques than simple string matching.

Professor Mark Keane.

Contact:
Dr Nick Kushmerick,
Assistant Director,
Smart Media Institute,
Department of Computer Science, UCD;
E-mail: nick@ucd.ie

Hypertext navigation on the world wide web

The huge increase in the amount of information on the Internet has not been matched by improvements in retrieval performance. Much of this information is in hypertext documents, which are chunks of information containing a network of links to text, graphics, images, sound, video or other digital data. These documents can be read in a traditional "linear" fashion, from start to finish, or can be "browsed" in a "non-linear" fashion, allowing the user to "move" from one chunk of information to another by following links.

The goal of the INTENTS (INTElligent Navigational Tools for Hypertext DocumentS) project is the design of a suite of intelligent knowledge-based tools to assist in the construction, navigation and management of hypertext documents. In order to facilitate intelligent navigation of a document corpus (e.g. the World Wide Web), a model of each user is constructed. This user model is then used to provide a navigational aid to the user on his or her

Thomas MacGreevy.

"journey" through the corpus. In order to encode information about the content of documents, languages which describe metadata ("data about data") are being investigated, including XML (eXtensible Markup Language) and RDF (Resource Description Framework).

INTENTS is funded by the Advanced Software Technologies Initiative (ASTI), which is funded by Enterprise Ireland's Software Programme in Advanced Technology (PAT) and the National Software Directorate, and is part of the CoSEI, the Computer Science and English Initiative, a inter-disciplinary initiative in UCD. One set of documents used in the INTENTS project is provided by a sister project within CoSEI, the Thomas MacGreevy Archive, which is providing in hypertext form part of a chronology of Thomas MacGreevy, poet, art critic and former Director of the National Gallery.

Contact: John Dunnion,
Smart Media Institute,
Department of Computer Science,
University College Dublin;
E-mail: john@kavanagh.ucd.ie

Cortical software reuse: a model of large-scale cortical computation

The Cognitive & Computational Neuroscience Centre (CCNC) is an inter-disciplinary initiative established at UCD to carry out research at the interface between Computer Science and the Neurosciences. This Basic Research project funded by Enterprise Ireland makes use of dynamical systems theory to understand brain function both at a neural and cognitive level.

At the neural level, a number of researchers have argued that the dynamical properties of firing neurons may have a central role to play in explaining how the brain computes. For example, one important dynamical feature is the process of synchronised oscillation of neuronal firing patterns. It has been shown that clusters of neurons at simultaneously stimulated sites in the cortex and elsewhere in the brain can exhibit synchronous oscillations over distances comparable to the size of the cortex. Recently, it has been shown that synchrony appears at multiple scales, from small pools of locally connected neurons to sites on opposite sides of the brain. It is hypothesised that synchronised firing plays a central role in "binding" together disparate features of sensory input (e.g. colour and shape), allowing us to perceive objects in the world as coherent wholes, with

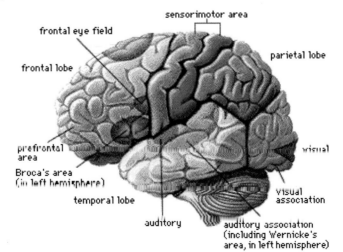

the correct features associated with the appropriate objects.

We are developing a theory called Cortical Software Re-Use (CSRU) which aims to account for a range of neuro-computational and cognitive-developmental phenomena. The central concept of the theory, that of "software re-use", is borrowed from the field of software engineering. Put simply, it states that dynamical neural processes from the sensory-motor areas of the brain provide the computational building blocks for higher level functions up to and including those involved in cognition and language. The aim of our research effort is to provide a rigorous mathematical foundation, implement it as a computational model, and test it against neuroscientific data.

Contact: Dr Ronan Reilly, CCNC, Department of Computer Science, UCD; E-mail: ronan.reilly@ucd.ie

1. Department of Computer Science, UCD
2. Department of Mathematical Physics, UCD
3. Department of Psychology, UCD

UNIVERSITY OF LIMERICK

Academia-Industry research links at the University of Limerick

The University of Limerick (UL), under its President, Dr Roger G.H. Downer, recognises research as a central element in its institutional mandate and aspires to an institutional ethos that is characterised by enquiry and a continuing quest for new knowledge. It has been eminently successful in attracting research funding from industry, the European Union, Irish National Agencies and other private and public funding sources. UL is now poised to build upon the solid foundation that was established during its formative years.

The University has contributed to meeting Ireland's growing need for special expertise and leadership in new technologies and believes that it must play an active role in developing new ideas and being at the cutting edge in R&D.

Research Mission and Organisation

Research is central to the mission of the University and is closely linked to the academic structure at College level. The University is organised into six Colleges - Business, Education, Engineering, Humanities, Informatics & Electronics, Science - each having an Assistant Dean for Research who has responsibility for the stimulation and co-ordination of research activity in the College. The Associate Vice-President for Research has responsibility for the overall co-ordination of the University's research activities, including recruitment of postgraduate students, promotion of research awards, seed funding schemes, approval and monitoring of research contracts, encouragement of inter-disciplinary research, advice to Faculty and Staff, the University's European Liaison Office and University/Industry research linkages. This approach to the organisation of research is regarded as being on a par with best practice in Europe.

Research Strengths and Priorities

The University is fully committed to the Programmes in Advanced Technology (PATs), which were established to strengthen Ireland's indigenous capability in selected niche areas of S&T which would impact on innovation and international competitiveness. Currently, UL houses four Centres associated with three of the PATs:
- **AMT Ireland** - Electronics Manufacturing
- **Materials Ireland** - Industrial Materials Technology
- **PEI Technologies**
  - Thermofluids
  - Circuits and Systems Design.

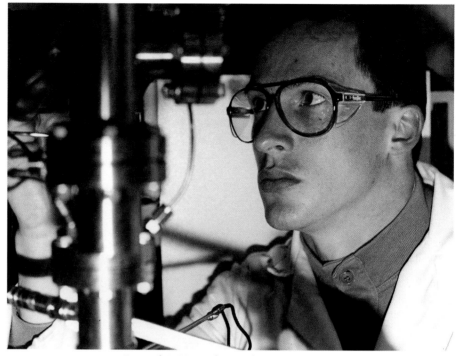

Research in Materials & Surface Science at UL.

Research in Bioelectronics/Biomedical Engineering at UL.

The University has recently undertaken an audit of its research capability, and its strengths lie in the following science and technology areas:

- **Materials and Surface Science**
  - Science of Structural Materials
  - Active Materials
  - Catalysis
  - Interface Science
- **Communications and Electronics**
  - Communications Networks & Security
  - Communications Signal Processing
  - Circuits and Systems
  - Sensors and Associated Microelectronic systems
- **Bioscience and Technology**
  - Food Technology & Microbiological Safety
  - Molecular Biochemistry
  - Biomedical Sciences & Engineering
  - Human and Clinical Sciences
  - Environmental quality
- **Informatics**
  - Applied Mathematical Sciences
  - Computational Intelligence
  - Interactive Media
  - Language Localisation
  - Software Re-engineering

- Innovation, Manufacturing and Design
- Aeronautical Engineering

The University also houses a Centre for Biomedical and Environmental Sensor Technology (BEST) which is a joint venture with Dublin City University, Queen's University, Belfast and the University of Ulster, funded by the International Fund for Ireland.

Industry - University Linkages

The University of Limerick has been involved with industry from its inception, and operates the largest Co-operative Education Programme (where students spend up to eight months in an industrial/commercial environment as part of their academic programmes) in Europe. During the placement, students carry out in-house research projects which can be continued once back at the University during the student's final year.

The link between technical progress and sustainable economic development and employment is now acknowledged, and universities can assist by entering into partnership with companies in order that industry can exploit their technological expertise and facilities. Over the last few years, UL has established the framework to facilitate appropriate levels of co-operation between academics, and flexibility in meeting the diverse needs of industry.

European Union Industrial Research at UL

The European Union Science & Technology Framework programmes continue to be important for funding major research projects. The programmes are industrially led and involve collaboration between companies and universities from different member states. The benefits for industry in these types of partnership are clear. A recent study shows that, under the EU Science Framework Programme: Industrial and Materials Technologies, research projects carried out by companies in collaboration with Universities led to greater competitiveness than projects involving companies acting without University support for their R&D.

Faculty at the University of Limerick have been extremely successful in attracting funding under these European programmes - placed seventh overall among European Universities in total funding under Industrial and Materials Technologies sub-programme in both the 3rd and 4th Framework Programmes (1990-1998).

EU Projects involving industrial partners include work on Inorganic Membrane Reactors, Catalysts, Aeromechanical Design of Turbine Blades, Laminar Flow on Aircraft, Aircraft Crash Survivability, Advanced Ceramics for Wear Resistance, High Temperature Corrosion, Speciality Glasses, Glass-ceramics for Dental Applications, Biomaterials for Hip Joints, Mobile Communications, Computer-supported Co-operative Work, Ergonomics, etc.

University - Industry Applied Research Scheme

This scheme, which is administered by Enterprise Ireland, is designed to encourage applied research projects which have direct industrial and commercial application. Some recent examples of these types of projects at UL include the following:

Professor Phil Burton is working with Silicon and Software Systems of Clonskeagh, Dublin, on Hardware/Software Codesign for "Systems on a Chip". Professor Burton also has projects with Analog Devices, Limerick. Dr Vincent Casey is working with Abatis Medical Technologies Limited, Limerick, on a Tourniquet Force Sensor for Intravenous Regional Anaesthesia. Mr Colin Piercy is working with AEG Servo Systems Limited on Development of an Automated Magnet Bonding System. Dr Tom Sorenson is working with Sun Microsystems on Prediction of Radiated Emissions in computer workstations. Dr Khalil Arshak is working with BMS Ireland, Limerick, on the Development of a Thick Film/ASIC Strain Gauge Sensor System. Dr Colin Birkinshaw is working with Elan Corporation Research Institute, Dublin, on Characterisation of Cyanoacrylate Polymers in a Drug Release Application. Dr Michael Pomeroy and Professor Stuart Hampshire are working with Harris Ireland on two projects concerned with ceramic varistor devices. Other topic areas include:

- Radio Frequency Scale Abatement
- High Speed Machining of Composites
- Manufacturing Systems Planning using Simulation
- Advanced Control Networks for Automotive Applications
- Advanced Semiconductor Device Fabrication.

These are just a few examples of industrially-related research projects at UL.

The University Technology and Enterprise Development Unit – Campus Companies

This specialised unit within the Office of Research was established to support and develop the innovative activity emanating from the University, and its brief includes:

- The Innovation Process
- Linkage Opportunities between Researchers and Industry
- Technology Transfer and Diffusion
- Campus Companies
- Intellectual Property Rights/Patents.

The Campus Company programme, initiated in 1994, is promoted and operated in close co-operation with the Shannon Development Company – the Regional Development Agency. Of the current Campus Company pipeline of 37 projects, 17 are trading: the other 20 are at varying stages along the development cycle. Companies trading include Piercom and ZPM Europe.

Innovation Board

The Innovation Board, a joint initiative between the University of Limerick and Shannon Development, has as its focus: *Innovation through the commercialisation of Knowledge and Technology*.

The Board meets monthly and is co-chaired by the Associate Vice-President Research of UL and the Director of the National Technological Park, Limerick, on which UL is located. The work of the Board has been to find mechanisms for accelerating the technology commercialisation process. Projects would typically:

- be concepts emanating from the University's research activity
- be commercialisable within a two year time scale
- would require a commercial champion.

A number of feasibility studies are in progress.

Atlantic University Alliance

The Atlantic University Alliance, launched in May 1999, is a co-operative initiative between the three universities on the Atlantic seaboard of Ireland, comprising University College, Cork (UCC), the National University of Ireland, Galway (NUIG) and the University of Limerick (UL), which provides an integrated approach to assist the economies of the Southern, Western and Shannon regions. It will facilitate the effective transfer and commercialisation of technology between the universities and industry through applied interdisciplinary programmes.

The three universities have recently completed a technology needs and resources study of the three regions. This was funded by Enterprise Ireland, and a programme to promote Research, Innovation and Technology within Enterprises (RITE) has been formulated from this study. This will harness the collective strengths and resources of UCC, NUIG and UL to facilitate innovation within companies and to meet the technology and management training requirements of industry, especially indigenous industry within the three regions.

The Future

The continuing success of University of Limerick researchers in industrially led, EU and National programmes ensures a pipeline for new ventures well into the future. The decades to come will bring new alliances with industry, new sources of funding, and new areas for research. UL intends to be to the forefront of developments. The University accepts that a strong focus on research is critical to sustaining and enhancing the interest and commitment of partners, to the generation of new ideas, to the training of young researchers, to the improvement of teaching, and to the further enrichment of our knowledge base.

Contact: Professor Stuart Hampshire, Associate Vice-President Research, UL; E-mail: stuart.hampshire@ul.ie

Bioscience and Technology at UL

The Bioscience and Technology grouping at the University of Limerick has core competencies within four key emerging research areas:
- Food Sciences
- Molecular Biochemistry
- Biomedical Sciences
- Environmental Quality.

Each of the areas has specialist "niche" interests which are quite unique in the Irish context. Furthermore, the departmental and collegiate structures at UL promote interdisciplinary research, since they bring together researchers from different fields.

Food Sciences

The research in Food Sciences is focussed on two areas where the group holds competitive positions internationally – Minimally Processed Fresh Foods and Food Ingredients. While most of the work involves fundamental science, it generally addresses practical industrial opportunities/problems in areas such as innovation, food safety and product quality. In the case of minimally processed foods, current work relates to the safety and quality of cut ready-to-use fresh produce packaged within preserving gas atmospheres. Sub-areas include the following:
- modelling gas equilibria within packages
- gas-enzyme interactions
- technology-pathogen effects, including microbial stress reactions
- competition between pathogens and spoilage organisms
- chemical and organoleptic (sensory) effects

Paula Burke working on food microbiology research at the College of Science, University of Limerick.

Work on food ingredients includes:
- protein functionality in foods (e.g. whipping/foaming properties)
- enzymatic modification of functional properties
- proteins from novel plant sources
- production and characterisation of nutraceuticals, e.g. blood pressure reducing peptides, peptides for enhanced mineral bioavailability and peptides which stimulate the immune system.

Molecular Biochemistry Group

The unifying theme within this group is the study, from a basic and applied perspective, of a number of niche classes of proteins, including antibodies, thermostable proteins and enzymes. Common goals link the research projects, including protein folding, protein stabilisation, protein interaction and the application of these topics, for example, in preventing the aggregation of biopharmaceuticals and the production of stable biosensor devices. Projects include the development of environmental and food biosensors, antibody engineering, polymerase chain reaction (PCR) probe development for food pathogens, and enzyme additives for animal feeds.

Some of this work in this group has immediate potential for application. One of the chief characteristics of enzymes is their requirement for an aqueous environment. Many applications such as biosensors or organic synthesis require a non-aqueous environment. Research at UL is focussed on important classes of enzymes in non-aqueous environments.

Rapid identification techniques for food pathogens are of increasing importance in ensuring food safety. One research group is developing DNA probe technology for the rapid identification of potential pathogens. When implanted materials or novel biomaterials are studied using animal cell culture techniques, little work has been done in biomedical research at the molecular level. This group is investigating the use of specific protein and mRNA induction as a means of diagnosing biocompatibility.

Barry McGrath, PhD postgraduate student, using automated DNA sequence analysis for sequencing sites of integration of bacterial conjugative transposons, at the College of Science, University of Limerick.

Biomedical Sciences Group

The Biomedical Sciences Group is an inter-collegiate alliance of researchers in the human sciences, biomedicine and biomedical engineering. There is important collaboration with the Vascular Surgery Group and Clinical Age Assessment Unit at the Regional Hospital, Limerick.

Michelle Kirby, BSc Sports & Exercise Science Degree student, measures the iso-kinetic muscle torque of the leg extensor muscles around the knee of Niamh Spratt, BSc Physical Education Degree student.

Research themes of the group focus on a multi-disciplinary approach to the problems of ageing, age-related disease and dysfunction. The group believes that the study and resolution of the problems of ageing requires such a multi-disciplinary approach: a synergy of experienced and innovative scientists, engineers and clinicians. Current research concentrates in three areas:
- Vascular function in ageing
- Musculoskeletal function in ageing
- Age-related metabolic disorders.

The Biomedical Sciences Group has developed an interdisciplinary research group with expertise in vascular medicine, blood flow biomechanics, cardiac physiology, muscle function, endothelial cell behaviour, and blood pharmacology and ageing. Another group is researching age-related metabolic disorders, such as insulin resistance.

Environmental Quality Group

This group involves interdisciplinary study of the environment and environmental management. The main areas of research are: sustainable management of a high quality environment, and fundamental research on various aspects of biodiversity.

The work of the group encompasses topics in applied ecology, soil science and environmental management. Current projects include:
- biodiversity in relation to management
- heavy metal concentrations in soil water and bioindicator invertebrates in contaminated sites
- chemical and physical properties of organic wastes in relation to land disposal
- policy issues in relation to sustainability.

Impact on teaching

Teaching and research at the University of Limerick are closely interlinked. Undergraduate courses in the area of Bioscience and Technology include
- BSc in Environmental Science
- BSc in Industrial Biochemistry
- BSc/Dip in Equine Science
- BSc in Food Technology
- BSc in Sports and Exercise Science
- BSc (Ed.) in Science Teaching
- BSc in Physical Education
- BSc in Nursing.

Research carried out by the associated faculty falls into the groupings in which a common theme is the molecular nature of biological and biomedical processes.

Contact: Dr John Breen, Assistant Dean for Research, College of Science, University of Limerick; E-mail: john.breen@ul.ie

Materials and Surface Science Institute

The Higher Education Authority in 1998 awarded £750,000 to the University of Limerick under its Research Programme for Third Level Institutions. The University is planning to support the formation of integrated multi-disciplinary Research Institutes based on carefully selected research themes. Over two-thirds of the HEA funding allowed the establishment of the Materials and Surface Science Institute (MSSI) in which more than twenty full-time academics from different Colleges across the University have come together with a view to co-ordinating their activities within a shared infrastructural context that allows co-operation between different departments and research centres. The Director of the Institute is Professor Kieran Hodnett, one of Ireland's most cited chemists and an international figure in Heterogeneous Catalysis.

Ireland can boast a very considerable success story in several advanced technological fields, for example, in electronics, pharmaceuticals and more recently in health-care products. At a time when more and more emphasis is being placed word-wide on the environment in which we live, the University of Limerick has made a strategic decision to take a major initiative to educate Ireland's young population in all modern aspects of the many new materials which will have impact on them, and also in their use in many and varied applications such as transportation, corrosion protection, health-care, the detection and treatment of environmentally hazardous effluents, as structural components or as catalysts for new and improved chemical processes.

Common to all these themes is a sound knowledge of the relationships between the bulk and surface properties *(chemistry and physics in two dimensions)* of the materials in question: to this end, the MSSI will have as a central theme the study of the surface properties of new materials, with a particular emphasis on how such properties affect their end usage (resistance to corrosion, catalytic properties, selectivities as sensor components, etc.).

The objectives of the MSSI are to educate scientists at postgraduate level in four key areas:
- Science of Structural Materials
- Active Materials
- Catalysis
- Interface Science.

The four themes are related to a number of research applications areas, currently the subject of work by Institute members. Each academic applies his/her research expertise in one or more of these fields to industrially related problems, in such areas as biomedical materials, catalysis, electronic materials (semi-conductors, dielectrics etc.), aerospace and aeroengine materials and magnetic materials. Research into thin films, thick films and coatings for such industrial applications is also conducted. The relationships between themes and research application areas are illustrated in the *Figure*. This is intended to be illustrative rather than comprehensive and many more synergies exist than are illustrated.

Over the past five years members of the MSSI have authored more than 170 refereed journal papers. Institute members have also been involved in writing and editing six books and authoring some 11 book chapters. In addition, some 14 patents have been registered. 84 completed postgraduate degrees have been supervised by Institute members since 1994. Currently, 88 postgraduates students are pursuing research programmes supervised by Institute members. This research activity has been underpinned by research contracts won by individual researchers. MSSI members also play a significant part in the specialist education of undergraduates.

MSSI members are highly successful in winning research funding. Over the last five years, joint income in excess of £7 million has been realised. Much of this funding has come from European Programmes, in particular the Joule Programme and the Industrial & Materials Technologies Programme in which the University was placed seventh out of all European universities under both the 3rd (1990-94) and 4th (1994-98) Science Framework programmes. A substantial number of the projects involve industrial leadership, and this ensures an industrially related focus to the research.

The Materials and Surface Science Institute will carry out both fundamental and applied research. The graduates will have wide experience of many different sub-disciplines and of modern state-of-the-art experimental techniques, and will be employable in a wide range of industries. Important linkages already exist with other European centres of excellence as well as with many leading US and Japanese laboratories. Because of this international profile, it is expected that the Institute will help to stimulate the establishment of additional employment in high-tech industries in the country that specialise in areas such as environmental monitoring, effluent treatment and modern materials in the electronic, health-care and engineering sectors.

Contact: Professor Kieran Hodnett,
Director MSSI, University of Limerick; E-mail: kieran.hodnett@ul.ie

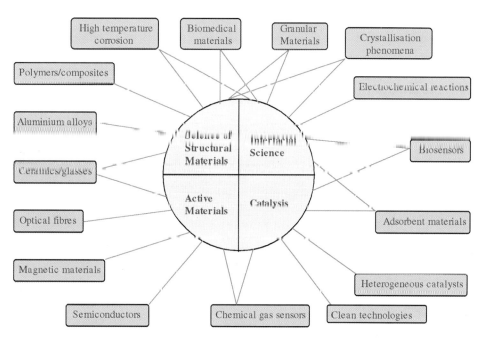

MSSI: core themes and current research applications areas.

LiteFoot: a smart dance floor or a new musical instrument

This article describes the development of LiteFoot, an interactive floor space that tracks dancers' steps, converting human motion into sound and images. The system can also record steps for further analysis for use in dance research programmes, choreographic experimentation and training.

Introduction

In early 1997 a team of researchers at the University of Limerick started to discuss the prospects of recording traditional Irish dance electronically. We realised that an electronic recording device could also be made to add to, or enhance, performance through sound and visual representations of dance. If such a system could be created, it would potentially allow researchers in ethnochoreology to record and analyse the subtle movements of master dancers as well as allowing modern dancers to explore possibilities with a dancer making his or her own music in real-time. We initiated a project, first reviewing existing systems and technologies. We found that existing systems were either expensive, invasive or not fulfilling the requirements.

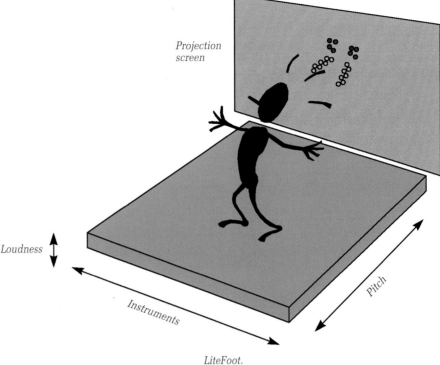

LiteFoot.

Designing LiteFoot

A number of issues had to be covered, both users' requirements and technical requirements versus constraints such as budget and a definite deadline – UL's 25th anniversary. The following requirements were identified as key elements for this design:

- The floor should be able to respond at a rate corresponding to at least 30 steps per second.
- The floor should have a reasonable spatial resolution (44 millimetres)
- The floor should be able to track multiple feet and dancers.

The LiteFoot prototype is a 1.76 meter square and 10 centimetres high floor element, filled with a matrix of 1,936 optical proximity sensors. When you stand on the floor the spatial locations of your feet are detected. An accelerometer detects the total impact force of the feet, providing a third dimension. Arbitrary mappings between the data coming from the floor and representations can be defined by the user. The visual representation has, so far, been a direct mapping of location to displayed groups of pixels, with the colour controlled by the impact force. For the auditory representation, the incoming data-stream controls the MIDI synthesizer of a standard sound card, with various musical scales in one dimension (X) and various sets of "instruments" in another dimension (Y). The impact force has been mapped to, for example, loudness (Z).

LiteFoot in Action

The LiteFoot premiere was in September 1997 in a performance in the University Concert Hall at UL. Both traditional Irish dance and improvised modern dance were performed, and several members of the audience reported the performance to be aesthetically and perceptually engaging. In January 1998 LiteFoot appeared in *The Late Late Show* as an example of new art and technology. Since then it has been demonstrated several times, including an international workshop on dance at the University of Limerick (*Trath na gCos*) and in an interactive exhibition in Limerick City (*Infusion* – a National Review of Live Art).

Continued Research

With the existing LiteFoot prototype, we can now evaluate its use to define the requirements for a second generation of LiteFoot. Some of our students, working on their final year projects, have extended our software libraries. There are many possibilities to be explored, for example, to use other kinds of mappings. Instead of direct mappings, orchestrations and sound, music or video sequences with parametric control could be mapped to areas of the floor. When several persons are active on the floor simultaneously, the level of "collaborative harmony" could be mapped to tonal harmony and temporal structures. As a play space, LiteFoot seems to be highly engaging, and there are also a number of possibilities to be evaluated with, for example, disabled people who might be able to extend their action range through training with different forms of auditory and visual feedback.

Acknowledgements

The LiteFoot team is an expanding interdisciplinary group and it is almost impossible to list all contributors to this project. The core of the group, so far, is Dr Niall Griffith (Department of Computer Science and Information Systems - CSIS), Dr Liam Bannon (Interaction Design Centre/CSIS) and Dr Catherine Foley (Irish World Music Centre). I also would like to thank Prof. Kevin Ryan for his financial support and the Department of CSIS for financial support and the loan of equipment during the development.

---

*Contact: Mikael Fernström,
Interaction Design Centre,
Department of Computer Science
and Information Systems,
University of Limerick.*

Communications and electronics research at UL

Electronic Engineering was one of the new core disciplines that was introduced when the National Institute for Higher Education (the forerunner of the University of Limerick) was established in 1972. As the University began to develop its research activities during the 1980s, research in the Department of Electronic and Computer Engineering (ECE) built on this base with particular emphasis on Computers, Telecommunications, Circuit Design and Microelectronics.

The success of these research activities was acknowledged nationally when the University was invited to become founder members of three programmes in advanced technologies: AMT (Microelectronics/Electronics Manufacturing), PEI (Circuit Design/Distributed Control) and Teltec (Mobile Communications/Networks). The ECE faculty have considerable expertise in control networks, integrated circuits, microelectronics, sensors, mobile/wireless communications, optoelectronics, protocols for communications networks and network security. Over 110 research Masters and Ph.D. students have graduated to date and the group has produced over 200 journal and conference publications in the last five years.

Because of the range of topics and the number of faculty involved, the activities are divided into four theme areas:

(i) Communications Networks and Security

This theme is concerned with improving service provisioning in wired and wireless systems, bringing together algorithms, architectures and associated hardware and software components. The major focus is on the next generation mobile systems, known in Europe as the Universal Mobile Telecommunications Systems (UMTS), and activities based on wireless Internet.

Communications security research is focussed on the creation of verifiably secure communication protocols and the verification of existing cryptographic-based secure protocols for use in the Internet, electronic commerce and mobile communications.

(ii) Communications Signal Processing

The focus of the research on mobile and wireless communications is on signal processing aspects and their hardware implementation. Transmitter and receiver technologies are being studied, including bandwidth efficient modulation and coding, equalisation, detection and synchronisation techniques.

The optoelectronics research is focussed on design and implementation of optoelectronic devices and optical communications networks.

(iii) Circuits and Systems

This theme includes integrated circuit design, with particular emphasis on mixed signal techniques, embedded system design and control for automation.

The focal point is real-time control and instrumentation, with a particular interest in connecting many microcomputers together via a network to achieve a significant level of automation.

(iv) Sensors and Systems

This theme is concerned with the research and development of a wide range of sensors and sensing technologies - e.g. thick film, polymers, optical fibres and their associated microelectronic systems for a wide range of applications such as gas sensing and particle concentration measurement in fluids.

The ECE Department intends to further strengthen its research base in the future by continuing to attract quality researchers and postgraduate students and by actively pursuing a variety of collaborative funding opportunities.

Contact: cyril.burkley@ul.ie

On-line radiometric analysis of peat

We're all familiar with the stylish TV ads that Bord na Mona have to promote their briquettes, and we're possibly envious also: blazing fire, attractive company... But spare a thought for the poor guy who must clean out the grate the morning after. All that ash!!!

That ash is the mineral matter which remains after the organic component of peat has been burned. It is one of two components of peat, along with moisture content, that those in Bord na Mona take seriously to ensure the production of high quality briquettes.

Ash levels must be closely monitored "on-line" (during production) and kept to a minimum. Peat contains anywhere from 2-12% ash; 4% is the ideal quantity because:

- Excess ash results in poor quality, bad burning fuel.
- Ash wears out factory components quickly, leading to excess factory downtime and raised costs.

With the assistance of the physics department in the University of Limerick, Bord na Mona now employs a novel system for the instantaneous analysis of ash on moving conveyors. This non-invasive system, which provides reliable, instantaneous data is based on the physical principle of gamma ray attenuation through matter. Using gamma radiation from two different sources - Caesium 137(Cs137) as a hard source (662 keV) and Americium 241 (Am241) as a soft source (~60 keV), and taking into account the fact that the mass attenuation coefficient μ is a measure of the amount of mineral matter in the peat, one can determine a percentage ash value.

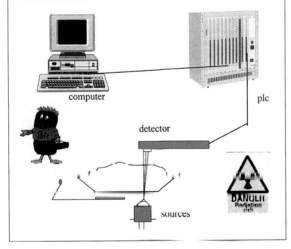

The radiation sources are housed in a stainless steel and lead container, and the beams are collimated as they exit the casing underneath the conveyor belt. The radiation passes through the belt and the peat it carries, and enters a scintillation detector located directly above the belt and source. The beam of gamma photons attenuated by the peat is then detected in the scintillation crystal whose light energies are converted into photoelectrons. These photoelectrons are amplified by a photomultiplier/amplifier assembly, and the Am241 and Cs137 pulses are separated by an analyser card.

These two streams of pulses carry different information regarding the peat. Radiation from Cs137, being the more energetic, is only attenuated by the height of the peat, whereas that from Am241 is attenuated both by the height and the mineral content. This fact allows us to extrapolate a value for ash attenuation and thus a percentage.

Data analysis is performed in a programmable controller by a complex STEP5 program, and linked to two slave PCs which give an instantaneous ash value for each wagon of peat being tipped into the factory.

Contact: Leah Wallace, Department of Physics, University of Limerick; E-mail: lmw@gemini.physics.ul.ie

Air turbine wave energy plants

Wave Energy is one of the major sources of renewable energy and is pollutant free. The conversion of energy from ocean waves has attracted considerable attention in UK, Ireland, Portugal, Japan, India and Korea.

The principal of a wave conversion system based on an oscillating water column is depicted in the *Figure*. The energy in the ocean wave is hydraulic and is contained by the oscillating motion of water. A chamber known as the device converts the oscillating water column into an oscillating air column. The pneumatic energy from the oscillating air column is converted into mechanical energy by a unique turbine known as Wells turbine, invented by Professor Wells, former Professor of Civil Engineering at The Queen's University of Belfast (QUB). The fundamental research and development and design methodology of the Wells turbine by the School of Aeronautical Engineering at QUB was a key component of the 75 kW Islay power plant in Scotland, built by the QUB team. The power plant station is the first wave power station to be built to an operational stage successfully.

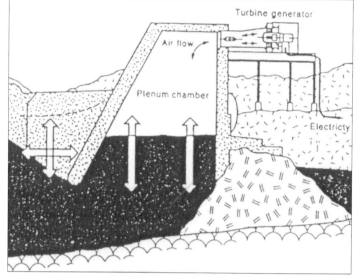

A wave conversion system based on an oscillating water column.

The Department of Mechanical Engineering at the University of Limerick (UL) and the School of Aeronautical Engineering at QUB are currently active in research, development and design of the next generation of wave power stations to be built in Portugal and Ireland. The focus is on modern computational, experimental and manufacturing methods for the cost-effective design of Air Turbines for wave energy conversion. The groups at UL and QUB are closely working with the Marine Institute at Dublin for the first wave power station to be built in Ireland. Ireland enjoys one of Europe's best wave climates, with near shore resource of 50 TW hrs. The Marine Institute has developed a plan for a 1-2 mW pilot plant to be operational by the year 2000 with a total installed capacity of 25 mW by year 2010.

*For further information contact: Dr A. Thakker,
Wave Energy Research Team, University of Limerick,
Department of Aeronautical & Mechanical Engineering, Limerick.*

Pore development in cold drawn PET fibres

This research was carried out at the Fibre Development Laboratory and the Analytical Research Centre of Asahi Kasei, Nobeoka City, Miyazaki Prefecture, Japan, and in the Department of Materials Science and Technology, University of Limerick, under the Asahi Scholarship and Study Programme.

Poly(ethylene terephthalate) (PET) is a polyester used in the textile industry. When fibres of this material are stretched or cold drawn, they thin down at a point, resulting in the formation of a neck. Drawing the material to many times its original length can result in a silvery or lustrous appearance. This change in appearance results from the formation of voids or pores within the fibre structure.

The mechanism of the development of pores within the material was investigated by carrying out cold drawing on monofilament PET fibres. Fibres were cold drawn to different draw ratios (the draw ratio is defined as the ratio of the final fibre length to the original fibre length). The formation of pores was observed using scanning electron microscopy (SEM). The pores were seen to develop initially at the fibre centre. As the draw ratio was increased, the distribution of pores was spread over the entire cross-section of the fibres.

Properties such as fibre density and fibre diameter were considered to be pore-sensitive, as their relationship with the draw ratio was altered at the point where the fibres became porous. The change in the relationship between these properties and the draw ratio allowed the identification of a critical draw ratio for pore formation.

The neck region of a fibre experiences a triaxial state of stress during the cold drawing process. Draw ratio is directly related to the force applied during the drawing process. As the draw ratio was changed, the shape of the neck was altered. The force applied to the fibre during the drawing process was resolved into components perpendicular and parallel to the neck.

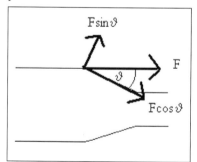

Figure 2. Schematic of the resolved forces associated with the neck during cold drawing. F is the force applied to the fibre during the drawing process. ϑ is the neck angle.

$F\sin\vartheta$ is an outward-acting force. As the critical draw ratio is approached, the value of $F\sin\vartheta$ is significantly increased relative to the value of $F\cos\vartheta$. This increase in the outward-acting force causes the material to fail at discontinuities in the structure, resulting in the formation of voids. *Figure 1* clearly shows that pores only form after the material passes through the neck region, during which it experiences the force $F\sin\vartheta$.

Figure 1. SEM micrograph of a longitudinal section through the neck region of a porous PET fibre.

Contact: Katherine Wallace & Martin Buggy, Department of Materials Science and Technology, University of Limerick.

Fundamental motor skill development

Fundamental motor skills such as hopping, jumping, skipping, kicking, throwing, catching and striking are prerequisites to the learning of sport specific skills such as those of basketball, football, gymnastics, tennis, badminton etc. Sport specific skills are comprised of fundamental skills and variations of them. It is very difficult to obtain proficiency in sport skills unless the prerequisite fundamental skills are present.

The fundamental skill phase of development begins in early childhood at about two to three years, and individuals have the potential to be fully proficient in most of them by about six years. Previous research has found that, while there is a considerable genetic influence in the early stages, complete development of the fundamental skills is very much dependent on environmental influences in the form of practice, learning and teaching. This latter fact is often not recognised, with even some professional educators assuming that such skills will emerge automatically.

Sport and exercise scientists at the University of Limerick have found that children are leaving primary school deficient in fundamental motor skills. These findings provide evidence for concerns expressed in the government strategy plan for Irish sport regarding the skill levels of Irish children.

With respect to the fundamental skill of throwing, researchers at Limerick have also found that, when teaching interventions are applied, children of four to six years can acquire full proficiency. This finding is consistent with similar research carried out in the USA. It was also found that, when exposed to appropriate teaching, older children of 12 years can catch up to achieve full development. Work is continuing whereby the research has been extended to include post primary pupils – and other fundamental skills are being examined.

A study is also about to begin which will examine the relationship between fundamental motor skills and sport specific skills from both applied and theoretical perspectives.

The focus on fundamental skills has implications not only for the development of highly skilled sports people, but also for health. Many modern day diseases are due in at least part to lack of physical activity. There is evidence to show that people are more likely to take up or continue participation in sports if they have adequate degrees of skill. A wide repertoire of fundamental skills obtained in childhood would make more specific skills easier to acquire in adolescence and adulthood.

Typical video still used in researching the skill of throwing.

Contact: Dr P.J. Smyth,
Department of Physical Education
& Sports Science,
University of Limerick;
E-mail: pj.smyth@ul.ie

Equine science research: technical developments for an indigenous industry

Ireland is the largest producer of thoroughbred horses in Europe. While production standards in Ireland are generally superb, technical barriers to the growth and development of the industry remain. The Equine Research Unit at UL performs applied and basic research designed to overcome these technical barriers.

Equine nutrition trials use urine and faecal collection techniques to determine digestibility of dietary components

Behavioural profiles in stabled horses
Repetitive behaviours, or "stereotypies", are common in stabled animals. Sterotypies are really coping behaviours that help the animal to reduce stress. Stereotypies are of concern as they can be damaging, detract from performance, and may indicate a need for environmental enrichment. Research at UL is addressing the incidence of these stereotypies in stabled horses and will identify environmental and husbandry factors that might reduce their incidence.

Developmental Orthopaedic Disease (DOD)
DOD is the name given to a number of disorders in the growing joints and surrounding cartilage in the limbs of young thoroughbreds. The incidence of DOD has increased recently. Symptoms develop from disturbances in the ossification process. DOD causes athletic impairment and affects the saleability of the animal and profitability. Genetics, nutrition, environment and trauma contribute to DOD. Studies at UL focus on defining the incidence of DOD in farms in the region and identifying nutritional and growth parameters associated with its development.

Sugar beet levels in equine diets
Sugar beet pulp (SBP) is a by-product of the sugar industry. While SBP is used as an ingredient in horse feeds, the Horse Industry has been slow to include significant levels of sugar beet pulp in diets because of concerns about the development of "choke", arising from impaction of the oesophagus when poorly hydrated SBP is fed. UL research shows that diets containing over 40% sugar beet pulp do not adversely affect health or performance.

Hoof quality in Irish thoroughbred horses
The athletic ability of the horse determines its value. Hoof diseases cause significant loss. Studies at UL are providing basic data on hoof quality and identifying the role of nutritional supplements such as biotin, zinc and methionine on hoof growth, wear and quality. These studies are necessary because of the unique combination of high rainfall and ill defined mineral intake that many Irish horses are subjected to. This research forms part of a programme of studies on hoof quality aimed at developing better shoeing materials in the horse.

Contact: sean.arkins@ul.ie

Molecular technology in action – a tale of two ß-Lactams

Many pathogenic (disease causing) bacteria produce an enzyme which is exceptionally efficient at breaking the four-membered, ß-lactam, ring in penicillin and other closely related structures. This reaction renders these compounds useless as antibiotics, and so the presence of these ß-lactamase enzymes – so called because they cut the ß-lactam ring – is a primary defence that many bacteria use to resist the effects of antibiotics.

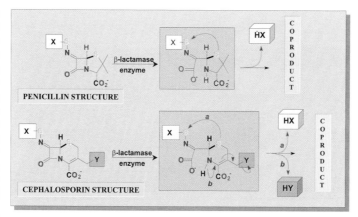

A short time ago we reported on the preparation of the first penicillin-based structure that behaves as a ß-lactamase-dependent prodrug. In this structure, cleavage of the ß-lactam ring triggers the very rapid release of a well-defined part of the side chain (see Figure). The unique value of this pattern of reaction lies in its potential to exploit the presence of ß-lactamase enzymes in pathogenic bacteria. Our prototypic structure is readily cleaved by ß-lactamase enzymes like other penicillins, but this one can carry a sting in its tail – literally. Each time that critical four-membered ring is cleaved, a unique fragment X is released, and this can be designed to be fatal to the bacterium – but only when it is cleaved off: this is the essence of what our current research is about.

Another organic chemical which is widely used as an antibiotic is known as a cephalosporin: like penicillin it too has a ß-lactam ring. Very recently we were able to show that the latent, or hidden, reactivity pattern that we had incorporating into the penicillin structure also worked in cephalosporins. But here there is an added bonus. An in-built reaction pattern of cephalosporins, which results in the release of component Y (see Figure), occurs sequentially with the release of our side chain component X. Our prototypic cephalosporin is thus the first ß-lactamase-dependent dual-release prodrug. The dual-release feature should bring a singular enhancement to the cytotoxic capacity of cephalosporins in their usage as the prodrug component in the anticancer therapy known as antibody-directed enzyme prodrug therapy (ADEPT). Derivatives of this dual-release prototype have potential application in combating the problem of antibiotic-resistant bacteria also.

For further information contact:
Timothy Smyth; E-mail: timothy.smyth@ul.ie

Centre for Applied Mathematical Sciences, UL

In June 1998 the Society for Industrial and Applied Mathematics (SIAM) conducted the workshop "**Uncertainty Management and Assessment**" and, according to the December 1998 issue of the SIAM news, the workshop report "**argues for the need to understand and, to the extent possible, to explicitly represent and quantify all sources of uncertainty in models and simulation**". Meeting this challenge will inevitably require increased collaboration between applied mathematicians and statisticians.

The Centre for Applied Mathematical Sciences (CAMS) was formed in 1998 and is a research centre comprised of applied mathematicians and statisticians working under the major unifying theme of mathematical modelling. CAMS has identified the analysis of uncertainty in models of physical processes as a major research theme. The members of CAMS fall naturally into two groups – the **Applied Mathematics** group and the **Applied Statistics** group.

The Applied Mathematics group has research interests in fluid mechanics and transport phenomena, geophysical flows, industrial flows, singular perturbation problems and convection-diffusion problems. Current topics include modelling the large scale ocean circulation, baroclinic instability in the ocean, flow over bottom topography, dynamics of strong eddies, stability of large scale atmospheric waves, modelling the thin viscous flow which occurs during the coating of substrates, mechanisms for episodic subduction

Left to right: Professor P.F. Hodnett, CAMS; Professor A. Acrivos, CCNY; Professor M. Wallace, CAMS.

on Venus, and finite difference and adaptive computational meshes.

The Applied Statistics group is engaged in research in geostatistics, survival analysis, Bayesian smoothing, epidemiology, meta-analysis, spatial data, hierarchical modelling, image analysis, reliability and process control. Current topics include manpower planning in the Irish Public Service, sequential analysis applied to reliability testing, Bayesian methods for meta-analysis of epidemiological studies, frailty models for repeat accident data in Dublin Bus, Markov chain models for caries clinical trial data, Bayesian methods for smoothing disease incidence rates over time and space, and the analysis of gait patterns from sensory motion detectors.

The recent appointment of Andrew Fowler, Senior Research Fellow at Corpus Christi College, Oxford, as Adjunct Professor at UL is a major boost for CAMS. Fowler is a leading world figure in Applied Mathematics and will visit regularly with a view to developing research links. A Visitors' Programme involving visits to UL by leading researchers in the fields of Applied Mathematics and Statistics has been initiated. A recent visitor was Professor A. Acrivos who is Director of the Benjamin Levich Institute for Physico-Chemical Hydrodynamics at City College, City University of New York (CCNY) and former editor of *Physics of Fluids*.

Contact: Professor Don Barry; E-mail: don.barry@ul.ie

Data mining made easy

MINE*it* – OUR PROFILE:

Hundreds of millions of pounds are spent every year by companies of all sizes on market research performed by second parties. Ironically, much of these costs are related to the surveying, collection, and analysis of data *already* possessed by the client company. The data mining solution produced by **MINE*it*** will allow companies to discover associations and sequences "hidden" within their internal databases. With this package, companies can discover previously untapped information sources while minimising costs and maintaining complete confidentiality.

MINE*it* is a new IT solutions company specialising in the marketing and provision of data mining and business intelligence services and products. Originating as a spin-off company from the Northern Ireland Knowledge Engineering Laboratory *(see box)*, **MINE*it*** has an unparalleled foundation in data mining technologies and methods. In addition, partnerships with both LPS (Dublin) and SPSS (Chicago) have further strengthened **MINE*it*'s** abilities to deliver the benefits of data mining to companies interested in strategically positioning themselves with their customers, competition, and in the ever-growing e-commerce industry.

MINE*it* – OUR CUSTOMERS:

E-COMMERCE - Commerce conducted on the internet is growing at a tremendous rate, and bringing with it new opportunities to market one's products and services. **MINE*it*** offers a variety of solutions to companies involved in e-commerce. For example, the **MINE*it*** data mining solution provides a means for the host company to build a variety of detailed browsing prediction models based on each individual visitor's unique buying history and navigational behaviour while online. E commerce customers continuously generate digital trails while surfing the net, which contain valuable marketing information, including which sites were visited, the frequency of visits, and records of recent purchases. With this knowledge, companies can divide customers into various groups with similar interests and purchasing patterns, and then personalise all online advertisements and emphasise products that complement these unique characteristics. This electronic version of direct marketing gives the company an opportunity to increase overall sales, as web designers can optimise their sites for maximum commercial impact. Due to the number of sites where potential customers can purchase a given product, it is vital to improve customer satisfaction at every opportunity to ensure customer attraction and retention. Included within the business intelligence package produced by **MINE*it*** is the Customer Relationship Management (CRM) tool. CRM will help companies to capitalise upon opportunities for cross selling their products and to improve upon existing customer service systems. The data mining package offered by **MINE*it*** is beneficial to all e-commerce businesses as it can analyse customer-to-business sites as well as business-to-business sites.

FINANCE – Data mining technology can be especially useful in financial companies where detailed customer databases can be analysed, revealing highly profitable information. Information derived in this process can be used for a variety of purposes including: customer retention analysis; payment or default analysis; fraud detection; and an analysis of the potential for cross-selling certain products and services. Fraud detection would be considered among the most useful of these tools in the financial services sector. By way of analysing customer databases, the **MINE*it*** data mining package can identify trends and patterns that indicate damaging behaviour such as: account delinquency; credit card fraud; and account transaction fraud. Cross-sales analysis is also highly useful in the financial sector. The **MINE*it*** package can indicate what products and services would be most successfully paired for sale to identified groups.

TELECOMMUNICATIONS – Due to increased deregulation and the consequential increase in service providers, higher than average customer turnover rates have become reality for companies in the telecommunications industry. Over the last decade, customer retention has become equally as important as customer attraction. The **MINE*it*** solution package includes a powerful churn management tool that can predict which customers are more likely to switch to another carrier based on the profiles of those who have left in the past. As a result, companies with this information can target a defined group of potentially lost customers with appropriate marketing campaigns and special offers. Other such uses of data mining technology within the telecommunications industry include: fraud detection; cross-sales analysis; and network traffic management analysis.

MINE*it* also provides business intelligence solutions in both the PUBLIC and RETAIL sectors. For public sector companies, the **MINE*it*** solution can provide a variety of predictive models, including land value prediction, and also has fraud detection capabilities. In the retail sector, data mining activities can provide detailed analysis of: sales productivity; exposure to risk; and churn.

MINE*it* – OUR SOLUTION:

The **MINE*it*** business intelligence package offers our client-companies the opportunity to gain a better understanding of their customers. Our clients gain the ability to analyse and predict customer behaviour. Such information can be used to not only benefit the company, but also its customers, through improvements in both customer service, and product offerings. **MINE*it*** can provide the solutions your company needs to gain a true competitive edge.

Contact: MINE*it* Software Ltd,
University of Ulster, Newtownabbey,
Co. Antrim BT37 0QB Northern Ireland;
Tel: (01232) 368875;
Fax: (01232) 366068;
Email: info@mineit.com;
Web: www.mineit.com

NIKEL

Since its inception in 1992 at the University of Ulster, NIKEL has gained an international reputation for data mining research and technology deployment. NIKEL's primary focus is to equip companies in Northern Ireland with the technical knowledge to use advanced software in their strategic and operational plans, consequently helping these companies to gain an international competitive advantage.

Current areas of research at NIKEL include: KBS systems, medical informatics, business process engineering, geographical information systems, multimedia, and planning & scheduling.

To obtain further information regarding NIKEL, please visit:
www.nikel.ulst.ac.uk

University of Ulster's Applied Research Centre - ARC

1999 saw the official launch of the Applied Research Centre (ARC) in conjunction with the new multi-million Nortel Networks Research Programme, which is a joint collaboration between Nortel Networks, the University of Ulster, and Queen's University.

ARC is based in the Faculty of Informatics on the Jordanstown Campus of the University of Ulster. The group has emerged over the last few years in presence both academically and industrially. Its roots lie in the several collaborative projects with Nortel Networks, the High Performance Computing Group, and the Medical Informatics and Telehealth Group.

The group specialises in Hybrid Artificial Intelligence Systems, Data Warehousing, Data Visualisation and High Performance Computing. Its research is planned and designed in a generic nature that can be applied to many domains, such as medical diagnostics, geological surveying and financial forecasting/prediction. This approach enables fresh techniques to be introduced to the telecommunications domain, and indeed techniques from the telecommunications arena to be introduced to the other research fields.

(Left to right) Dr Edwin Curran, Ms Mary Shapcott, Dr Roger Johnson (Nortel Networks), Dr Kenny Adamson, Mr Roy Sterritt, Dr Alfons Schuster.

The Partnership

The Faculty of Informatics has undertaken research collaboration with Nortel Networks for many years now. This research has been additionally funded by successful project applications to EU (Stride), the Engineering and Physical Sciences Research Council (EPSRC) and the Department of Trade & Industry (DTI) (AIKMS, TCS programmes), Industrial Research and Technology Unit (IRTU) (Start) and the Department of Education (NI) (DPhil & CAST). The Northern Ireland Telecommunications Engineering Centre (NITEC), one of Nortel Networks's global R&D sites, is based only a few miles from the University. Both local Universities provide high calibre graduates to the company.

ARC sees that its competitive advantage lies in the fact that it is not its aim to focus on any one specialised area such as data mining or even telecommunications. This ensures it is flexible, dynamic and in a position to harness the best and most relevant of techniques for the application of the research. It is its aim to develop fundamental pure research, apply this with a solid application, and develop embedded research in local companies to assist in developing a high-tech economy with long-term stability.

Extracting Cause and Effect Relationships from Complex Systems

The Faculty of Informatics, UU Jordanstown, in 1995 was successful in obtaining funding from a highly prestigious EPSRC/DTI research programme. They beat top-class applications from the best Universities around the UK to belong to the "Architectures for Integrated Knowledge-Manipulation Systems" (AIKMS) programme.

The team at UU developed a system that extracts and manages large sets of complex, non-deterministic live data from equipment within Nortel Networks. They added "artificial intelligence", the ability for the computer to model human decision making, by combining Evidential Reasoning – a technique to measure evidence based on the probability of interconnecting events occurring; and Evolutionary Computing – an optimising technique emulating the Darwinian process of natural selection – "survival of the fittest". This was implemented on a multi-processor computer platform to speed-up the mathematically complex process.

The system or Knowledge Discovery architecture developed accepts live management data from the Synchronous Digital Hierarchy (SDH) telecommunications equipment. The SDH is the international standard for the new generation of broadband

Nortel Networks' Monkstown Visitors Centre depicting the operation of the management of their High Speed Global Networks.

network and telecommunication services. The increased bandwidth makes possible sophisticated services such as video on demand, ISDN and ATM data transfer, and video conferencing.

The management of this level of sophistication becomes more difficult, particularly when a fault in the network occurs. Under extreme fault conditions a tremendous amount of management data can disrupt the network and flood the network manager, making it difficult to identify the cause of the disruption.

The "NetExtract" architecture could be implemented to mine the incoming data and predict with increasing accuracy the underlying fault.

Testing of Complex Systems

The GARNET project (1997-1999), partially funded by IRTU, has assisted Nortel Networks in automating the testing of their high capacity broadband transmission and switching equipment.

> Their 10 gigabit per second transport products provide the equivalent of 1000 paperback novels per second down a fibre thinner than a human hair.

In our view, the automated testing of this equipment will give Nortel Networks a competitive advantage in a fiercely competitive industry. It should reduce the expense and time to market of their products, while freeing up specialised engineers for further design effort.

Two major lines of investigation were undertaken in GARNET. The first was the simulation and modelling of these transmission systems as processes for execution on a machine with multiple processors. Sixty-four of these processors can fit into a machine the size of a typical PC, which has only one processor. These parallel machines are characterised by operating independently on their assigned instructions.

This approach offers the opportunity for Nortel Networks to test multiple configurations of large scale networks before customer installation. This offers a potential solution to the practical and logistical problems encountered in the test lab with the actual equipment as a "captive office".

The second line was the continuation of the Artificial Intelligence research, with its adaptation to provide an assurance level for auto-testing.

The live data from an over-night automated test run at Nortel Networks is fed into the Knowledge Discovery architecture. The visualisation of the mined data provides the engineer with the human assurance that the high capacity equipment behaved within the norms during the automated tests.

This approach offers the advantages that several weeks' worth of manual testing can be performed automatically in one night, and a day's worth of manual data correlation can be automatically mined in a few minutes.

Into the New Millennium

1999 also saw the start of a new collaborative research programme between Nortel Networks, Queen's University and the University of Ulster, supported by IRTU funding, addressing a top down network design approach to telecommunication systems. This forms part of the strategic development in increasing the telecommunication capability within the universities in Northern Ireland. Twenty additional posts are being created to undertake the applied research. The work will focus on:

- Reducing the time to market by increasing the amount of engineering reuse
- Improving testing efficiency in new product developments & manufacture
- Addressing how network wide services can be effectively provided
- AI and Knowledge Engineering.

This new programme will enable the Faculty of Informatics to continue to apply its innovative and fundamental research to the telecommunications domain. It is becoming increasingly renowned for its research in Data Mining, Knowledge-Based Systems and AI. These areas will be extended in the new research, along with investigation into Temporal Probabilistic Networks, Temporal Databases, Distortion and Geometrical User Interfaces and TMN (Telecommunication Management Network) research.

The group's primary interest is Soft Computing (SC). Lotfi Zadeh from the University of California at Berkley explains that soft computing differs from conventional (hard) computing in that, unlike hard computing, it is tolerant of imprecision, uncertainty and partial truth. In effect, the role model for soft computing is the human mind. The guiding principle of soft computing is to exploit the tolerance for imprecision, uncertainty and partial truth to achieve tractability, robustness and a low solution cost.

Currently the principal areas that make up SC are neural network theory (NN or ANN – artificial neural networks), fuzzy logic (FL), and probabilistic reasoning (PR), with the latter incorporating belief networks, genetic algorithms, chaos theory and sections of learning theory. What is important to note is that SC is not a melange of FL, NN and PR. Rather, it is a partnership in which each of the partners contributes a distinct methodology for addressing problems in its domain. In this perspective, the principal contributions of FL, NN and PR are complementary rather than competitive.

This project provides the opportunity to harness these approaches in a hybrid fashion for the network management and testing arenas in the telecommunications domain and, at the same time, enables the staff involved to engage in strategically important research.

For further information about ARC or the Nortel Networks-UU collaborative projects, contact Dr Kenny Adamson (Director); Tel: 01232-368163; E-mail: k.adamson@ulst.ac.uk; Arc web site: www.ulst.ac.uk/arc

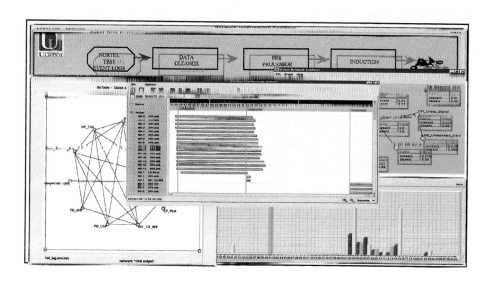

Screen shots of some of the Visualisation User Interfaces developed by ARC to complement the Knowledge Discovery Architecture.

UNIVERSITY OF ULSTER — PHIL EAMES

Centre for Sustainable Technologies (CST)

The Centre for Sustainable Technologies undertakes both fundamental and applied research in the areas of Solar Energy and Energy in Buildings, River Hydraulics and Hydropower, and Sustainable Materials and Processes. The centre was formed in 1998 to more accurately represent the expanding range of activities undertaken within the centre for Performance Research On the Built Environment (PROBE).

The majority of research undertaken is performed in collaboration with other university research groups, research centres, industry or government, with successful projects undertaken within Ireland, the UK, Europe and World wide. Recent collaborators have included, BP Solar, Pilkington and HR Wallingford in the UK, TNO in the Netherlands, TFM and PSA in Spain and the University of Patras in Greece, L.N Mithila University in India, and the University of Sydney in Australia. In addition to an extensive portfolio of research funded by the Engineering and Physical Sciences Research Council and the European Union Framework programmes, research is also undertaken for both industry and government. Major research areas at present include:

- The design, development and production of high-efficiency, non-imaging integrated photovoltaic facade cladding elements, that give similar performance to present systems at a significantly reduced cost.
- The development and testing of advanced high performance evacuated glazing that has both good visual transmittance and a mid-plane heat transfer coefficient in the range of 0.4 to 0.6W/M2 K.
- The processing of locally-grown timber with regard to both drying and machining to enhance its value by producing a better quality product.
- Experimental and theoretical investigations into mobile bed meandering compound channels.

A purpose built vacuum system for producing experimental samples of evacuated glazing with dimensions of up to 0.5 by 0.5 m.

An extensive range of experimental research facilities have been established. These include:
- Both line-axis and planer solar simulators for testing photovoltaic and solar thermal systems under controlled well-characterised conditions.
- An illuminated calorimeter and window weather test equipment to enable the evaluation of glazing system performance.
- An infra-red thermography system to enable the non-intrusive measurement of surface temperature for both small scale experiments and large scale building envelopes.
- Vacuum systems for the fabrication of evacuated glazing samples, the development of spectrally selective glass coatings and for vacuum casting.
- Small scale instrumented timber drying kiln and controlled timber machining facilities.
- A medium scale river flood channel facility is situated at the University technology unit in Carrickfergus to enable models of river hydrology to be developed.
- Outdoor test facilities for solar thermal and photovoltaic systems are located adjacent to an instrumented research building which has a low heat loss fabric and triple glazing.

A range of optical and thermofluid models utilising both ray-trace and finite volume/element methods have been developed in-house, to enable detailed spatial and temporal predictions of temperatures and fluid flows within solar energy and building/building facade systems to be made. Extensive computing facilities at present include a network of Sun ULTRA computers and a Sun 4 processor HPC 450 machine.

Contact: School of the Built Environment, University of Ulster, Newtownabbey, Co. Antrim BT37 OQB, Northern Ireland; Fax 01232-368239; E-mail: pc.eames@ulst.ac.uk; Web site: //www.engj.ulst.ac.uk/SCOBE/CST

UNIVERSITY OF ULSTER — JIM SHIELDS

Fire safety engineering research and technology transfer

The Fire Safety Engineering Research and Technology Centre (Fire SERT) at the University of Ulster is an internationally acknowledged leader in the field of fire safety science and engineering. Fire SERT provides a wide range of research, educational, product development, testing and consultancy services to an increasing number of clients. The reputation of Fire SERT is founded on its research portfolio and taught fire safety engineering programmes, and is the only centre in Ireland offering such a comprehensive range of fire safety related services.

Ongoing research programmes include: the behaviour of glazing systems exposed to real enclosure fires; interactive behaviours of common wall and ceiling linings in enclosure fires; effect of exposure of structural steel to different thermal environments; and aspects of human performance in fire emergencies. The output of Fire SERT's research is disseminated to a wide range of beneficiaries including: British Standards Institution, International Standards Organisation, Government Bodies, Fire Scientists and Fire Safety Engineers. Collaborative programmes with other institutions are such that Fire SERT's work has immediate application in the development of the next generation of fire and evacuation simulation models.

In June 1999 Fire SERT will be joined by Professor Vladimir Molkov, formerly Head of Fire Modelling and Fire Safety in Buildings, All-Russian Research Institute for Fire Protection, in Moscow. His addition to Fire SERT's staff marks the beginning of the next phase in the development of new research initiatives and services to industry

Fire SERT's international reputation in the field of Fire Safety Engineering is founded on its research portfolio and its taught fire safety engineering programmes. Fire SERT's MSc Fire Safety Engineering course first introduced in 1990 attracts students from all over the world to study at the University of Ulster. New developments in relation to the MSc Fire Safety Engineering include the introduction of additional pathways for students wishing to specialise in Health and Safety, Fire Safety Management, Fire Simulation and Safety Design. It is intended that these additional options will be available from October 1999 onwards.

Work is already in hand to develop this

Predicting the behaviour of Glazing Systems in real enclosure fires.

capability of offering these and other programmes by way of block release and/or distance learning, and it is hoped that progress will be sufficient to allow the first enrolment of students in September 2000. For all at Fire SERT there is much to do and look forward to.

Further information on any of the above can be obtained from:
Professor Jim Shields,
School of Built Environment,
University of Ulster, Newtownabbey,
Co. Antrim BT37 0QB, Northern Ireland;
Tel: 01232-368702; Fax: 01232-368700;
E-mail: tj.shields@ulst.ac.uk

UNIVERSITY OF ULSTER   JIM CURRAN

Mission Statement
To create centres of excellence for innovation, technology transfer and enterprise that optimise access to the R&D resources and services at the University of Ulster.

Strategic Background
Success in today's fast moving global economy depends critically on exploiting the knowledge, skills and ideas of the science base to be in the vanguard of technologically advanced economies. To participate in and take advantage of the new knowledge-driven economy, the University of Ulster established UUTECH Limited, its wholly owned umbrella organisation specifically charged with the responsibility for technology transfer and exploiting the innovative research emanating from its internationally recognised research centres. It serves as a formal businesslike interface enhancing the University's close relationship with the local business and industrial sectors by providing a strategic focus for technology transfer activities that make a vital contribution to the diversification of the local economy.

UUTECH Remit
To take a stake in start-up companies, to exploit Intellectual Property Rights (IPR) by patenting and licensing, to manage and develop campus incubator facilities and consultancy activities.

Intellectual Property Rights and Technology Brokering
UUTECH gives additional impetus to the protection, management and exploitation of the University's IPR through patenting and licensing, and has sole responsibility for negotiating these matters for the University.

Technology Transfer
UUTECH also has responsibility for the University's technology and knowledge transfer activities, such as the Teaching Company Scheme, the Manufacturing Partnership, and the consultancy activities of the University. Various supported initiatives also sustain and encourage entrepreneurial innovation. The overall vision is to establish NI as an internationally recognised world class region for commercialisation of innovative technologies.

Venturing
UUTECH negotiates and holds the University's stake in high technology spin off companies which emerge from the University's research and knowledge base. Joint venturing with existing companies also bolsters high technology commercial development. UUTECH is a partner in the University Challenge Fund (NI) Limited, and sources suitable innovative high technology ventures that can achieve value-added growth by accessing this seed funding.

Campus Incubation
UUTECH manages Campus Business Incubator Centres being developed to nurture high technology start-up companies. These facilities are aligned with the respective campus-specific research strengths: with Coleraine focusing on the life, health and environmental sciences; Magee on software; Jordanstown on engineering and informatics; and the new Springvale campus on medical technologies. Services available at the Incubator Centres include:
- Business start-up assistance, support and mentoring.
- Office and laboratory accommodation; access to University technical expertise.
- Campus amenities; IT and educational infrastructure.

For further information contact: Jim Curran - UUTECH Manager; Tel: 01232-368019; Fax: 01232-366802; E-mail: jp.curran@ulst.ac.uk

---

UNIVERSITY OF ULSTER   DAVID MCILVEEN-WRIGHT

The Northern Ireland Centre for Energy Research & Technology

Sustainable development is important for the future of all companies and is crucial for the future of mankind. In essence, it is about getting the maximum benefit from limited resources and producing the minimum environmental impact.

The Northern Ireland Centre for Energy Research & Technology (NICERT) was formed with assistance from the European Development Fund, through the Industrial Research & Development Unit's Technology Development Programme, with the remit of promoting sustainable development within Northern Ireland and the Republic of Ireland, and identifying the opportunities which this creates.

NICERT has considerable expertise in helping industry to realise its potential in energy R&D or to optimise its use of energy. It can help with energy audits, identifying opportunities for energy reduction, and the technical, environmental and economic analysis of energy saving projects or proposals. This analysis can look in detail at one part of the process or consider the total plant system. Of particular interest are schemes aimed at the utilisation of biomass or waste streams for the generation of heat or power.

Research staff from NICERT using the latest computer simulation techniques to evaluate projects.

NICERT can assess agricultural or industrial wastes/biomass and perform feasibility studies to determine their potential as a fuel. This will identify the most promising routes for matching the energy recovered from the waste with a company's energy demands.

NICERT has advanced laboratory facilities, which include gas analysis equipment and a range of portable analysis and data logging equipment for collecting data on site. NICERT has one of Europe's most advanced laboratories for developing and evaluating refrigeration and heat pump equipment using new replacement refrigerants and lubricants. NICERT has recently made exciting developments in high temperature heat pumps for heat recovery applications, in the smart control of refrigeration equipment, and in the

Research work in progress on a current project.

industrial application of high temperature glide refrigeration and heat recovery plant.

The aims of NICERT are to support the development of a renewable energy industry and to foster the application of energy management and the rational use of energy. For these aims to be realised, there needs to be a greater awareness of energy-related issues, and the underpinning technologies and practices that will influence future development. Industry must realise that R&D is essential for survival, and NICERT is here to help provide an R&D base for industry to tap into.

Contact: Dr David McIlveen-Wright, Business Manager, NICERT, University of Ulster, Cromore Road, Coleraine BT52 1SA; Tel: 01265-324477; Fax: 01265-324900; E-mail: dr.mcilveen-wright@ulst.ac.uk

Environmental toxicology at the UU – a new initiative

The World Health Organisation has confirmed that "a considerable proportion of the European population is exposed to environmental factors to an extent which is considered to pose a threat to some aspect of human health". Environmental Toxicology is concerned with the effects of chemical contaminants on ecological systems and organisms, both large and small. Of over six million known chemicals, detailed toxicity tests have only been performed on a few hundred.

Environmental Toxicology research within the School of Environmental Studies at the University of Ulster has built upon existing strengths in the aquatic science and analytical chemistry areas. Current work aims to improve our understanding of the behaviour and effects of contaminants on organisms and the environment, along with the development of methods to improve monitoring and effluent quality.

Industrial collaboration is integral to this research group, and successful partnerships have already been formed, including Randox Laboratories Limited (the leading British manufacturer of clinical chemistry reagents, environmental diagnostic kits, and quality control sera) and Northern Ireland Electricity plc. Government agencies also form a key part in funding Environmental Toxicology research at UU, with grants obtained from the Natural Environment Research Council and the Department of the Environment.

Present research projects include: impact of sulphur dioxide emitted from power stations on soils; development of a rapid toxicity assay; development of environmental quality standards and biomarkers; trace metal toxicity testing; ecological assessment of lakes; phosphorus loss to fresh-water; and pesticide risk assessment for freshwaters.

On the educational side, a new MSc in Ecotoxicology has been recently introduced at UU. This course fulfils the demand for trained personnel in the environmental regulatory agencies, in companies subject to such regulation, and those involved in providing support services such as monitoring and consultancy. Research opportunities also exist for companies or individuals with an interest in Environmental Toxicology, and it is hoped that Environmental Toxicology research at UU will continue to flourish as we enter the new Millennium.

Toxicological effects of contaminants are being investigated in atmospheric, terrestrial and aquatic systems.

Contact: Dr Brian Rippey,
Environmental Toxicology & Chemistry,
University of Ulster, Coleraine, Northern Ireland, BT52 1SA;
Tel: +44 (0) 1265-324085; Fax: +44 (0) 1265-324911;
E-mail: BHRT.Rippey@ulst.ac.uk;
http://www.ulst.ac.uk/faculty/science/nnru/envtoxchem.html

The Northern Ireland Centre for Diet and Health (NICHE)

The Northern Ireland Centre for Diet and Health (NICHE) is an enthusiastic multi-disciplinary research team within the School of Biomedical Sciences at the University of Ulster, Coleraine. Its remit is to conduct high quality research programmes on the complex links between dietary habits, nutrient status and aetiology of chronic diseases, and to provide research support and advice for industry for developing new food products, particularly in the area of functional foods. Scientists from the group have developed novel methods for assessing nutrient status with regard to micronutrients, and have developed innovative techniques for the rapid assessment of, for example, antioxidant levels in food and body fluids. The group is also recognised as being one of the leaders in developing methods for assessing dietary intake.

The main research topics include health benefits of antioxidant micronutrients (particularly carotenoids, vitamin C, vitamin E and copper); the influence of n-3 polyunsaturated fatty acids, vitamin A, and carotenoids on immune function; effects of n-3 fatty acids on heart disease and stroke; and the influence of folate on serum homocysteine, a risk factor for heart disease. Studies are also being conducted on probiotics, oligosaccharides, complex carbohydrates, phytoestrogens and flavonoids in relation to cancer prevention, and on the influence of disease on nutritional status. Obesity is a major area of research that focuses on dietary, socio-economic and pyscho-social factors, and the related area of effects of macronutrients on appetite and satiety.

Professor Ian Rowland, Director of NICHE.

The breadth of expertise at NICHE in the areas of human volunteer studies, diet and health research, in vitro systems, and consumer response to foods, enables us to provide a high quality, comprehensive service to the food industry for the development and evaluation of new food products with health promoting properties.

NICHE has built up an extensive network of collaborations with leading universities and research institutes throughout the world, and works with major food companies including Kellogg's, Coca-Cola, Milupa-Nutricia, Hoffman-La Roche and Golden Vale.

NICHE is supported by the European Regional Development Fund through the Industrial Research and Technology Unit's Technical Development Programme.

Contact: E-mail: dietandhealth@ulst.ac.uk

UNIVERSITY OF ULSTER — ROGER ANDERSON, CHRIS HUDSON, DECLAN MCKEEFRY & KATHRYN SAUNDERS

Vision Science looks to the future in Coleraine

The Vision Science Research Group at the University of Ulster in Coleraine has acquired substantial funding in the form of two large grants from the Wellcome Trust and another from the British Diabetic Association to sustain a programme of research investigating the mechanisms which underlie visual processing, and to improve understanding of how disease processes impair visual function. The Group, which has recently moved into refurbished laboratory space, is an established part of the research effort within the School of Biomedical Sciences which gained a 5* grade at the last UK Research Assessment Exercise.

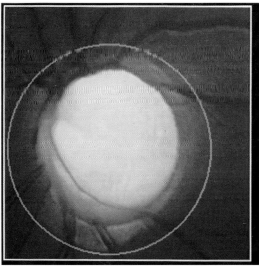
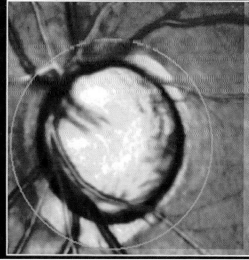

Scanning laser section through the optic nerve head in Glaucoma.

Glaucoma

Glaucoma is one special area of study. This is a degenerative ocular condition affecting the optic nerve head, and it accounts for a large proportion of registered blindness in the UK and Ireland. The mechanisms of damage are still poorly understood, but progression involves damage to the axons of the retinal ganglion cells which form the optic nerve head. Evidence that deficits of the short-wavelength (blue) sensitive visual pathway can be measured early in glaucomatous eye disease has led to the development of new tests to isolate and measure the sensitivity of the blue pathway in the visual system. The Group is at the forefront of the development of new techniques to non-invasively measure the density of blue-sensitive retinal ganglion cells by measurements of resolution for short-wavelength gratings in peripheral vision, in order to better detect the condition and monitor its progress. This project has been funded by the Wellcome Trust for two years and should lead to a better understanding of the mechanisms of damage in glaucoma.

Diabetes

Macular oedema, which affects the central area of the retina, is the leading cause of visual impairment and blindness in diabetics, and is another research priority. The Group has developed sensitive techniques, based on the sensitivity of the blue-sensitive pathway, for the improved monitoring of early visual loss in diabetics. Such changes can occur before significant loss of achromatic sensitivity, or loss of visual acuity. These techniques will be used to investigate the causes of the blue-sensitive loss. This involves monitoring the effect of the manipulation of blood glucose and blood oxygen levels on the short-wavelength mechanism and retinal capillary blood flow. Identification of the factors which impact upon the blue-sensitive mechanism will provide insight into the pathogenesis of diabetic macular oedema and the mechanism(s) which ultimately underlie sight loss. This work has received funding from the British Diabetic Association for three years.

Brain recording

Electro-diagnostic techniques are also utilised to examine visual function at the level of the brain's visual cortex. One aim is to ascertain whether motion-related activity can be generated in Visual Evoked Potentials (VEPs) by motion which is defined purely by change in colour, rather than by change in luminance. This is done by comparing the characteristics of VEPs generated by the onset of both luminance-defined and chromatically-defined motion, in order to gain an insight into the commonality, or otherwise, of the neural processes that subserve these different kinds of motion perception.

Paediatric problems

The Group's activities are not only confined to the investigation of age-related eye problems. Children's vision also has a special place in the research of the Group, particularly the assessment and amelioration of accommodative problems in children with cerebral palsy. This research aims to determine which technique is optimal for assessing accommodation in the child with cerebral palsy, which types of neurological impairment leading to cerebral palsy predispose to high refractive error and accommodative dysfunction, and whether the prescription of spectacles to correct for inadequate accommodation actually improves visual status.

A Wellcome Trust equipment grant has allowed studies like these and others to progress through the purchase of specialised equipment, including a Scanning Laser Tomographer, Blood Flowmeter, Visual Evoked Response Imaging Systems, and a Spectrophotometer.

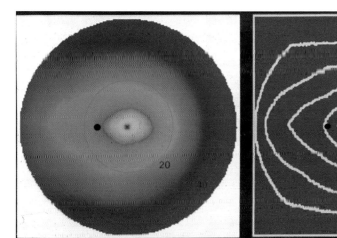

Iso-resolution contours across the retina.

Contact: Dr Roger S. Anderson,
Vision Science Research Group,
School of Biomedical Sciences,
University of Ulster at Coleraine,
Co. Londonderry, BT52 1SA, N. Ireland.

Probing molecular and cellular aspects of diabetes

Diabetes mellitus represents a complex disorder characterised by the dysfunction and destruction of the pancreatic beta cells. The continuing escalation in the incidence of the two main forms of this disorder (type 1 and type 2 diabetes mellitus) prompts the ongoing research into the molecular and cellular aspects underlying the diabetic syndrome. Such research offers the promise of improved treatments, together with the knowledge needed to cure or even prevent this chronic disorder.

The Diabetes Research Group in the School of Biomedical Sciences at the University of Ulster, Coleraine, has a prolific research profile focussing on five key areas:
- Bioengineering insulin-secreting cells for the future gene therapy of diabetes.
- Unravelling the complex mechanisms regulating pancreatic B-cell function.
- The role of glycation and other structural modification of biologically active peptides in the pathogenesis and potential treatment of the diabetic state.
- The discovery, targets and action of pharmacological and natural antidiabetic agents.
- The involvement of glucose toxicity and drug metabolising enzymes in pancreatic islet cell dysfunction and destruction.

This high-impact research contributed to the award of the highest grade of 5* in the 1996 Research Assessment Exercise, emphasising the Groups' international/national excellence. Since 1996 the Group has also generated over 50 full scientific papers and a further 50 published scientific communications from both national and international conferences.

Indeed, the novel and innovative research activities of the Diabetes Research Group are strengthened through an extensive network of international collaborations, achieved through the foundation and co-ordination of the Islet Study Group of the European Association for the Study of Diabetes, the Islet Research European Network, and the Islet Research Latin American Network.

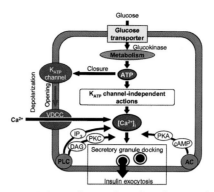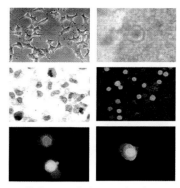

Left: Regulation of insulin release. Right: Focus on cellular morphology and induction of cell death in clonal pancreatic beta cell lines.

Further details on current activities and opportunities of working with the Diabetes Research Group are available at: http://www.ulst.ac.uk/faculty/science/diabetes/update/index.html

Contacts: Professor Peter R. Flatt or Dr Neville McClenaghan, School of Biomedical Sciences, University of Ulster, Coleraine BT52 1SA; Tel: (0) 1265-324491; Fax: (0) 1265-324965; E-mail: pr.flatt@ulst.ac.uk, nh.mcclenaghan@ulst.ac.uk

Biotechnology research at the University of Ulster

The main research projects of this group include:
- Production of microbial enzymes and biosensor systems for medical diagnosis;
- DNA probes for rapid detection of toxigenic bacteria;
- Ethanol production by thermotolerant yeasts;
- Biodegradation of textile dyestuffs effluent;
- Adsorption of radionuclides and metal ions by biomass;
- Drug targeting and analysis and degradation of antiulcer drugs;
- Biodegradation of xenobiotics and biosurfactant production.

A new addition to the work of the Biotechnology Research Group is Proteomics/Peptidomics. The activities of this research are centred towards molecular structural inventory of proteins and peptides in an array of cells, tissues and secretions. Our focus at the present time is on peptidic profiling of human neuroendocrine tumour tissues, neuroendocrine cell lines, and the complex dermal venoms of neotropical phyllomedusine frogs. The purpose of the work is directed towards novel peptide/protein discovery, with subsequent applications both clinically and commercially. This research involves collaboration with clinicians locally, nationally and internationally, the School of Biological Sciences in the University of Manchester, with whom we have recently established an Amphibian Research Centre, and a range of commerical companies.

This group's activities are reinforced by the recent establishment of a Centre of Innovation in Biotechnology. It is supported by the International Fund for Ireland and involves collaboration with both QUB and BioResearch Ireland.

An air-lift fermenter being used in Biomedical research.

Contact:
Professor Roger Marchant, Head,
School of Applied Biological
& Chemical Sciences,
University of Ulster, Coleraine,
Co. Londonderry BT52 1SA;
Tel: 01265-324450;
E-mail: r.marchant@ulst.ac.uk

SANDY STEACY & JOHN MCCLOSKEY

Investigating earthquakes

"In Ireland? Are you mad?" This is the most common reaction when people learn that scientists in the School of Environmental Studies at the University of Ulster are actively researching earthquake processes. "We don't have earthquakes here" generally follows. Although not completely true, Irish earthquakes are certainly not large enough to cause concern. None the less, earthquake physics is an important area of study in the Geophysics Research Group.

Most of the earthquake studies in Coleraine involve computer simulations. However we recently completed an eight-month field experiment in the Gulf of Corinth, Greece, and are at present analysing the data. The aim of the project was to search for repeating microearthquakes, clusters of very small earthquakes with almost identical locations, sizes, and seismograms. Repeating microearthquakes have been observed in the San Andreas fault zone in California, but the geology in Greece is much more complex and hence sequences of repeating earthquakes might also be more complex.

These repeating events are important because they may help us understand the recurrence behaviour of much larger earthquakes. Because the time between large earthquakes (the recurrence interval) at any location is typically hundreds to thousands of years, we have little data on how regularly they recur. In the few locations where the data are reasonably good, it appears that the recurrence interval varies widely from event to event. Since many estimations of seismic hazard

Damage to a reinforced concrete building from the 1986 Kalamta earthquake.

are based on the time since the previous event, and the recurrence interval of the expected event, a good understanding of the recurrence behaviour of large events is very important for seismic risk analyses.

Repeating microearthquakes may be small scale analogues to large earthquakes because they seem to involve the near total rupture of small faults in much the same way as big events involve the near total rupture of much larger faults. Since these very small events occur much more frequently than large, damaging earthquakes, they are ideal for investigating recurrence behaviour in great detail. If repeating earthquakes are identified in Greece, a more extensive field experiment will be carried out in hopes of understanding the recurrence behaviour of microearthquakes in a complex geologic environment.

---

*The project was funded by the UK Natural Environment Research Council and was carried out in collaboration with scientists from the Dublin Institute for Advanced Studies, University College Dublin, and the Earthquake Planning and Protection Organisation in Athens.
For more information on the Geophysics Research Group and its studies on earthquake physics, fluid flow in fractured porous media, and aeolian sand transport, contact:
Dr John McCloskey (head of group) or Dr Sandy Steacy;
Tel: +44 (0) 1265-324401;
E-mail: j.mccloskey@ulst.ac.uk, s.steacy@ulst.ac.uk*

COLIN BREEN & RORY QUINN

Maritime archaeology

The Centre for Maritime Archaeology (CMA), housed in the Coastal Research Group at the University of Ulster, was formed in February 1999 and officially launched by the Receiver of Wreck on 26 April 1999. The Centre is jointly funded by the University and by the Department of the Environment (Northern Ireland). It is currently staffed by two lecturers, one in maritime archaeology and the second in marine archaeological geophysics, as well as by three Research staff from DoE's coastal archaeology unit. The Centre is engaged in a variety of research projects including: large-scale mapping of the seabed off Northern Ireland using a variety of geophysical and marine survey techniques; an ongoing foreshore survey of Strangford Lough, Co. Down, and a maritime landscape study of Bantry Bay, West Cork. Excavation on a number of wreck sites, including *La Surveillante*, a French frigate scuttled in 1797, are also central to the research programme.

500 kHz sonograph of the Oregon wreck site, Northern Ireland. The remains of Oregon (1946) lie in gullies in Silurian greywackes off Wilson's Point in Belfast Lough. Divers are currently ground-truthing the geophysical data.

The *La Surveillante* research project in Bantry Bay is unique in Irish archaeology in that it draws together for a the first time a truly multi-disciplinary group, combining researchers from various backgrounds, including maritime archaeology, marine geophysics, marine geology, and oceanography. Work to date has included imaging the wreck site using high-resolution geophysics, archaeological survey and excavation on the wreck, and environmental sampling within the site's environs. This work has shown that the wreck is well preserved in a low energy environment, with portions of the lower part of the hull buried in two metres of sediment.

The Centre is well equipped with a suite of high-resolution marine geophysical equipment including side-scan sonar, magnetometer and a Chirp sub-bottom profiler, supported by a Geographical Positioning System (GPS) unit. The Centre is also equipped with diving and other marine survey gear.

A postgraduate diploma/MSc course in Maritime Archaeology will be run from October 1999 within the School of Environmental Studies at the University. The course is aimed at graduates from an archaeological and earth-science background who wish to pursue a career in this area, or who would like to develop their interest in this field.

*Contact: Colin Breen; Fax: 01265-324911;
E-mail: cp.breen@ulst.ac.uk;*

YOUTH SCIENCE FEATURE   CHARLES MOLLAN

In praise of our Young Scientists

It is very satisfying, once again, to acknowledge the sponsors who have made this expanded Youth Science Feature in *The Irish Scientist 1999 Year Book* possible. The major sponsors were the STI Awareness Programme (administered by Forfás), Esat Telecom, and Samton Limited. In addition, most of the sponsors of special prizes at the Esat Telecom Young Scientist and Technology Exhibition (YSTE) also kindly agreed to sponsor articles by their winners:

An Bord Glas
Dublin Institute of Technology
The Food Safety Authority of Ireland
The Geological Survey of Ireland
The Information Society Commission
The Institute of Chemistry of Ireland
The Institute of Petroleum
The Institute of Physics (Irish Branch)
The Institution of Engineers of Ireland
Intel Ireland Limited
Irish Business & Employers' Confederation
The Irish Science Teachers' Association
The Marine Institute
The National Rehabilitation Board
The Patents Office
Shaw Scientific Limited
The Teachers' Union of Ireland.

Left: Andrina Moore, Event & Sponsorship Manager at Esat Telecom, organiser of YSTE. Exhibition 2000 will take place at the RDS from January 11 - 15.

All the young people invited to contribute articles did so – and our best thanks are due to them for this additional effort. Hopefully the reports of their work will encourage others to take an interest in science, and will encourage all our scientists of the future to explain to the general public what they have been doing.

The Feature begins with an article describing the encouragement which a commercial company is giving to a local school. Many thanks to Yamanouchi Ireland for sponsoring this article. More and more companies are giving this kind of support to the science teachers and pupils in schools, and it is hoped that such activities can not only continue but expand – thus encouraging more of our talented young people to choose scientific and technical careers.

YAMANOUCHI IRELAND & COOLMINE COMMUNITY SCHOOL   JIM MCCARTHY

An industry-school links programme

The IBEC Business and Education Links Programme was started in October 1995. This Programme was set up to improve the image of, and focus on, business in the education system. It was also designed to assist students in making the transition from second-level school to the world of work or further education. At present, there are over 150 companies and 220 schools involved in the Programme - in which there is usually a link established between a company and a school. It is intended that a link will be offered to all second level schools by 2002. In 1998, this type of link was established between Yamanouchi Ireland and Coolmine Community School.

Initially, a number of meetings took place in the school between Yamanouchi representatives and Heads of the Science Department in the School. Following this, the two Coolmine teachers, Phil Solan and Jim McCarthy, visited Yamanouchi and were shown around the plant. A programme for the 1998-1999 school year was agreed. The overall objective was to increase the number of students who study chemistry in the school and the number who subsequently study chemistry at third level. In January 1999, Siobhan Briggs, a Yamanouchi senior development chemist, visited the school to address three groups of third year and transition year students about their subject choices for the Leaving Certificate (LC). Her purpose was to promote the science options, and particularly chemistry. She highlighted the career choices available to students who have a chemistry qualification.

In March 1999, two groups of LC chemistry students from Coolmine visited the Yamanouchi plant on different days. The plant tour helped

Yamanouchi Ireland Company Limited, Damastown, Mulhuddart, Dublin 15.

to make the industrial chemistry content of the LC chemistry course more real to the students. The role of a chemist and the career choices available to students who have a chemistry qualification were highlighted. This was done by means of short outlines given by a number of Yamanouchi employees of their educational background (with particular reference to chemistry), work experience and the nature of their current work. These employees included chemists working in the areas of quality control, research and development, health and safety, manufacturing, and environmental management.

Other benefits also accrued to the school as a result of this programme. Yamanouchi provided twenty-five laboratory coats for use by students in Senior Cycle chemistry practical classes. These have proved particularly useful, as there are two fifth year and two sixth year chemistry classes in Coolmine. The Company also provided a number of Periodic Tables for use in the school.

It is hoped to develop and extend this programme in the 1999-2000 school year. It is intended to have further plant tours for students, and further talks to students about LC science options. A visit to the plant by all the science teachers is planned, and Yamanouchi staff may be involved in "mock" interviews for students in the school. When the new LC chemistry syllabus (which has a greater emphasis on industrial chemistry) is implemented in September 2000, this link with a pharmaceutical company will be of even greater benefit to the chemistry students in Coolmine.

SCOIL MHUIRE GAN SMÁL, BLARNEY, CO. CORK   SARAH FLANNERY

Cryptography – a new algorithm versus the RSA

My project dealt with cryptography – *the science of secrecy* – which is the study of methods to conceal information so that only the intended recipient can read the information. Today cryptography has applications in sending secure e-mail, storing encrypted files and bank accounts, protecting credit card information when buying on the net, digital signing and much more.

My project investigated a new public-key cryptographic system called the Cayley-Purser Algorithm. My examination of the algorithm, which uses matrices over a ring, proceeds by comparing it with the celebrated and long established RSA public-key system. The latter was introduced by three Massachusetts Institute of Technology students, Rivest, Shamir and Adleman in 1977. The Cayley-Purser Algorithm is based on ideas given to me by Dr M. Purser. I programmed it and the RSA with the mathematical package MATHEMATICA in order to run extensive comparative tests on each of them. I found that the running time is of an order of magnitude faster than the RSA.

I needed a basic knowledge of Number Theory and Group Theory, two branches of Higher Mathematics, in order to explain the mathematics behind both algorithms and to establish through a series of mathematical proofs the security of the Cayley-Purser Algorithm.

*Sarah Flannery, who entered her project in the Senior Individual Section in the Chemical, Physical & Mathematical Sciences Category at the ESAT Young Scientist and Technology Exhibition in January 1999, won the top prize – **Young Scientist of the Year 1999**. Her teacher was Mr Sean Foley.*

THIS ARTICLE WAS SPONSORED BY SAMTON LIMITED.

ABBEY VOCATIONAL SCHOOL, DONEGAL TOWN   DAVID FOLAN

Wild mushroom – *Hydnum repandum*

My project studied the edible wild mushroom *H. repandum* and its relationship with the Birch tree. The work was carried out over two years and was divided into two parts: laboratory studies of the fungus, and an environmental study of its natural habitat.

The laboratory work included aseptic isolation of the fungus from collected mushrooms, optimising growth parameters (pH, heat, etc.) and attempts to cultivate the mushroom artificially.

The environmental study examined the conditions at an area known to produce mushrooms. These included tree tapping and the collection of sap daily over the growing season, surveying other species in the forest, and a long-term soil-sampling program.

When the results of both parts of the project were brought together, they showed that there is a link between the trees' autumnal sap and the triggering of the fungus to produce mushrooms. From these results it can be concluded that the tree provides the chemical or physical trigger or the energy required to produce mushrooms.

In the future, I hope to be able to use this information to develop a commercial growing technique: to achieve this, a much more detailed study would need to be carried out.

David Folan with his teacher, Ms Caroline Feeney.

David studies Hydnum repandum *in in the forest.*

*David Folan, who entered his project in the Senior Individual Section in the Biological & Ecological Sciences Category at the Esat Telecom Young Scientist and Technology Exhibition in January 1999, won one of the top prizes – the **Runner up Individual Award**. His teacher was Ms Caroline Feeney.*

THIS ARTICLE WAS SPONSORED BY SAMTON LIMITED.

AQUINAS DIOCESAN GRAMMAR SCHOOL, RAVENHILL ROAD, BELFAST — CHRISTINE MCCRUDDEN & CAITRIONA LAGAN

Parasites and propulsion – or when the experts get it wrong

Our project was an investigation into the behaviour of a marine parasite called *Cryptocotyle lingua*. This trematode infects a variety of coastal birds and has a complex life cycle involving periwinkles, fish and the definitive hosts which are birds such as gulls.

It was decided to use a larval stage of *C. lingua* because it is easy to obtain and does not infect humans. In particular, we were interested in the responses of the parasite to environmental signals and in how it locates a fish host. In order to study its behaviour, we used a novel approach whereby the responses of the parasite were recorded on computer via a video camera. The images were then printed for analysis. In our preliminary experiments, we found that the parasites were attracted to light. We used this aspect of their behaviour to study their swimming rate to light of varying intensifies and wavelengths. We also quantified their behaviour when offered a choice between light of different wavelengths.

Our investigation then progressed to looking at the responses of the parasites in a vertically mounted observation chamber. By using this set up, we able to count the number of parasites swimming, sinking through the water column, or just resting on the bottom. We then became interested in the response of the parasites to shadow stimuli, and showed that they swim faster, longer and over a greater distance. It had been suggested in a number of scientific papers that this shadow response increased the chances of the parasite infecting a host. However, we discovered that they swim too slowly to contact a fish in this way. In addition, other papers indicated that most *C. lingua* infections were found on the dorsal surface of the fish. This pattern of infection is not consistent with contact between parasite and host occurring as result of a shadow. If this were the case, most infections would be found on the ventral surface. We are of the opinion that, in this instance, the experts may have got it wrong!

Based on our results, we concluded that the parasite was well adapted to life in the sea and that it had evolved sophisticated behaviour which increased its chances of completing its life cycle. We think that our approach to investigating the responses of *C. lingua* could be developed to study parasites of medical or veterinary importance.

*Christine McCrudden & Caitriona Lagan, who entered their project in the Intermediate Group Section in the Biological & Ecological Sciences Category at the Esat Telecom Young Scientist and Technology Exhibition in January 1999, won one of the top prizes – the **Best Group Award**. Their teacher was Dr Joseph Rea.*

THIS ARTICLE WAS SPONSORED BY SAMTON LIMITED.

COLÁISTE SPIORAID NAOIMH, BISHOPSTOWN, CORK — HUGH HURLEY & CONOR MEEHAN

A modified water barometer

The first ever water barometer was built by the Italian scientist Gasparo Berti in 1641. It was, however, two stories high and therefore of no practical use. Nowadays water barometers need to be at least 34 feet high to work properly. For our project, we went about showing that a water barometer of manageable dimensions could be accurately built.

To do this we had to devise some way of decreasing the height of water in the barometer. We managed this by replacing the vacuum at the top with a fixed volume of gas. The pressure of this gas pushing down on the water column counter-acts atmospheric pressure and keeps the water height to a minimum.

We built a rough model of the barometer to find out what volume of gas we needed to enclose at the top. Too much gas would force the water down too low and, if there were too little, then the water height would still be high. When we had found the correct volume needed, we went ahead and built the real thing.

The barometer needed to be calibrated and it was this calibration that required the most attention in our work. We wrote a computer programme based on work done by UCC physicist, Dr Michel Vandyck, to do a statistical analysis of water height, gas temperature and atmospheric pressure. We got atmospheric pressure readings from the MET Eireann station at Cork Airport.

When it was calibrated, we then checked the accuracy of our barometer readings against readings from MET Eireann. The results we got were extremely accurate, up to 99.8% accurate in fact. We drew graphs to examine the sensitivity of the water column height to changes in both pressure and temperature. Finally, at normal room temperatures, we discovered that our barometer is up to five times more sensitive than the usual mercury barometer.

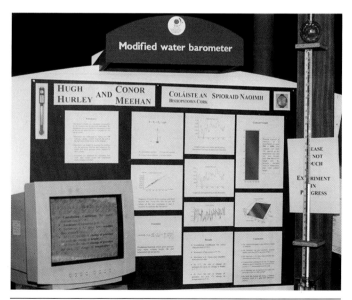

*Hugh Hurley & Conor Meehan, who entered their project in the Junior Group Section in the Chemistry, Physical & Mathematical Sciences Category at the Esat Telecom Young Scientist and Technology Exhibition in January 1999, won one of the top prizes – the **Runner-Up Group Award**. They also won a Special Award presented by the Institute of Physics (Irish Branch). Their teacher was Mr Dan Sweeney.*

THIS ARTICLE WAS SPONSORED BY THE INSTITUTE OF PHYSICS (IRISH BRANCH).

POBALSCOIL GHAOTH DOBHAIR, DOIRÍ BEAGA, LEITIR CEANAINN, TÍR CHONAILL   POL MAC SUIBHNE

Hydration practices in sport

Is e uisce an comhabhair is fluirsí san chorp. Cé go bhfuil se tabhachtach ni ghlacfí e mar cothaitheach.

Afach, tá sé riachtanach fá choinne,
1. Taisteáil (Transportation),
2. Teocht an choirp (Body temperature),
3. Dílea (Digestion),
4. Bealú na n-altanna (Lubrication of joints).

Caitear cibé 'uisce a chailltear ón chorp a bheith ionghlacthe arís ar mhaithe slainte an choirp. Cailltear cuid mhór uisce i rith aclaíocht go h-áirithe agus caitear an t-uisce seo a bheith ionghlacthe arís go h-éifeachtach chun an duine a chosaint. Níl morán eolas san am i láthair faoin ionglacthe agus cailluint léachtanna i rith na cleachtaithe d'imreoirí i spóirteanna fhoirne (team sports). Is é seo an fath a ghlac mé an tionscnamh seo.

There were three distinct parts to my project.

Section 1 - A player was monitored during two simulated football matches.
A. With proper hydration practises,
B. Without proper hydration practises.

The purpose of this section was to monitor the effects of dehydration on work performance. This was done by monitoring heart rate, blood pressure, fitness level, weight and water intake at regular intervals.

Section 2 - A series of five training sessions in which a group of students were subject to various stimuli and given knowledge to determine the most effective and most practical method for teams to follow for adequate hydration at training, as well as during competitive games.

Section 3 - I surveyed over 100 secondary school sports teachers and various other sports trainers in Co. Donegal. The purpose was to accumulate as much knowledge as possible on, and to assess the current hydration habits of, sports trainers and teams throughout the county.

Results:
1. Performance level can be impaired by as much as 48% by as little as 2% dehydration.
2. Proper and effective hydration practises are as follows:
   - One litre of water should be consumed one hour before game or training session,
   - Water should be consumed during and after game/training session,
   - Water should be made easily accessible i.e. at various points around the game/training session area,
   - Knowledge on the importance of fluid intake increases consumption,
   - Players should be regularly encouraged to consume water,
   - An isotonic flavoured drink can increase consumption.
3. The survey showed that 95% of sports trainers do not follow proper hydration practices.

I hope that the results and success of my project will highlight the need for proper hydration practises in sport and can be used to show how to conduct proper hydration practises.

Buiochas - Do mo mhuinteoir Eolaíochta, Michael O Giobuin, mo chomh scolairi, muinteoiri agus oibrithe na scoile.

Pol Mac Suibhne won first prize in the Senior Individual Section in the Biological & Ecological Sciences Category at the Esat Telecom Young Scientist and Technology Exhibition in January 1999. He also won a Special Award presented by Bord na Gaeilge. His teacher was Mr Michael O Giobuin.

THIS ARTICLE WAS SPONSORED BY SAMTON LIMITED.

SANDFORD PARK SCHOOL, RANELAGH, DUBLIN 6   EDWARD ABRAHAMSON, TOM WARD & MÁNUS DE BARRA

Vegetricity – the power of the future

Waste and pollution are two major problems which endanger not only our environment but also our future. Our project was an attempt to dispose of common household waste in a beneficial yet environmentally friendly fashion. To this end, we availed of existing biotechnology i.e. the microbial fuel cell, and we created a waste disposal system that produces electricity. (A microbial fuel cell is a device which produces electrical currents from glucose.)

The experiments had several main strands. The first was to determine whether or not vegetables and fruit could be used in the microbial fuel cell as a power source (instead of pure glucose) and to see if different foods gave different power outputs. It then struck us that our food samples might yield more glucose if they were treated with enzymes (such as amylase, maltase and pectinase) and so we carried out several experiments to see if enzyme pre-treatment affected glucose content.

The next stage was to attempt to make the process continuous. The advantage of this would be that if ever the microbial fuel cell were used in the home, one would not have to continually empty it out. To solve this problem, we immobilised both enzymes and yeast in sodium alginate beads and tested them to see if they were still effective in their immobilised state. They were. The organic juices could simply be dripped through these beads and the reactions would occur, producing power.

The last part of the project was to find an environmentally friendly way of disposing of the waste products of the cell - carbon dioxide and alcohol. We came up with the idea of using these waste products (which are slightly acidic) to neutralise the bases in the household's water system – e.g. cleaning agents.

We believe the microbial fuel cell has potential to generate power from household vegetable waste: however there is much more development work to be done. We did achieve a great deal of success in our experiments and we hope that our work has demonstrated this potential and will inspire others to develop our idea commercially. Perhaps adding the day's vegetable waste to the microbial fuel cell will someday be as much a part of the night time routine as switching off the lights and putting the cat out.

See our web page:
http://members.theglobe.com/eamonn_a/vegetricity.htm

(From left) Edward, Tom & Mánus at their stand.

Edward Abrahamson, Tom Ward & Mánus de Barra won first prize in the Senior Group Section in the Biological & Ecological Sciences Category at the Esat Telecom Young Scientist and Technology Exhibition in January 1999. Their teacher was Dr Alison Graham.

THIS ARTICLE WAS SPONSORED BY SAMTON LIMITED.

COLÁISTE RÁITHÍN, BRÍ CUALANN, CO. WICKLOW — DAITHÍ MAC SÍTHIGH

"Gaeilgeoirí nó Déagóirí" – studying the students who attend Irish medium schools

Around two thousand students in Dublin alone attend all-Irish secondary schools or Gaelcholáistí. These schools, such as Coláiste Eoin, Coláiste Iosagán, Coláiste Cillian, and Coláiste Ráithín in Co. Wicklow, provide a full education through the medium of Irish. My project set out to study the differences between the students from those schools and students from traditional English-medium schools.

I studied those students under headings such as sport, music, cultural awareness, etc. I also compared their religious and social backgrounds. From 800 returned questionnaires, gathered from seven schools in two different areas, I compiled comparative results between "similar" groups of students (i.e. age, sex, family income) from the Irish-language and English-medium schools.

My results indicated that, although the Catholic population of the Irish-medium schools were more likely to attend Mass every week, the proportion of students from minority religious backgrounds was also higher in those same schools. As with many of the other results, these "mixed messages" showed that the Irish-medium schools were not a homogenous group. However, it was very clear that the Irish-language schools attracted far more students from higher-income families. Virtually no students in these schools came from families where both parents were unemployed, but the other schools in the same areas all had many of these students.

Daithí Mac Síthigh with Mr Mícheál O Muircheartaigh, Cathaoirleach of Bord na Gaelige.

Other interesting results included the finding that, even though both groups of students knew who won the Premiership in England, the Irish-medium students knew far more about the Gaelic football and hurling competitions. On the other hand, the students from the English-medium schools showed more of an interest in traditional music than their "Gaelic" counterparts. Why is this? I felt, after studying the overall results, that the students in these schools were asserting their identity in a way that the Irish-medium students already could by attending an Irish-language school.

To conclude, I found that the perceived middle-class Irish learner still exists, but yet they still retain the characteristics of typical teenagers, like watching *Friends* and following English soccer teams.

Daithí Mac Síthigh won first prize in the Senior Individual Section in the Social & Behavioural Sciences Category at the Esat Telecom Young Scientist and Technology Exhibition in January 1999. He also won a Special Award presented by Bord na Gaelige. His teacher was Mr Terry Mulcahy.

THIS ARTICLE WAS SPONSORED BY SAMTON LIMITED.

MOUNT MERCY COLLEGE, MODEL FARM ROAD, CORK — RHIAN FITZGERALD & ELAINE DEWHURST

Food borne diseases – are you at risk?

In our project we dealt with the steadily increasing incidence of food borne diseases, e.g. *Salmonella, E-coli 0157:H7* and *Listeria*. We tried to explain the reasons for this increase.

We surveyed 480 students aged 16 to 18 years old from all over Ireland, to discover their level of knowledge concerning food borne diseases. In order to get a good geographical spread, we surveyed six urban schools (two boys, two girls and two mixed schools) and six rural schools (two boys, two girls and two mixed). We surveyed students from Cork, Dublin, Kerry, Monaghan, Kildare, Galway and Tyrone.

Our results were startling and interesting. Seventy five percent of those surveyed thought that antibiotics were the best cure for food borne diseases. This is not correct. Only 16% knew that fruit and vegetables could transmit food borne diseases. Only 3.92% knew that Septicaemia was a symptom of some types of food borne diseases.

On the basis of our survey results we compiled a factsheet which we believe will help improve the knowledge of our peers concerning food borne diseases. We have distributed copies of our factsheet to the schools we surveyed.

We designed a new type of domestic fridge after discovering that a high proportion of cases of food borne diseases are caused by temperature mismanagement and cross contamination. We completed an experiment in which we investigated the maximum and minimum temperature level in various domestic fridges. We discovered that each fridge's minimum temperature level was above the recommended 0-4°C. We recorded temperature ranges as great as 8°C in the course of a single day.

Above: Rhian Fitzgerald and Elaine Dewhurst

Our fridge has a glass door so that the its contents can be viewed without opening it, thus reducing the likelihood of heat exchange. It also contains drip trays to reduce the likelihood of cross contamination; the drips collected are then channelled to a central collection cup, which can easily be removed and cleaned. As we were concerned about the fridge temperature being too high, we fitted a digital temperature display panel. We also fitted a light that would flash if the fridge temperature went above 8°C and a warning buzzer if the temperature went above 12°C.

We believe that our results illustrate that more education is needed to prevent food borne diseases becoming more common. The education students receive in second level schools is not enough. Only 9% of those surveyed had received any information concerning food borne diseases.

As a result of winning the Food Safety Authority of Ireland Special Award, our project was displayed at the Authority's headquarters in Dublin. A brief resume of our project was also posted on the Authority's website.

Rhian Fitzgerald and Elaine Dewhurst won first prize in the Senior Group Section of the Social & Behavioural Sciences Category at the Esat Telecom Young Scientist and Technology Exhibition in January 1999. They also won a Special Award presented by the Food Safety Authority of Ireland. Their teacher was Ms Mary Cowhig.

THIS ARTICLE WAS SPONSORED BY THE FOOD SAFETY AUTHORITY OF IRELAND.

SCOIL MHUIRE GAN SMÁL, BLARNEY, CO. CORK — VINCENT FOLEY

Curve representation of digitised objects

My project is based on finding an alternative method to represent, and enlarge, digital images. The digital images that I am aiming at are small internet images, and small, low quality surveillance camera images.

The problem with digital images is that they are not generally scalable. So if the size of an image needs to be changed, some data has to be removed to make it smaller, or the existing data has to be stretched to make it bigger. This leads to a blocky appearance in the case of enlarging, which can be seen regularly in the program "Crimeline", when a small image is taken from a security camera, and stretched to a larger size.

The current solutions to this problem are a compromise between blockyness and blurring. This obviously leads to a lot of distortion.

The aims of my project are to enlarge a small digital image without introducing significant distortion.

To do this, I have written a new computer graphics program in the Delphi programming language, that uses a system of curved polygons to represent the areas of colour in the source image. The main steps that my program performs in enlarging an image are:

1. Copy colour data of the source image to an array
2. Find areas of similar colour in the image
3. Find boundaries of each colour area
4. Smooth jagged edges in boundaries
5. Detect "corners" in boundaries, and regenerate them
6. Draw polygons with modified boundaries to output image
7. Blend together overlapping polygons
8. Fill spaces between polygons.

The results I have achieved with diagrams have been very successful. Good, clean images are created with almost all detail at sharp points restored. With pictures, I have achieved reasonable results. Though not as good as I would like, the image quality at large enlargement factors still far surpasses what is available in commercial computer graphics programs.

I have found that the method of curved polygons is generally very effective as a means of enlarging many images.

Vincent Foley won first prize in the Senior Individual Section in the Chemistry, Physical & Mathematical Sciences Category at the Esat Telecom Young Scientist and Technology Exhibition in January 1999. He also won the Intel Excellence Award. His teacher was Mr Sean Foley.

THIS ARTICLE WAS SPONSORED BY INTEL IRELAND LIMITED.

OUR LADY'S COLLEGE, GREENHILLS, DROGHEDA, CO. LOUTH — AOIFE DILLON & ANN MARIE CAMPBELL

Slimeocity

Slimeocity was the title of our project and, as you have guessed, it involved the creation of slime. We decided to make slime because we wanted to have a **really** fun project and yet still learn about chemistry.

Slime is made by mixing polyvinyl alcohol – a polymer – and sodium tetraborate – a crosslinker. The crosslinker basically links all of the polyvinyl alcohol chains together to create a viscos-elastic gel.

After we had made the slime, we carried out a number of tests on it. We would get a perfectly grungy piece of slime and see what would happen if we changed conditions such as pH and temperature, or varied the concentrations or the amounts of the chemicals used. We discovered that these all had an effect on the slime's viscosity. For example, if temperature increased, viscosity decreased and, if pH decreased, so did the viscosity of the slime.

Another aspect of our project was to see if we could find another crosslinker instead of sodium tetraborate which would also join the chains of polyvinyl alcohol together to make slime. After using a great number of other chemicals, including sodium silicate and aluminium sulphate, we discovered that copper sulphate worked quite well.

Finally we decided that it would be interesting if we compared three different types of slime:
a) Slime purchased in a shop (e.g. toy slime),
b) Slime made using bought polyvinyl alcohol, and
c) Slime made using polyvinyl alcohol which we prepared ourselves.

To make polyvinyl alcohol was a difficult process. Firstly we had to get vinyl acetate and polymerise it into polyvinyl acetate. This then had to be hydrolysed to form polyvinyl alcohol. During this process we even made glue! We think that the glue we made is similar to P.V.A. glue used in art classes. So here was one practical application of our project!

We had an excellent time in the R.D.S. - especially the slime fight (but shh - don't tell anyone!!).

Aoife Dillon & Ann Marie Campbell won first prize in the Senior Group Section in the Chemistry, Physical & Mathematical Sciences Category at the Esat Telecom Young Scientist and Technology Exhibition in January 1999. They also won a Special Award presented by the Institute of Chemistry of Ireland/Royal Society of Chemistry. Their teacher was Ms Martina Coyle.

THIS ARTICLE WAS SPONSORED BY THE INSTITUTE OF CHEMISTRY OF IRELAND.

KILDYSART VOCATIONAL SCHOOL, CO. CLARE — BRIAN MURPHY, EUGENE GINNANE & ANNE O'DONOGHUE

Parental involvement in primary education

An article in the Education Section of the Sunday Times, entitled *"Hostages to their children's homework"*, provided the inspiration for a young scientist project done by three students of Kildysart Vocational School. In the article it was claimed that parents are taking an increasingly unhealthy interest in their children's school and homework and that some parents were over ambitious. However the article did not quantify or prove that such changes have or are occurring. The Project tried to quantify the changes in Co. Clare and to find out were the results significant scientifically.

The students decided to do multistage sampling of 5th/6th class students in both rural and urban Co. Clare. There are 125 National Schools in Co. Clare, and the county was divided up into 10 x 10 mile blocks for the first stage of random sampling. Tables of random digits were used to select blocks and schools.

The team with Minister for Education & Science, Mr Micheál Martin.

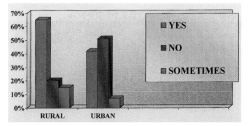

Left: An example of the results – The response to the question: *Do you help your son/daughter with their homework?*

The parents of these students were given a questionnaire, which was then databased. Analysis and cross-analysis between questions also was done. However, the results would be meaningless unless a certain confidence could be built into the results, so significance testing of results was done. The chi Squared test, Chi test with Yates correction, and the Z statistic were used for such.

Some of the key results are as follows: Fathers would be expected to have less involvement in helping out at homework but, whilst 90% of mothers help out, 56% of fathers now help out with homework. Parents were also cross analysed in relation to their own educational attainment levels and the responses they give. Of parents who have attained Inter Cert, 48% help out, whilst of those parents who have attained Leaving Cert or Third Level education, 55% help their children.

Using the Z statistic at the 95% confidence levels proved that 80-89% of all the sample frame population would have replied in a similar fashion.

Brian Murphy, Eugene Ginnane & Anne O'Donoghue won second prize in the Junior Group Section in the Social & Behavioural Sciences Category at the Esat Telecom Young Scientist and Technology Exhibition in January 1999. They also won a Special Award presented by the Irish Statistical Association. Their teacher was Mr Leo O'Donoghue.

THIS ARTICLE WAS SPONSORED BY SAMTON LIMITED.

CONVENT OF MERCY, ROSCOMMON — BRIANA HEGARTY, MELANIE CONNOR & CAROL KEIGHER

Operation deceleration: an inquiry into speeding – who speeds and why?

As we are all in transition year this year, we thought it would be the perfect opportunity to enter the Esat Young Scientist and Technology Exhibition. As there have been so many deaths on our road due to speeding and no reasons why available, we decided to base our project on the reasons why people speed and what type of people speed.

We conducted surveys through face to face interviews in our local town of Roscommon to achieve the relevant information. We surveyed a total of 100 drivers – 50 male and 50 female. As we used a face to face surveying method, 100% of the surveys were returned and we had total control over our target population, which was a great advantage to us.

We also conducted tests to show the effects of music on people's driving habits. This was done by playing a tape, while travelling in the car with the subject, and noting the change in speed by the unsuspecting driver.

In order to analyse our information, we used a computer stats program – Systat (s.p.s.s.) which was available to us in the school. We then analysed our information with the help of our science teacher Mr Harlow. In order to evaluate our results we used t-tests, Correlation Coefficient techniques and chi-square tests.

The following are our findings:
1. Less female drivers speed than male drivers.
2. More males think the speed limit is too low than females.
3. Fast music is inclined to make people drive faster, while slow music has the opposite effect.
4. People who drive cars with large engines are more prone to speeding than people with small car engines.

From these results we conclude the following:
1. Female drivers are less prone to speed than male drivers.
2. There is a difference in the attitudes of male drivers and female drivers towards the speed limit.
3. Music has a significant effect on people's speed.
4. There is a connection between increased engine size and speeding.

Briana Hegarty, Melanie Connor & Carol Keigher entered their project in the Intermediate Group Section in the Social & Behavioural Sciences Category at the Esat Telecom Young Scientist and Technology Exhibition in January 1999. They won a Special award presented by the Irish Statistical Association. Their teachers were Mr Patrick Harlow & Ms Stella O'Gara.

THIS ARTICLE WAS SPONSORED BY SAMTON LIMITED.

Coláiste Eoin, Bothar Stigh Lorgan, Co. Dublin — Tadhg O'Broin

Alzheimers – cabhair teicneolaiocht

What I intended to do in my project for the Esat Young Scientist & Tecnology Exhibition 1999 was to design and build an alarm system which would alert a minder of a person suffering from Alzheimers when the patient got out of bed.

One of the most distressing things for carers is that they cannot get a restful night's sleep because the Alzheimers patient will quite often get out of bed during the night and possibly wander out of the house. This could be extremely dangerous for the patient.

As part of my research for the project, I obtained a list of the equipment available on the market. I found that what was available was expensive and in some cases too sophisticated. My priority therefore was to come up with a device that was both Inexpensive and easily portable so that it could be used in different situations.

It was of paramount importance that the patient would not be aware of, or frightened by, an alarm. At first I made a wired system but, with further research, I decided that a radio (wireless) type would be better. The wired system would be effective, but it would mean that wires would have to be laid from room to room and would not have the portability of the radio system.

The system I built consisted of a pressure sensor which would

Left: Tadhg O Broin with Dr Arthur O'Reilly, Chief Executive NRB and Mr Donie O'Shea, Head of Disability Technology NRB.

activate a radio signal transmitter and a receiver which would emit a sound to alert the carer. The sensor was incorporated into a bedside mat so that, when the patient stepped out of bed, he/she would activate the transmitter, sending the signal to the carer who would have the receiver close by. The receiver could be carried on a belt like a mobile phone.

The cost of the components at full retail price came to slightly less than £IR30.00. This compared with prices of several hundred pounds for other alarm systems.

The final piece of research entailed having the device tested. This was done in a nursing home, and the Matron said it was most effective and would be of great help to carers of Alzheimers patients.

Tadhg O'Broin, whose project was carried out through the medium of the Irish language, entered his project in the Intermediate Individual Section of the Chemistry, Physical and Mathematical Sciences Category of the Esat Telecom Young Scientist and Technology Exhibition in January 1999. He won a Special Award presented by the National Rehabilitation Board. His teacher was Mr Ciaran O Cualain.

This article was sponsored by the National Rehabilitation Board.

Coleraine Academical Institution, Co. Derry — Paul Kissick

Programming and robotics for primary schools

The way in which programming is currently taught in primary schools is through the use of a small, robotic buggy, which is able to take commands, in the form of Logo, and convert these into movements. Logo is a very basic programming language, which uses simple commands such as "Forward 60" and "Right 90". This can be taught using computers, but being able to see what the children have done takes a long time.

This is why the robotic buggy, currently used, is such a good idea. Or is it? The problem with existing products is the necessity to wait for the children to actually write the program on the device itself. This takes up valuable time with children having to program and watch the program, using the same device.

I proposed the idea of having the programming and viewing as completely separate devices. There is a separate programmer, used with software designed for use with a PC, and a separate buggy. The only thing linking the two devices is a cartridge. The cartridge contains the program, written by the children on the PC. The cartridge can then be plugged into the buggy, and the program run.

The software was designed for a PC, because I already had experience with writing programs, and found this a lot easier than for use with an Apple Mac, or BBC Micro. The interface was designed to be as user-friendly as possible. The keyboard is virtually unused. The mouse, instead, is used. This is due to children finding it easier to use the mouse than the keyboard. And, instead of moving in ridiculously small or large distances, the distances and rotations were standardised into steps of 10cm and 15 degrees, respectively. The maximum allowed per operation was limited to 100cm and 90 degrees. From experimentation, the maximum number of operations was limited to 36, ample for any program.

From testing at a local primary school, I found that the children learnt how to use the software very quickly, and were actually able to teach each other how to use it.

Paul Kissick won second prize in the Senior Individual Section in the Chemistry, Physical & Mathematical Sciences Category at the Esat Telecom Young Scientist and Technology Exhibition in January 1999. He also won Special Awards presented by the Institution of Engineers of Ireland and Dublin Institute of Technology. His teacher was Mr Gareth Clarke.

This article was sponsored by the Institution of Engineers of Ireland and Dublin Institute of Technology.

SUTTON PARK SCHOOL, SUTTON, CO. DUBLIN — GABRIEL BIETZ & JOSHUA BIETZ

Spotlight on *Elodea*

The aim of this project was to investigate the effects of changing environmental conditions on the production and activity of photosynthetic pigments in *Elodea Canadensis* (Canadian pond weed).

We grew the pond weed under different wavelengths of light (blue, red, green and white) for several months, and compared the results using several different experimental methods.

Using a photomicroscope, we compared the appearance of the cells, including the numbers and distribution of chloroplasts. We investigated the types of pigments each plant produced, using thin layer chromatography. We then tested the reducing power of the pigments by investigating their ability to reduce a solution of D.C.P.I.P. This was done using a colorimeter. Possible changes in absorbance peaks were investigated by analysing the chlorophyll in a U.V. Spectrophotometer. Finally, the gross primary productivity of each of the plants was compared. This was done by growing samples of the plants in the dark for 48 hours, and then exposing them to strong light, and measuring the rate of increase in dissolved Oxygen in the surrounding water.

The results of these investigations showed that the plants grown under blue light only developed poorly, and were not as efficient as those grown under white light. Those grown in red light only thrived the best. The plants grown in green light only died. Two samples were grown at each of the wavelengths, and the results were consistent.

The results from the plants grown in blue light surprised us, as there is an absorbance peak at the blue end of the spectrum. We assume there must be a specific need for the longer wavelengths of light for maximum efficiency. Another interesting observation was that the plants grown in red light consistently developed thicker cell walls than those grown in blue light. This may be to facilitate the absorbance of the shorter blue wavelengths.

We really enjoyed competing in the Esat Telecom Young Scientist competition, and found the experience very valuable. We would like to thank as well as commend our teacher, Ms Walshe, for all of her assistance, encouragement, and support. Finally we would like to thank Sutton Park School for all of their help and the use of their facilities.

Joshua & Gabriel at the Exhibition.

Gabriel & Joshua Bietz entered their project in the Senior Group Section in the Biological & Ecological Sciences Category at the Esat Telecom Young Scientist and Technology Exhibition in January 1999. They won a Special Award presented by the Irish Professors of Botany. Their teacher was Ms Anna Walshe.

THIS ARTICLE WAS SPONSORED BY THE IRISH PROFESSORS OF BOTANY.

ST FRANCIS COLLEGE, ROCHESTOWN, CO. CORK — DAVID O'DRISCOLL, ALAN CREMIN & STEPHEN FLANAGAN

Natural food colourings extracted from carrots

Our project deals with the extraction of the natural colour pigment, Carotene, from carrots and its use as a food colorant. We took Food Science as an Applied Science for the Junior Certificate and, while studying food additives, we became curious whether there were natural alternatives to the synthetic additives dealt with in our course. During our search for natural additives, we found that a local company, Quest International, was involved in the production of natural food colourings.

At Quest we found that many food products owe their colour to a natural food colouring called Carotene. This carotene is currently extracted from palm fruit in Indonesia for use in the world food industry. Carotene is also present in carrots, but is not extracted from them as a food colouring at present. We decided to attempt this extraction ourselves.

We ran a series of extraction trials to determine the best method of extracting the carotene from the carrots and settled on an Acetone extraction. Once we had extracted the carotene from the carrots, we set about using it to colour food. We successfully coloured samples of both lemonade and ice-cream with our extracted carotene. We compared our samples with samples of lemonade and ice-cream coloured with the existing synthetic and natural carotene food colours on the market and found that our colour was up to the same standard as these existing products.

We then completed a feasibility study to determine whether it would be commercially possible to produce our natural colour in Ireland and compete against those products already on the market. We found that our colour would be nearly ten times more expensive to produce than its synthetic counterparts. However we believe that with further study, and the use of carrot varieties with higher colour content, we could dramatically reduce the cost of our colour and greatly improve its commercial feasibility. We also believe that the health benefits, especially for the heart, attached with using carrots as a source of carotene would make food products coloured with our colour more attractive in the eyes of the consumer.

We have achieved our goal of finding a natural alternative to at least one synthetic food additive and have gone one step further to actually produce it ourselves.

Stephen, Alan & David discuss their carrot extracts.

David O'Driscoll, Alan Cremin & Stephen Flanagan entered their project in the Senior Group Section in the Biological & Ecological Sciences Category at the Esat Telecom Young Scientist and Technology Exhibition in January 1999. They won a Special Award presented by An Bord Glas. Their teacher was Ms Olive Sexton.

THIS ARTICLE WAS SPONSORED BY AN BORD GLAS.

SALERNO SECONDARY SCHOOL, GALWAY · NIAMH MCKEOWN

Starfish, predators of shellfish

Starfish predation on shellfish is a very serious problem, particularly in Lough Swilly, where they are estimated to eat 1500 tonnes of mussels each year. This project has given some very basic information on the types of starfish involved and the density that can be tolerated in a shellfish farm. Once the population gets out of control, significant effort is required to reduce the population. With the extent of the losses this causes an important industry, perhaps this basic research provides a good starting point on the most effective controls.

Over four hundred and fifty starfish were collected from five sites where shellfish farming is carried out. Four different species of starfish were found, but two of these do not cause a difficulty for the shell fishermen. The two serious predators are *Asterias rubens* the common starfish and *Marthasterias glacialis* the spiny starfish.

The common starfish was found at all five locations including a site at Killary Harbour with a very low salinity level. The spiny starfish was found at three of the sites. The starfish found in each area were all of the same size - i.e. no spread of sizes from juveniles to adults. Where the common starfish and the spiny starfish were found at the same site,

Clare Hanrahan, Esat Telecom, with Niamh McKeown.

the spiny starfish had an average size greater than the common starfish, again with only a slight variation in size of all the specimens examined. This would indicate that the different species can survive on the same site without competing for the same size prey.

Due to the slow movement of adult starfish, two other methods of controlling the population are available: provide a sacrificial area in the mussel farm that the starfish gather in and clear an area of 100 metres between this sacrificial area and the mussel beds of any food. The mussel beds could be combed using starfish mops to prevent any build-up of starfish, or traps such as lobster pots could be used to control the starfish. This method has worked for the fishermen at Clarinbridge but has not yet been tried at Lough Swilly.

Niamh McKeown won joint first prize in the Junior Individual Section in the Biological & Ecological Sciences Category at the Esat Telecom Young Scientist and Technology Exhibition in January 1999. She also won a Special Award presented by the Marine Institute. Her teacher was Ms Carmel Donlon.

THIS ARTICLE WAS SPONSORED BY THE MARINE INSTITUTE.

BUNDORAN VOCATIONAL SCHOOL, CO. DONEGAL · MICHELLE CONDRON, ANDREA MCGOWAN & ELAINE DORAN

A microbiological analysis of Bundoran's bathing waters

Bundoran's beauty is its greatest attribute. The location of the beach, under 100m from the town, ensures that thousands of visitors flock to Bundoran each year. This influx of tourists during the summer leaves Bundoran open to use and abuse. Our primary aim was to carry out a microbiological analysis of the sea water, determining its quality, isolating the source of any pollution and providing recommendations for its elimination.

We decided on two different sampling sites. The first was located at the centre of Main Beach. Our second site was a stream which enters the beach after flowing through the town. During its course the stream collects litter and animal faeces which led us to believe it may be contaminated.

We obtained a copy of the criteria for the Blue Flag status and carried out our tests to comply with Blue Flag guidelines.

Our first job was to detect how many coliforms, if any, 100ml of each sample contained. This is called a presumptive test and determines whether or not the water is safe for bathing. The standard of the seawater was generally very good, with only 43 coliforms per 100ml. The stream was not, containing 1,100 coliforms per 100ml, being in breech of the Blue Flag guidelines.

As this test did not specify which coliforms were present, we decided to test for *E-coli*, quite a topical and widely feared bacteria. We had to be extremely careful working with a pathogen such as this. E.M.B. agar was used to determine if *E-coli* was present. The colonies of bacteria on each plate revealed a green metallic sheen, typical of *E-coli*. Further tests could have been carried out to confirm this – however, time and money were restricted luxuries, and we also felt it would be too dangerous, especially since experiments were carried out in our school lab.

We feel that, even allowing for dilution, the stream must have

Elaine, Michelle & Andrea.

a considerable effect on the bathing water quality. Although it is diverted during the summer months, it still represents a hazard to winter surfers and swimmers.

We discovered that for towns with a population of under 2,000, such as Bundoran, it is legal to pump raw sewage into the sea. Unfortunately Bundoran's huge influx of tourists during the summer has not been taken into consideration. We were informed that there are plans for a £50 million scheme to be put into action by the year 2005 for a sewage treatment plant for the whole of Donegal Bay. We recommend that the proposed plant should be installed and operational as soon as possible.

We also recommend that the stream is diverted all year round to ensure high quality recreational amenities in Bundoran in the future.

Michelle Condron, Andrea McGowan & Elaine Doran entered their project in the Intermediate Group Section in the Biological & Ecological Sciences Category at the Esat Telecom Young Scientist and Technology Exhibition in January 1999. They won a Special Award presented by the Marine Institute. Their teacher was Ms Jacqui Dillon.

THIS ARTICLE WAS SPONSORED BY THE MARINE INSTITUTE.

LORETO COLLEGE, ST STEPHEN'S GREEN, DUBLIN — SARAH DEENY & DENISE TYNAN

Science thrills without the bills – a website for National Schools

In 1991 an international survey was carried out to investigate educational progress in maths and science. Although the survey can only be taken as a sample of Irish education at a given moment, the results are still disturbing. Among other things it showed that Science received scant attention in the primary school classroom. When Denise and I saw these statistics, particularly in relation to the amount of time and the way in which science is taught, we were understandably shocked. In a world where science is becoming more and more important to future careers, it seemed completely irrational to us that, although a full science course is provided at secondary level, little or none is provided at primary level. Some might argue that chemistry and physics are both taught at junior cycle level and that there is little need for it to be taught at primary level. However, with no sound base, the more complex subjects are often lost or badly understood by some students.

Others might argue that it is hard to teach science in an average national school classroom as it would lack the necessary facilities. With this in mind, we decided to make a website as a source of information for national school teachers who would like to teach practical science in a classroom situation. We especially decided to try and show that simple chemistry, and in particular physics, can be taught well in a classroom situation, as in the survey it was revealed that this area received a scant amount of attention.

Therefore, in selecting the data for the website, we needed to select subjects that fitted the following criteria. All experiments had to have materials that were cheap and readily available, they had to be safe and suitable for children aged 8+: however perhaps the most important factor was that they had to be practical in nature as we believe that this is the best way to teach science.

In constructing our website we used HTML Hyper Text Mark-up Language: we used HTML to make the basics of our page and then switched to a HTML editor to "jazz up" our pages.

We believe that we have shown that the internet is a valid educational tool and source of information. Also we believe that the project has shown that science is easily taught in our national schools and that all sciences, not just biology and nature studies, can be taught cheaply and effectively. We hope that our project has helped to dispel any notions that science is out of reach of any primary school in this country.

Sarah Deeny & Denise Tynan won first prize in the Intermediate Group Section in the Chemistry, Physical & Mathematical Sciences Category at the Esat Telecom Young Scientist and Technology Exhibition in January 1999. They also won Special Awards presented by the Institute of Physics (Irish Branch) and the Information Society Commission. Their teacher was Ms Sheila Porter.

THIS ARTICLE WAS SPONSORED BY THE INSTITUTE OF PHYSICS (IRISH BRANCH) AND THE INFORMATION SOCIETY COMMISSION.

BELVEDERE COLLEGE, DUBLIN 1 — JOE FITZSIMONS

Low cost 3-D scanner: virtual reality?

Today, in the broadcast and computer industries, virtual reality and 3D models are widely used. The development of 3D scanners is what enables these models to be transposed. Unfortunately, these scanners are very expensive and aimed primarily at the professional market. I believed that there must be a more cost effective way of digitising objects, and I set out to design a scanner that would prove that there was.

I first focused on identifying an efficient and effective method of scanning all surfaces of a static object. Next I worked on a method for propelling the laser scanner head along a track, and I designed the tracks to allow its smooth operation without slippage, with one rail used to power the head and the other to send back the signal.

The holding mechanism was critical and needed to be designed to minimise the interference with the scanning process. I positioned the holding prongs in such a way so that the laser head, travelling along the arc, is always perpendicular to the arc and will always pass through the centre of the circle, and thus miss the supporting prongs.

I then integrated these elements into a final design and animated it on the computer screen. I called it "Rotoscan" because it rotated around the object to be scanned.

"Rotoscan" takes an original approach to 3D measurement based on the measurement of one length and two angles, rather than the more conventional approach based on the length of these sides and one angle. It can be further modified to scan colour as well as the shape of

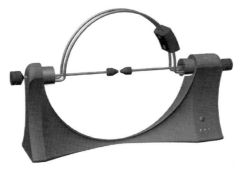

Illustration of the final design.

an object. It can also be modified to print intricate designs onto models or objects.

"Rotoscan" has many advantages over the models commercially available. An unexpected benefit was that, unlike other scanners on the market, it would be capable of generating not only polygon meshes but also NURBS (non uniform rational b-splines) which facilitate smoother curves and higher quality modes.

"Rotoscan" has many applications – including computer animation, modelling for computer games, industrial design, and architectural modelling. It also would be of great value to computer artists – amateur and professional.

The scanner I have designed would, I believe, be fast, flexible and easy to use and would meet the criteria for a scanner capable of high quality digitisation at low cost.

Joe Fitzsimons won third prize in the Senior Individual Section in the Chemical, Physical and Mathematical Sciences Category at of the Esat Telecom Young Scientist and Technology Exhibition in January 1999. He also won a Special Award presented by the Irish Business and Employers' Confederation (IBEC). His teacher was Mr Dermot Ryder.

Further details of Joe's design can be viewed on his web site, www.intelcities.com/knowledge_way/rotoscan/ or http2//www.rotoscan.com

THIS ARTICLE WAS SPONSORED BY IBEC.

OUR LADY'S COLLEGE, GREENHILLS, DROGHEDA, CO. LOUTH — KARIE MURRAY & GEMMA O'HALLORAN

Mop it up!

The main aim of our project was to make a cotton cloth which would absorb oil. At the outset of this research, we assumed that the untreated tea towels would pick up water more efficiently than they would pick up oil, and we wanted to see if we could change this by a chemical process.

Based on further research we decided on a method which we could use to convert the ordinary cotton tea-towel into an "oil loving cloth". We selected two fatty acids – Stearic Acid and Lauric Acid – which we wanted to react with the untreated cotton. The first step was to reflux the acid with Thionyl Chloride ($SOCl_2$) to begin the process of making the corresponding acid chloride. Excess $SOCl_2$ was then distilled off and the resulting acid chloride was reacted with the cotton using Toluene as a solvent and Pyridine as a catalyst. The resulting material was cleaned with Toluene and dried. It was then tested to see if there was any change in its efficiency at absorbing oil and water.

The first test carried out involved putting a sample of each type of cotton (stearoylated, lauroylated and untreated) into separate beakers of oil and water. The second test examined the ability of each type of cloth at absorbing oil and water by capillary action. The final experiment was to test the selectivity of the samples to see how effective they were at absorbing oil from a mixture of oil and water.

After the untreated cotton had been treated with either of the acid chlorides, the fibres of the cotton became very weak and brittle. However, in all experiments carried out on the stearoylated cotton, it always preferred to absorb oil to water. Treating cotton with Stearoyl Chloride made the cloth hydrophobic. When placed in separate samples of oil and water, the lauroylated cotton absorbed more water than oil. The untreated cotton generally absorbed almost equal amounts of oil and water, but when shredded it absorbed substantially more water than oil.

Next year we hope to use several more acids, namely Decanoic Acid, Myristic Acid and Oleic Acid, and different types of cotton, to investigate if the length of the Carbon chains will affect the strength or absorbency level of the cotton.

Karie Murray & Gemma O'Halloran won third prize in the Senior Group Section of the Chemistry, Physical & Mathematical Sciences Category at the Esat Telecom Young Scientist and Technology Exhibition in January 1999. They also won a Special Award presented by the Institute of Petroleum. Their teacher was Ms Geraldine Mulvihill.

THIS ARTICLE WAS SPONSORED BY THE INSTITUTE OF PETROLEUM.

BALINTEER COMMUNITY SCHOOL, DUBLIN 16 — STEPHEN McGUINNESS

Telephone home link

My interest in electronics began at the age of nine when I got a present of an electronics kit for my birthday. My interest developed more and more in following years. When I went into secondary school, my science teacher Mr Walsh suggested that, having such an interest in electronics, I should enter the Esat Telecom Young Scientist & Technology Exhibition in the RDS.

I called the Project "Telephone Home Link". My original idea was that people leaving their home for a period of time going on holidays/business trips etc. could phone up their home using a DTMF (dual tone multi frequency) phone and press phone buttons to operate household services and appliances. These might include lights, radio, heating, etc. I had a lot of trouble splitting the tones and, after much study and experimentation, it only worked properly on occasions due to tone drift.

I then came up with the idea of using a separate unit to produce tones. My project then consisting of two parts – one is a base unit and the other a remote unit. The base unit is kept at home and wired to the phone line. It is possible to connect ten different lights, heating and a radio. The remote unit is the part you carry around with you. It is a small box about the size of a small cassette player, and has three buttons. When each button is pressed, it creates a different tone through a loud speaker. What the users have to do is phone their house from either a mobile or land line phone and, when the phone is answered through a standard answering machine, the tone buttons are pressed on the remote unit to perform the desired tasks (e.g. turn on/off lights etc. at home).

The system is very useful as a security device, as it can be used to turn on/off lights at random, ensuring that any potential intruder would assume the house is occupied. Another very good use of the system is to turn on central heating in winter time and possibly prevent burst pipes due to a very cold snap.

I now hope to improve my present system whereby the user would be able to use the phone buttons directly, thereby eliminating the need for the remote unit and answering machine.

I would also hope to have a password system, therefore eliminating any possible unauthorised access to the system.

Stephen McGuinness entered his project in the Junior Individual Section in the Chemistry, Physical & Mathematical Sciences Category at the Esat Telecom Young Scientist and Technology Exhibition in January 1999. He won a Special Award presented by Shaw Scientific Limited. His teacher was Mr Barry Walsh.

THIS ARTICLE WAS SPONSORED BY SHAW SCIENTIFIC LIMITED.

Heavy metal ions in plants

Our interest in this area followed a newspaper article discussing the health benefits of eating watercress due to the large quantities of metal ions contained within the plant. We decided to investigate the uptake of certain heavy metal ions by plants to determine if there was any correlation between the concentration of metal ions in the hydroponic system and the retention of metal ions in the root and stem systems.

We used a hydroponic system – one in which plants are cultivated without the use of soil – as this allowed us to control the ions present and their concentrations.

Mung beans were grown in the hydroponic system with transition metal ion solutions, as these heavy metals are often found on landfill sites. An acid mixture was used to digest the plant material before it could be analysed in an atomic absorption spectrophotometer to determine the levels of metal ions in the plant remains. The analysis was carried out in the School of Chemistry, Queen's University, Belfast.

The first set of plants were grown in 500 parts per million (ppm) solutions of zinc, nickel and copper. However, after a couple of days, many of the plants showed possible signs of premature death. These high concentrations were too toxic for the plants. The second set of plants were grown in 250 ppm solutions of tin, nickel and copper.

In the second part of the investigation, no visible signs of death were observed, suggesting that the plants could tolerate the presence of tin, nickel and copper at moderate levels.

From the results it was observed that the percentage of the metal ion found in the root system was considerably greater than the percentage found in the stem. This suggested that the presence of high concentrations of the heavy metal ions disrupted the transport of minerals in the xylem of the plants. The most feasible explanation for this was that the metal ions were precipitated in the roots as insoluble salts and therefore further transportation up the xylem was prevented.

A possible application of our investigation is to use plants to remove metal ions from polluted soils, such as those of landfill sites. A subsequent article on *Tomorrow's World* (BBC TV) used plants to extract gold from soil.

Karen McCluskey prepares a heavy metal ion solution for the hydroponic system.

Ciara Greenan, Karen McCluskey & Julie Ellison entered their project in the Senior Group Section in the Biological & Ecological Sciences Category at the Esat Telecom Young Scientist and Technology Exhibition in January 1999. They won a Special award presented by the Institute of Biology. Their teacher was Dr M.R.J. Dorrity.

THIS ARTICLE WAS SPONSORED BY SAMTON LIMITED.

Decomposition in coniferous and deciduous woodland soil

When we set out to do our project we asked ourselves *"If dead leaves don't go to heaven, then where do they go?"*, and throughout the ten weeks of our project we found the answer.

Go out to your garden and take a fistful of soil. Look closely at it....... what do you see....... well all you see is mud, right? **Wrong,** what you don't realise is that in that soil there are about two million organisms living and breathing, but that isn't all they do. These organisms are what get rid of those rotten leaves in your garden. They eat them and turn them into a substance called humus. This substance then nourishes the soil which helps your flowers and plants grow. The breaking down of dead matter into humus is known as decomposition.

In our project we did a study on how fast the leaves took to be turned into humus. We did this in two different woodlands, deciduous and coniferous, and compared the speed of decomposition in each.

This is how it was done:
We picked an undisturbed area in each woodland and dug eight holes. In each hole we placed a piece of blotting paper, a partly decomposed leaf, a fresh leaf and a small piece of birchwood (an icepop stick). In the same holes we put all the items above wrapped in a piece of

Kathryn Prendergast & Eimear McCarthy.

netting and buried them. The netting was to stop the larger organisms such as earthworms, beetles etc. from decomposing the materials. This way we could measure the rate of decomposition by microorganisms – e.g. fungi and bacteria. Each item was weighed before it was buried. Every week we returned to the woodlands and dug up one hole until all the holes were empty. We weighed each item again and compared the weight from when they were buried.

We found that the decomposition was faster in the deciduous woodlands because (i) there were more organisms to decompose and (ii) there was a low amount of acid in the soil. We did tests to check these reasons. Organisms can decompose faster in a low acidic soil.

So now you know that dead leaves don't go to heaven but are recycled into humus to nourish your soil and that, when you look at your soil, you are looking at the little creatures that do this.

Eimear McCarthy & Kathryn Prendergast won first prize in the Junior Group Section in the Biological & Ecological Sciences Category at the Esat Telecom Young Scientist and Technology Exhibition in January 1999. They also won a Special Award presented by the Irish Science Teachers' Association. Their teacher was Ms Joan Faherty.

THIS ARTICLE WAS SPONSORED BY THE IRISH SCIENCE TEACHERS' ASSOCIATION.

TERENURE COLLEGE, TEMPLEOGUE, DUBLIN 6 — BRIAN LAWFORD, JOHN McCAMBRIDGE & JAMES O'BRIEN

The self-repairing wheel

The Self-Repairing Wheel utilises new and existing technologies, allowing quick development in the market place. The unit, which is installed in the wheel, has the ability to re-inflate the wheel whenever a puncture occurs. The Model on display is a working model for demonstration purposes only. Our new design has been constructed to increase both convenience and safety for drivers and passengers as it removes not only the inconvenience associated with a wheel puncture but also the safety hazards a stranded car can cause, especially on our motorways and many country roads.

Drivers today not only desire but also require safety. When the safety potential of our unit is fully realised, it will save both tyres and rims from being damaged. The added safety will prevent road accidents and thus road deaths, something wanted by all.

One of our system's most vital components is a liquid rubber / compressed air solution, an existing product used throughout the market. Our unit consists of a stainless steel container mounted inside the wheel rim. This container has two valves: one is used to release the solution upon puncture, the other's function being the replenishment of solution in the container. The first is automated and triggered when the tyre contacts the switch built into the valve, thus opening it and releasing the solution. The Self-Repairing Wheel works in the following manner.

When a puncture occurs in a wheel, air begins to escape. This causes the tyre to flatten and press on the valve. When open, this valve allows the solution to enter the wheel from the container present in the wheel. The compressed air/liquid rubber solution automatically tries to exit the tyre via the puncture, where the rubber solution solidifies and blocks the hole and the compressed air inflates the tyre.

Our product increases car and driver safety, hopefully lowering road accidents, and provides a safer drive for all involved, as well as being able to be replenished as many times as needed.

Brian Lawford, John McCambridge & James O'Brien entered their project in the Junior Group Section in the Chemistry, Physical & Mathematical Sciences Category at the Esat Telecom Young Scientist and Technology Exhibition in January 1999. They won a Special Award presented by The Patents Office. Their teacher was Mr Thomas Hughes.

THIS ARTICLE WAS SPONSORED BY THE PATENTS OFFICE.

ST JOSEPH'S COMMUNITY COLLEGE, KILKEE, CO. CLARE — DANIEL ROCHE, CONNOR ROONEY & JAMIE FITZGERALD

Cow pat power

In the first stage of our project to produce alternative electricity using energy from cow manure, we have a digester sitting in a water bath set to 44°C. The digester is a five litre vessel containing four litres of cow manure with an air space above it to allow for frothing during the fermentation process. The digester provides an anaerobic environment in which the methane producing bacteria (methanogens) can work.

The digester has an outlet pipe to a gas storage vessel where the biogas is collected by the displacement of water. We use this method because methane forms an explosive mixture with air.

The methane can be used in two ways. One method is to burn it directly to heat water or to provide lighting. The other involves converting the chemical energy of the methane into electricity. We chose the second.

Because of the small quantities of gas produced by our digester, we decided to use a steam engine to generate steam to turn a small motor which then lights an LED.

The Law of Conservation of Energy states that: "Energy cannot be created or destroyed - it is converted from one form into another". But in the conversion heat is lost, so there cannot be a total conversion to another form of energy. It is inefficient from an energy point of view to use a steam engine to generate power, as a lot of energy is lost to heat, and due to friction due to the mechanical workings of the engine. A lot of energy is consumed before any energy is converted because water has a very high specific heat capacity and a very high latent heat.

We were not allowed to bring naked flames and explosive or toxic materials to the exhibition so we made a model of a biogas producing unit. Our model could be scaled up to meet some of the power needs of a farm, an average household or a small community.

Conclusions

On a farm with dairy cows a lot of water is used every day to wash the animals in preparation for milking and to wash the milking machine. If the energy from the manure produced by these cows was harnessed, the farmer could cut down on detergent and have a more hygienic and efficient cleaning system.

In our research we found that this process is used in third world countries like China and India. We think that the project has a lot of potential and could be developed much further.

Jamie, Conor & Daniel.

Daniel Roche, Connor Rooney & Jamie FitzGerald won second prize in the Junior Group Section in the Chemistry, Physical & Mathematical Sciences Category at the Esat Telecom Young Scientist and Technology Exhibition in January 1999. They also won a Special Award presented by The Teachers' Union of Ireland. Their teacher was Ms Angela Rahill.

THIS ARTICLE WAS SPONSORED BY THE TEACHERS' UNION OF IRELAND.

COLÁISTE RIS, DUN DEALGAN, CO. LOUTH — THOMAS GERNON

Geomorphology of river valleys in Louth

For three years I have entered prize-winning projects at the Young Scientists' Exhibition and this year I concentrated on the geomorphological effects of glacial ice on river flow and beach formation in Co. Louth. I found that the ice sheet had caused several major rivers to be dammed. The resulting lakes broke through and the floods brought loads of debris to the coast. I also found that beaches (now raised) were formed of large pebbles which were too large to be carried by present day rivers. They could have been caused by such floods.

My laboratory work with Ordnance Survey maps and numerous fieldwork digs helped me formulate a county wide picture of moraine damming and river capture. I studied one river in detail, the Mattock, and found it had been diverted to a new course about 20 km south of its original course. Capture and flooding had scoured much material from its valley(s) and brought to the coast quantities of large pebbles.

I studied one raised beach at Port, Co. Louth. It seemed to comprise small pebbles at the surface, and larger pebbles mostly 1 metre below the surface. The raised beaches had clearly been built by the waves of the sea, but the debris was provided by rivers which brought it down from inland locations. The water flow which brought down the larger pebbles to the (now) raised beaches could have come from flooding after moraine dams were breached. These larger pebbles >17 cm. in diameter were brought to the (raised) beach 6,000 ybp (years before present) by strong floods.

Thomas Gernon receives his award from the Director of the Geological Survey of Ireland, Dr Peadar McArdle, in the presence of Minister for Education and Science, Mr Micheál Martin, and Mr Sean Corkery, Chief Operating Officer, Esat Telecom.

The present-day seabed was once dry land which had normal river valleys on its surface. Sea level has risen and the valleys have disappeared. The most interesting aspect of my project was to examine the (as far as I know) unresearched topic of rivers which cross a tidal mud-flat area such as Dundalk Bay. I developed a hypothesis which stated that the foreshore mud forms in laminar concentric shells under and around the present foreshore river, which probably follows the deep valley which existed during the Midlandian Glaciation (exposed due to a drop in sea level). Even though the muddy foreshore is inundated by tides twice daily, the river still retains its original course.

Thomas Gernon entered his project in the Senior Individual Section in the Chemistry, Physical & Mathematical Sciences Category at the Esat Telecom Young Scientists' Exhibition in January 1999. He won a Special Award presented by the Geological Survey of Ireland. His teacher was Mr Sydney Peck.

THIS ARTICLE WAS SPONSORED BY THE GEOLOGICAL SURVEY OF IRELAND.

GAEL CHOLÁISTE AN CHLÁIR, ENNIS CO. CLARE — SÉAMUS Ó ROIDEACHÁIN

Mixing dynamics in Killone Lake

This is a project about the mixing dynamics in a small (20ha.), low outflow lake in Co. Clare (Killone). The surrounding rocks are Namurian shale and Carboniferous limestone with a watershed of 500ha. To find out how water and pollution move about in the lake, I needed to investigate the mixing agents - which are wind, inflow/outflow, stratification, evaporation, diffusion, underwater currents, seiche (tide), solar convection and rain.

I talked to local people and fishermen about the lake's recent history. I gathered maps and drew a more accurate diagram of the lake. I found out the depth contour lines using a boat and a measuring tape. I transferred all this onto a computer drawing file. From this I made both fibreglass and Medite scale models at 667:1. I visited the Geological Survey of Ireland to find out more about the lake's geological history.

The best website was about Lake Tegel in Berlin because it explained how wind-mixing worked. This study was very significant as it gave me a good idea of how to approach wind-mixing in Killone Lake.

I had access to computer modelling facilities in the engineering department at the University of Limerick. The computer-

modelling programme displayed simulations of how wind over the surface could affect the sediment at the bottom of the lake in that a stiff breeze can mix the water just as much as a strong storm. In the scale model I used dyes to show underwater currents and a video camera to track the mixing generated by a fan. These simulations agreed with the results of underwater float experiments.

I learned that Killone Lake is quite deep at 18m with a further 13m of mud below this. It was a large waterfall thousands of years ago but now only 100 litres per second pass through. I also learned how the water mixes in it. The lake can get a serious algal bloom during summer. Phosphates dissolve out of the mud during warm periods in the summer, as there is a large flat area in the middle of the lake at between 6 and 10m deep. They then mix with the ever-present nitrates and trigger the bloom. Steady wind is the main mixing agent and not seiche as suspected. Cleaning the lake to less than 1% of current levels of pollution would take about five years if only clean water went into the lake from now.

Séamus Ó Roideacháin (Jim Redington) won joint first prize in the Junior Individual Section in the Chemistry, Physical & Mathematical Sciences Category at the Esat Telecom Young Scientist and Technology Exhibition in January 1999. He also won a Special Award presented by the Dublin Institute of Technology. His teacher was Ms Mary Masterson.

THIS ARTICLE WAS SPONSORED BY DUBLIN INSTITUTE OF TECHNOLOGY.